FIG. 1. Robley Dunglison in his prime. Portrait painted about 1848 and presented as a token of esteem and friendship to Dr. Robert M. Patterson. Courtesy of Dr. William B. Bean (a descendant of Dr. Patterson), who obtained it from his cousin, Father Odey Patterson.

TRANSACTIONS

OF THE

AMERICAN PHILOSOPHICAL SOCIETY

HELD AT PHILADELPHIA
FOR PROMOTING USEFUL KNOWLEDGE

NEW SERIES—VOLUME 53, PART 8
1963

THE AUTOBIOGRAPHICAL ANA OF ROBLEY DUNGLISON, M.D.

Edited with Notes and Introduction by
SAMUEL X. RADBILL, M.D.

THE AMERICAN PHILOSOPHICAL SOCIETY
INDEPENDENCE SQUARE
PHILADELPHIA 6

December, 1963

To

WALTON BROOKS McDANIEL 2d
Curator of Library Historical Collections
of the
College of Physicians of Philadelphia

This work is inscribed as a
Token of High Esteem

Library of Congress Catalog
Card Number 63–22636

THE AUTOBIOGRAPHICAL ANA OF ROBLEY DUNGLISON, M.D.

Edited with Notes and Introduction by

SAMUEL X. RADBILL, M.D.

CONTENTS

INTRODUCTION

Robley Dunglison, distinguished American medical educator,[1] was brought to this country from England in 1825 to serve as the Professor of Medicine at the newly founded University of Virginia. While there he was personal physician to Thomas Jefferson, James Madison, and James Monroe and was called into consultation in the treatment of Andrew Jackson.[2] Leaving Virginia in 1833, he taught for three years at the University of Maryland, and then for the rest of his life at Jefferson Medical College of Philadelphia.

Dunglison's *Ana*, prepared in Philadelphia, about 1852, were designed to preserve personal recollections and records thought to be of particular interest to his family. To the original eight holograph volumes he added supplementary notes from time to time. Thus the work is not a diary, or even an autobiography in the usual sense. Seven of these volumes were presented to the College of Physicians of Philadelphia in 1904 by Mrs. Violette Fisher Dunglison, widow of Dr. Dunglison's son, Richard, and the eighth, with the supplementary notes, was given to the College in 1933 by its then recently retired librarian, Mr. Charles Perry Fisher.

[1] S. X. Radbill, Robley Dunglison, M.D., 1798–1869: American Medical Educator, *Journal of Medical Education* 34: 84–94, 1959.

[2] S. X. Radbill, Dr. Robley Dunglison and Jefferson, *Transaction & Studies of the College of Physicians of Philadelphia*, 4th ser., 27: 40–44, 1959.

The decision to publish the *Ana* was based on a number of considerations. In the first place it is evident that Dunglison himself hoped for eventual publication. But more than this, what he has written has achieved historical importance after lying fallow for over a hundred years. Furthermore, his personal recollections of four American presidents, as well as intimate anecdotes concerning a great number of other prominent men of the nineteenth century, promise to be of great interest to general historians as well as to those concerned with biography. Much of what he records here about medical schools with which he was connected directly and indirectly is practically unknown, while significant aspects of his own work, particularly in the field of medical literature, are here publicly presented for the first time. Significant, too, is the extended account of his participation in the experiments of William Beaumont on the physiology of gastric digestion.

The Dunglison manuscripts have not been generally known to scholars. Dr. Samuel D. Gross used them in the preparation of a biographical sketch of Dr. Dunglison not long after his death.[3] Henry S. Randall consulted the *Ana* while preparing the biography of Thomas Jefferson during Dunglison's lifetime,[4] and Robert C. McLean used them in writing a biography of George Tucker as recently as 1961. But nearly all others who have mentioned Dunglison's memoranda seem to have been unaware of the existence of the originals.

Reminiscences such as the *Ana* contain may be colored by personal bias; with this in mind, Dunglison, who kept copies of much of his correspondence, and retained other records of his many activities, backed up much of his statements with documentary proof. This adds to the historical value of these memorabilia. There is a serious dearth of personal documentation such as this. The account of his training and professional activities before leaving England for America reveals the difficult path in the way of an aspirant to a specialized career in medical teaching and writing. He exerted a decided influence upon the form and direction of medical education in the United States, and three medical schools with which he was associated have endured. To their lasting success he contributed no small part.

During his teaching career at Jefferson Medical College, he attracted many students from all parts of

[3] *Transactions of the College of Physicians of Philadelphia*, n.s., 4: 294–313, 1874.

[4] See Appendix I.

3

the country, especially from the southern states. His writings were held in high esteem, and probably no American medical author was more prolific. Admired and respected by a wide circle of friends, he was celebrated for his conviviality as a host and gay participant at social gatherings. He held office in many medical and scientific organizations and was esteemed at scientific meetings for his diplomacy and erudition. By the time he wrote these memoirs, he was in his prime and one of the great national medical figures of his day.[5] As corresponding secretary, he was one of the most useful and active members of the American Philosophical Society, honorably known in Europe as well as in America. A rapid and discriminating reader with a natural bent for foreign language, he was indefatigable in collecting information which he incorporated into his lectures and then gathered into basic textbooks on physiology, therapeutics and materia medica, hygiene, medical jurisprudence, medical education and ethics, and, above all, into his comprehensive dictionary, which received the greatest public acclaim, brought him financial rewards, and of which he was justifiably most proud. By his own confession, he preferred teaching to practice or research. While his medical work covered practically all the nonsurgical specialties of today, even including psychiatry, and the medical conditions of children, his actual practice was limited to attendance on the four presidents mentioned; he officiated at the birth of the son of at least one of his colleagues (Thomas Addis Emmet); was called to treat some other friends and colleagues, and conducted teaching clinics at the schools and hospitals with which he was associated. He knew the medical facts of his day and knew them well; and he knew well how to impart this knowledge to his students, and, through his writings, to the medical world at large. For this he was properly famous. But with little experience in private practice, he lamented the fact that intelligent people will confide in the lowest empiric rather than in the most learned and accomplished physician.

As an innovator he has several "firsts" to his credit. Limited by Thomas Jefferson to consultation practice, he was the first full-time professor of medicine in this country. Likewise in conformity with the desire of Jefferson, he was the first to include formally and clearly a series of lectures upon the history of medicine within the medical curriculum in the United States. His syllabus on medical jurisprudence, while not the first on this subject in the country, was one of the first academic publications at the University of Virginia. His textbook on human physiology, published in 1832, was the first comprehensive treatise on physiology by an American author and earned for him the title of "Father of American Physiology." A year later, the

dictionary was published, again a "first," for it was the first indigenous work of its kind in the United States. For this his students nicknamed him the "walking dictionary." When, at his own request, he was appointed Professor of Hygiene at the University of Maryland in 1833, he became, as it were, the first professor of preventive medicine and public health in the nation. True to his penchant for publishing what he taught, he promptly put his lectures into print, and although Shadrach Ricketson's book on the subject in 1806 preceded his by more than a quarter of a century, Dunglison's book on hygiene was really the first formal textbook of hygiene in this country.

Robley Dunglison took a lively interest in the experimentation of William Beaumont on the physiology of gastric digestion, and was briefly an active participant in this basic research, quickly recognizing its scientific value. Although this constitutes nearly the sum total of his scientific research, it was not insignificant, for he not only influenced Beaumont indirectly but he himself performed some of the experiments on the juice, outlined chemical examinations which were performed by his friend, Emmet, who was a more expert chemist, and sent all the results to Beaumont. Specific suggestions regarding detailed experiments to be performed by Beaumont, which must have taken a good deal of careful thought on the part of Dunglison, set an orderly course for Beaumont to follow, and when Beaumont's book is compared with the report Dunglison prepared for the *American Journal of the Medical Sciences* (which was withheld from publication in deference to the wishes of Beaumont), it seems evident that Dunglison had much to do with the definitive work of Beaumont. Indeed, of all the reputable scientists to whom Beaumont applied, Dunglison was the only one who came to his aid, and were it not for the fact that Beaumont was induced by his cousin Samuel Beaumont to publish his work by himself, it is highly probable that Dunglison would have been associated with him in its publication.

Robley Dunglison also played an important role in the improvement of the care of the insane poor of Pennsylvania, and his efforts on behalf of the education of the blind are noteworthy, too. With his friend Chapin, a form of raised type for the use of the blind was developed, and there is a memoir of Dunglison in this raised print in the library of the College of Physicians of Philadelphia. His activities at the Philadelphia General Hospital were beneficial to patients and students, and his account of the struggle between Medical Board and Board of Guardians is accurate and factual, and one of the few first-hand accounts, by a participant, of an affair which terminated all teaching functions at the hospital for a decade.

PORTRAITS OF DUNGLISON

In the William N. Bradley Collection of Reproductions of Portraits of Philadelphia Physicians in the Col-

---

[5] Richard H. Shryock, The College of Physicians in Historical Perspective, *Transactions & Studies of the College of Physicians of Philadelphia*, 4th ser., 27: 150–157, 1960.

lege of Physicians of Philadelphia, there are eight different likenesses of Dr. Robley Dunglison. Probably the earliest of his portraits in existence is a miniature in possession of the University of Virginia, which may have been painted for him (as was one similarly for his wife) just before leaving London for America.[6] Next, in point of time, is a silhouette which reveals him, rather youthful, in a high stiff collar.[7] Another is a charming full-length silhouette by Aug. Edouart in 1843, which shows him lecturing.[1] In 1847 F. C. Bruce painted a portrait of him which Dunglison then gave to his friend Robert M. Patterson; it is now in possession of Dr. William B. Bean.[8] A lithograph by A. H. Ritchie, from a daguerreotype by M. P. Simons was made probably ten years later.[9] Somewhat older in appearance, middle-aged and handsome, is the lithograph in an album of Jefferson Medical College portraits.[10] A phototype by F. Gutekunst portrays him at a more advanced age, gray-haired and bearded.[11] A similar full-length daguerreotype of about the same period is in possession of the College of Physicians of Philadelphia.[1] In 1868 Thomas Sully painted a portrait for the Musical Fund Society which is likewise now at the College of Physicians of Philadelphia.[2] Finally, there is a nice posthumous portrait done by S. B. Waugh in 1876 which hangs in the library of the Jefferson Medical College.[12]

EDITORIAL TREATMENT

In the original manuscript each page is numbered consecutively through the first seven volumes, but the text really begins on page 137. Volume VIII of the original manuscript was separately paginated by Dr. Dunglison; he drew up a table of contents and a bibliography, and listed other matters at the beginning of volume I, and the supplemental notes were not numbered at all. His table of contents at the beginning has been omitted, but the other lists will be found at the end, where they are presented as appendices. He wrote only on one side of each page, reserving the blank side of the page for later insertions, so that many of the odd-numbered pages are blank. Some of the manuscript volumes continue on to the next without any break in sentence or text. The editor has, therefore, separated the entire work himself into chapters with subheadings. Each page of the original manuscript has subject headings in the left-hand margin. For the sake

of economy, these have been omitted, but some have been used for subheadings. No changes have been made in spelling of words unless obviously mistakes on the part of Dr. Dunglison.

Deliberate emphasis has been given in the footnotes to medical schools, with occasional remarks regarding medical education and medical literature. An attempt has been made to identify as many as possible of the numerous individuals mentioned, and few of the medical personalities have been missed. As a rule, only a single identifying note is given for each person; this can be located from the index. In some instances, notations relating to men and events dealt with by Dunglison have been added from other sources for comparison with his point of view. An index which Dunglison compiled at the end of the seventh volume has been incorporated into a more comprehensive index prepared by the editor.

ACKNOWLEDGMENTS

Walton B. McDaniel 2d, Curator of Historical Collections at the College of Physicians of Philadelphia, has been my "alter ego" in editing this work, and the College willingly granted permission for its publication. To Richard H. Shryock, Librarian of the American Philosophical Society, I am indebted for friendly help and advice. Mr. John Alden Tifft, great grandson of Dr. Dunglison, not only gave me encouragement but furnished a number of family anecdotes. Dr. John M. Dorsey, enthusiastic editor of the *Jefferson-Dunglison Letters,* provided me with information for some of the footnotes, and Dr. William B. Bean granted permission to reproduce his portrait of Robley Dunglison. The Alderman Library of the University of Virginia not only permitted the use of their rich resources but have allowed me to reproduce (in Appendix I) original material which they possess pertaining to Dunglison. Mr. John Cook Wyllie, Librarian, and Miss Anne Freudenberg, Assistant in Manuscripts, facilitated for me the use of the division of rare books and manuscripts at the Alderman Library. Mrs. Elizabeth S. Adkins, Librarian of the Hospital of the University of Virginia, was a great help to me, as was Dr. Elizabeth G. McPherson at the reference department of the Manuscripts Division of the Library of Congress. My dear friends Dr. and Mrs. M. Robert Beckman, have given me assistance in editing, recording and typing, generously, without reward.

Finally, to my wife some acknowledgment must be made for her sympathy and equanimity which she always maintained throughout the harrowing period when books, papers, ashtrays and even spilled ink would have tried the patience of a saint. Without her enduring complaisance this work would never have reached completion.

This work was supported by a grant from the Penrose Fund of the American Philosophical Society.

---

[6] Andrew de Jarnette Hart, Jr., Thomas Jefferson's Influence on the Foundation of Medical Instruction at the University of Virginia, *Annals of Medical History,* n.s., **10**: front cover and page 98, 1938.

[7] Bradley Collection, College of Physicians of Philadelphia.

[8] *Virginia Medical Monthly* **87**: 676, 1960.

[9] *Annals of Medical History,* n.s., **10**: frontispiece, 1938.

[10] James F. Gayley, *History of Jefferson Medical College,* 1858.

[11] John M. Dorsey, *Jefferson-Dunglison Letters,* 1960.

[12] George M. Gould, *History of Jefferson Medical College* **1**: 120, 1904.

The following strictly private
autobiographical Ana

may afford   interest to those who are
dear to me, and may be hereafter a source
of   satisfaction in fixing the dates
of occurrences, where a question may arise.
They will also recall to the recollection of
one[+] for whose special use they are written,
the numerous happy events that have
transpired in which we have been concerned
together, and many facts, which, if mentioned
to her, may have made but a casual and
evanescent impression.
      [+] (now alas! no more.)

I. FORMATIVE YEARS

Birth and early education. Medical education and practice in London. Accepts professorship in the University of Virginia. Marriage. Society membership. Early literary contributions.

BIRTH AND EARLY EDUCATION

I was born in Keswick, Cumberland, England, on the 4th of January, 1798;[1] and was educated, in the first instance, to be a West India planter; my great uncle Joseph Robley, being an extensive landed proprietor and planter in Tobago and Saint Vincents, after whom, indeed, I was named;[2] and having promised to take charge of me. He died, however, when I was very young, and the idea of my going to the West Indies was gradually abandoned. A portion of my early education was obtained at Green Row Academy on the Solway,[3] then enjoying much reputation as a scholastic institution, under the competent management of Mr. Joseph Saul,[4] a man of undoubted ability and educational tact; yet, I am proud in being able to say—with a due sense, I hope, of what I have obtained from others, that I have been—not figuratively but literally—a "self made man"; for my mother's means were such, that it was necessary for her children to be actively occupied in their earnest endeavors for their own advancement.[5]

MEDICAL EDUCATION AND PRACTICE
IN LONDON

After this, I was somewhat undecided as to the profession I should adopt, whether law or medicine; but ultimately determined on the latter; passed a portion of my medical pupilage in Cumberland,[6] and the remainder in London, where I acted, for some time, as an assistant to Charles Thomas Haden, of Sloan Street, who became a most valued friend, but died, alas! too soon for science;[7] attended a course of lectures in the University of Edinburgh[8] and the Royal Infirmary there[9] and subsequently in London, attended lectures;[10] and at the École de Médecine and sundry private courses in Paris;[11] passed my examinations at the Royal College of Surgeons, and the Society of Apothecaries in London; and commenced the exercise of my profession in that city in the year 1819.[12] General practice was, however, very distasteful to me; and, after graduating, by examination, in the University of Erlangen, my inaugural dissertation being "De Neuralgia"[13]—I determined in 1823 to restrict myself to the practice, especially obstetrical, of medicine, which I pursued with every prospect of success, having formed an extensive acquaintance with many of the medical practitioners of London through the different medical societies to which I belonged, and of some of which I had been an officer and active member; and through the periodicals, in two

---

* Daybook, journal or memoranda.
[1] His father was William Dunglison and his mother, née Elizabeth Jackson, was a Robley on the maternal side (Henry Lonsdale, *Worthies of Cumberland* 6: 262, London, 1875).
[2] It seems to have been a custom of the northwest of England to receive in Baptism the mother's maiden name.
[3] In the Abbey Holme.
[4] A Quaker schoolmaster with pretensions to classical poetry.
[5] After the death of Robley Dunglison's father, his mother married James Atkinson. When her uncle, Joseph Robley, died, he left a legacy which enabled her to give her family a liberal education.
[6] Dunglison received the first part of his medical education in the surgery of Mr. Robinson, a friend of the family (J. Alden Tifft, personal communication). At seventeen, he was apprenticed to John Edmondson, a surgeon of repute in Keswick.
[7] Charles T. Haden (1786–1824), one of the first to adopt the use of Laennec's stethoscope in England, surgeon to the Chelsea and Brompton dispensary, transmitted a great taste for music to his assistant, Dunglison.

[8] The Library of the College of Physicians of Philadelphia possesses the student notebook with notes written by Dunglison at the clinics of James Home (1758–1842), Professor of Clinical Medicine at Edinburgh.
[9] Appended to Dunglison's student notes are clinical cases presented at the Royal Infirmary by Daniel Rutherford (1749–1819).
[10] Guy's, Saint Bartholomew's, Saint Thomas', Saint George's and other hospitals and a dozen or more private medical schools were available to medical students in London.
[11] Paris was at this time an advanced center of medical education and research, attracting medical students from England and America.
[12] The practice of medicine in and around London was controlled by the three medical corporations: The Royal College of Surgeons, The Royal College of Physicians and the Society of Apothecaries. The Apothecaries Act of 1815 placed almost all who wanted to practice medicine under the control of the Society of Apothecaries.
[13] Dunglison evidently obtained his diploma *in absentia* by submitting his thesis and a proper fee, a common practice at the time, since the degree in medicine apparently was not available in London. While in England he was an "Esq." Upon his arrival in America he was designated on the passenger list as "Dr. of Medicine," and he remained Dr. Dunglison ever after.

FIG. 2. Keswick.

QVOD FELIX FAVSTVMQVE ESSE IVBEAT

DEVS OPTIMVS MAXIMVS

SVB AVSPICIIS

AVGVSTISSIMI ET POTENTISSIMI REGIS ET DOMINI

DOMINI

MAXIMILIANI IOSEPHI

REGIS BAVARIAE

REGIS ET DOMINI NOSTRI LONGE CLEMENTISSIMI

EX DECRETO GRATIOSI MEDICORVM ORDINIS

IN ACADEMIA REGIA FRIDERICO-ALEXANDRINA ERLANGENSI

PRORECTORE MAGNIFICO

VIRO ILLVSTRI ET AMPLISSIMO

D. THEOPHILO ERNESTO AVGVSTO MEHMEL

AVGVSTISSIMO BAVARIAE REGI AB AVLAE CONSILIIS PROFESSORE PHILOSOPHIAE ET AESTHETICES PVBLICO ORDINARIO

BIBLIOTHECAE ACADEMICAE DIRECTORE

VIRO PRAENOBILISSIMO

ROBLEY DVNGLISON

ANGLO

MEDICINAE ET CHIRVRGIAE LICENTIATO

COLLEGII REGII CHIRVRGORVM LONDINENSIS SOCIO SOCIETATI MEDICAE AB EPISTOLIS AD EXTEROS DANDIS SOCIETATIS HVNTERIANAE
LONDINENSIS SOCIETATIS SCIENTIARVM REGIAE NANCIANAE LINNEANAE PHARMACEVTICAE PARISIENSIS PHYSICO-
MEDICAE ERLANGENSIS SODALI ANNALIVM MEDICORVM LONDINENSIVM EDITORI

POST ERVDITIONIS INSIGNIS SPECIMINA IN EXAMINIBVS DATA

NEC NON

POST EXHIBITAM DISSERTATIONEM INAVGVRALEM

DE NEVRALGIA

DOCTORIS MEDICINAE

GRADVM IVRA ET PRIVILEGIA

DIE IV. DECEMBRIS cIↃ IↃ ccc xxIII

RITE CONTVLIT

D. ADOLPHVS HENKE

CONSILIARIVS AVLICVS THERAPIAE CLINICES ET MEDICINAE PVBLICAE PROFESSOR PVBLICVS ORDINARIVS CLINICI MEDICI DIRECTOR

SOCIETATIS PHYSICO MEDICAE ERLANGENSIS H. T. DIRECTOR. ACADEM. CAESAR. LEOPOLD. CAROL. NATVRAE CVRIOSORVM SOCIETATIS MEDICAE HEIDELBERGENSIS
RHENANAE BONNENSIS BERNENSIS WETTERAVIENSIS ETC. SODALIS

ORDINIS MEDICI H. T. DECANVS ET PROMOTOR AD HVNC ACTVM LEGITIME CONSTITVTVS.

QVOD ITA FACTVM ESSE SOLEMNE HOC DIPLOMA SIGILLIS MAIORIBVS ET REGIAE LITERARVM VNIVERSITATIS ET ORDINIS MEDICI
NEC NON ORDINIS EIVSDEM DIRECTORIS ASSESSORVMQVE PORRO PRORECTORIS MAGNIFICI ET VNIVERSITATIS SECRETARII
AVTOGRAPHIS MVNITVM TESTATVR.

FIG. 3. Medical diploma from the University of Erlangen. The original is in the Historical Collections of the College of Physicians of Philadelphia.

of which I was engaged in the editorial department—the *London Medical Repository*,[14] in conjunction with my distinguished friend Dr. Copland; and the *Medical Intelligencer*,[15] which was a critical *resumé* of the articles contained in the different domestic and foreign journals—of both of which periodicals I shall speak hereafter.

From the commencement of my practice as a physician, I determined to devote myself largely to Obstetrics and the Diseases of Women and Children; and was made *Physician Accoucheur to the Eastern Dispensary*,[16] one of the most extensive charities of the kind in London. I determined, also, to teach these subjects; and actually gave instruction to a single pupil, at my lodgings. I decided on making known my intention publicly, and in the number of the *Medical Repository* for May, 1824, it was announced: "Dr. Dunglison, physician-accoucheur to the Eastern Dispensary, will commence a course of lectures on the Principles and Practice of Midwifery, etc., at the commencement of October next."

ACCEPTS PROFESSORSHIP IN THE UNIVERSITY OF VIRGINIA

All my plans, however, became overthrown in the autumn of 1824, by the arrival in London of Francis Walker Gilmer, Esqr,[17] who had been sent over by the Board of Visitors of the new University of Virginia of which Mr. Jefferson was Rector, to select professors for the same, and with full powers to appoint them.[18] His attention was perhaps first directed to me by Dr. Birkbeck,[19] the enlightened physician, philosopher and philanthropist, who may be regarded as the founder of those useful establishments, the Mechanic's Institutes, of which our own admirable Franklin Institute of Philadelphia[20] is a shining example. When the offer of the position was made to me it was a matter of serious deliberation on my own part, and on that of some of my

intelligent friends, and especially of Mr. Callaway,[21] the assistant Surgeon of Guy's Hospital, and of Mr. Leadam[21a] and others, whether I should avail myself of it. All agreed, that if I remained in England for a few years, my worldly affairs would be in a more prosperous condition than if I went to Virginia; one overwhelming reason, however, decided me. I was ardently attached to a daughter of Mr. Leadam, whom I could not expect to be able to marry if I remained in London, for years to come; whilst if I embraced the American offer I could do so immediately. As soon as Mr. Gilmer had offered the professorship for my acceptance, I called in Tooley Street at the residence of Mr. Leadam and asked his daughter Harriette if she would go to America. General Quiroga—a well known guerilla officer in Spain, and Major Dickson, who had served on the same side with him, were present, which prevented my entering into a full detail of the matter.[22] She at once assented, however, and I accepted the proposition of Mr. Gilmer, and entered into a covenant with him on the 28th of September 1824, of which the following is a copy:

This covenant entered into on the 28th day of September, in the year 1824, at London between Francis W. Gilmer, attorney in fact for the University of Virginia of the one part, and Robley Dunglison of the other part, witnesseth, that the said Gilmer, attorney in fact for the Rector and Visitors of the University of Virginia, doth hereby appoint the said Dunglison a professor in the said University and covenants with the said Dunglison, that he shall, as professor aforesaid, occupy one of the pavilions of the University free of rent; and shall receive from the said University as a salary for his services the sum of *fifteen hundred dollars*, money of the United States, *per annum*, payable semi-annually or quarterly, as the Rector and Visitors may prefer; and moreover the said Dunglison shall be entitled to receive as a tuition fee from each student who attends his class and no other, the sum of fifty dollars *per annum;* from each student attending his class and only one other, thirty dollars *per annum;* and from all other students attending his class, twenty-five dollars *per annum*, nor shall these fees be diminished for the space of five years during which this covenant shall last, without the consent of the said Dunglison; and the said Dunglison on his part covenants with the said Gilmer, that he will proceed with as

---

[14] First published in 1814, it continued for a number of years under various titles.

[15] *The Medical Intelligencer* ran to four volumes, 1820 to 1823.

[16] Founded 1782; located in White Chapel.

[17] Francis Walker Gilmer (1790–1826), childhood friend of Thomas Jefferson's grandchildren, practiced law in Virginia.

[18] Gilmer was instructed to seek men possessed of a "due degree of science, talent for instruction and correct habits and morals" and reported to Jefferson that Dr. Dunglison was "a very intelligent and laborious gentleman and a writer of considerable eminence in various medical and anatomical subjects." Dunglison was only twenty-seven years of age when he was engaged to teach at the University of Virginia. Perhaps Jefferson still had in mind "Hints Concerning Public Education," which Joseph Priestley sent him in 1800, advising that young men be engaged for the professorships at the University of Virginia. (See Edgar Fahs Smith, *Priestley in America 1794–1804*, pp. 117–122, Phila. 1920.)

[19] George Birkbeck (1776–1841), M.D., Edinburgh, active lecturer at the London Mechanics' Institutions founded in 1825, and at other similar institutions throughout Britain (Thomas Kelly, *George Birkbeck: Pioneer of Adult Education*, Liverpool, 1957).

[20] Dunglison was elected a member in 1837.

[21] Thomas Callaway (1791–1848), apprentice to Sir Astley Paston Cooper, became one of the busiest surgeons of London. When Dunglison was anxious to equip his anatomical theatre at the University of Virginia in 1825, Calloway purchased a good deal of the equipment for him in London (Jefferson Papers, University of Virginia, July 28, 1825).

[21a] John Leadam (1779–1845).

[22] Harriette Leadam (born in 1802) was considered a very nice looking girl at the time. There used to be a painted miniature of her with the stiff dark curls worn in those days. She was very popular, and it is said that at the time Dunglison was seeking her favor, a clergyman, Thomas Lee (or Lea), was also paying her attention. It happened that both suitors arrived on the same day to present their request to her father. One was ushered into the drawing room and the other into the consulting room, and Mr. Leadam had to go backwards and forwards between them and his daughter to ascertain their respective positions and his daughter's decision. It was a trying day (J. Alden Tifft, personal communication).

little delay as possible to the University of Virginia, and will there teach to the best of his ability, and with due diligence, Anatomy, Surgery, the History of the Progress and Theories of Medicine, Physiology, Materia Medica and Pharmacy; that he will suffer no waste to be committed in his tenement; that he will maintain the internal of his pavilion, the doors, windows and locks external, in as good repair and condition as he shall have received them; that he will conform to the rules and regulations of the visitors, etc., and to secure the faithful performance of every of the above covenants, the said Dunglison binds himself to the said rector and visitors of the University in the sum of five thousand dollars, money of the United States; and the said Gilmer binds the Rector and visitors of the University of Virginia aforesaid to the said Dunglison in the same penalty. In witness of all of which, the said parties have hereto set their hands and affixed their seals, on the day and year first above written.

L. S.

Mr. Gilmer was a man of unquestioned ability and well educated. His literary productions are highly creditable to him. He was appointed Law Professor at the University of Virginia, but did not live to undertake its duties. In his speech, he had much of the peculiarity of tone that distinguishes many of the Virginians, which struck unpleasantly upon the ear in the first instance, but this soon wore off, and he was popular with those with whom he became acquainted in England.

MARRIAGE

After I had entered into a covenant with him I immediately set about making my preparations for the voyage; and the most important preliminary step was to become married to my excellent wife, Harriette Leadam, daughter of John Leadam, Esquire, a medical practitioner of Tooley Street, Southwark, which joyful event took place on the 5th of October, 1824. I write this at the expiration of nearly twenty-eight years and can declare, that the union has been one of unmixed happiness to us both, and that she has ever proved to be one of the most amiable and affectionate of women. We were bound together at the church of St. Oliver, Tooley Street.

SOCIETY MEMBERSHIP

During my residence in London, I was an active member of two of the medical societies there—the old *London Medical Society*,[23] which held its meetings in Bolt Court, Fleet Street, and the *Hunterian Society*,[24] with which I was connected from its very foundation. Of the former, I was Secretary for foreign Correspondence and a member of the council; and of the latter a Member of the Council. Both these Societies passed

[23] The Medical Society of London was founded in 1773 by John Coakley Lettsom.
[24] The Hunterian Society, still flourishing, was founded in 1819, largely through the efforts of William Cooke, who played an important part in the early fortunes of the Society and wrote a biography of its first president, Sir William Blizzard (Sir D'Arcy Power, *British Medical Societies*, pp. 78–90, London, 1939).

very complimentary resolutions in my favour on my leaving England. In answer to my letter of resignation in the former, I received the following letter from Mr. James Field, the Registrar:

Medical Society House
Bolt Court, Fleet Street
Oct. 5, 1824.

Sir:
I beg leave to inform you, that the letter, which you did me the honour to address to me and in which you tendered to the Council of the Medical Society your resignation of the office of Secretary for foreign Correspondence, I laid before the Council at their meeting at the opening of the Session.
I am instructed to inform you, that the resignation was accepted by the Council.
I am directed also to lay before you the following unanimous Resolution of the Council.
RESOLVED, that the *best thanks* of the Council be presented to Dr. Dunglison for the zeal he evinced for the Society's welfare during the short period he held the office of Foreign Secretary, and for the very great benefit the society has experienced from the Doctor's official exertions. The members of the Council, while they cannot but deeply lament the resignation of Dr. Dunglison, sincerely hope, that the circumstances, which occasioned it, may prove advantageous to his interests and welfare.
I am further directed to acquaint you that your request to be ranked as a *Corresponding Member* during your absence was most willingly complied with by the council.
I have the honor to be, Sir,

Your obedient servant,
James Field, Registrar
Robley Dunglison, M.D.
Professor of Medicine
in the Univ. of Va. of
America, etc., etc.

The Hunterian Society passed resolutions equally complimentary to me; and suspended one of their by-laws in order to declare me at once a corresponding member.

On my visit to England in 1854, my sister gave me the following letter from Dr. W. Cooke of Trinity Square, London, which had been sent to Miss Leadam who, sometime after placed it in my sister's hand. It was never sent to me in America.

My dear Friend:
You will be aware that the dinner of the Hunterian Society occurred on the 10th Inst. It is probable, that amidst the new and important duties you have undertaken, the precise time of our assembling did not occur to your recollection but I can assure you, that you were borne in recollection by us. Having had the pleasure of your company on similar occasions, we could not fail to observe a deficiency in our efficient members, and in our good society. We had, too, a lively remembrance how much you have contributed to the interest of our professional meetings by the extent of scientific and practical information you have acquired—and which you were so ready to diffuse. I may truly say, that we felt that we have lost one of our most useful and most honorable supporters. I think you are aware that we made you "a corresponding member" and may we hope for the great pleasure of receiving some communications from you?

At our anniversary Festival your health was drunk with the greatest cordiality, and the warmest wishes were expressed for your health and prosperity—and this I was charged to communicate. In doing so, I have fulfilled my official duty. Allow me, then, individually to say, that I most sincerely hope, that you will be amply rewarded for the great sacrifices you have made in quitting England. May God spare your health and that of Mrs. D. May you prosper abundantly where you are, and where, I doubt not, you will be very useful in the diffusion of medical science, and may you, at length, be safely reinstated in England.

With best compliments to Mrs. Dunglison,
I remain, my dear Sir
Yours with the sincerest esteem,
W. Cooke.

39 Trinity Square
February 25th, 1825

Dr.—then Mr.—Cooke was secretary to the Hunterian Society.

As soon as this letter was put in my possession, I wrote from Keswick to Dr. Cooke mentioning that I had then seen it for the first time; and informing him, that, a year or two ago, I had written to the Hunterian Society, asking them to tell me what books of mine they had, and offering to forward to them those they had not, if they would tell me by what medium. To this letter I received the following reply.

39 Trinity Square
August 15, 1854

Dear Dr. Dunglison,

I do not know whether you are known at *Keswick*, or are only a passing visitor; but I was so glad to hear from you that I try to say so, although my note may not reach you for another 28 years. I heard, some time ago, that you were in England, and I quite hoped for the pleasure of a call. You find, that I have not been changeable as to residence. All has gone on steadily and comfortably with us, except, that Mrs. Cooke's health has greatly failed within the last two or three years. I am glad you have at length received the note of the Hunterian Society. There must have been delinquency somewhere. But whilst I suggest delinquency as to our letter to you, some blame must rest on us, as to your communication two years ago. I have long ceased to be secretary. In 1839, the Society presented me with a Salver, having on it a kind inscription, which, on some occasion, when you were partaking of our hospitality, I hoped to have shewn you. Since relinquishing the secretaryship, I have held the bag as treasurer. We should most gladly receive any of your publications, addressed to me or to the Society's rooms, Bloomfield St., Finsbury Circus. My successors in the office of Secretary have often changed, and I cannot tell whether, or not, your letter was received. From some of your friends you will no doubt have heard of the Society's continued prosperity. Since you resided in this neighborhood, it has become greatly altered for the worse, and were it not, that I never depended very much on the immediate residents, I must have sought another locality. In many quarters, the lapse of years since you left has occasioned great changes. Enjoying, as I generally do, a good state of health, I am too ready to lose sight of the years I have attained—and you, dear sir, whatever may be the energy you still enjoy, are creeping onward towards that which I have attained—"the evening of life"—may it be to each of us a bright and happy evening—and when we come to the close of life may it be in full preparation for that better life where there is perfect holiness and unalloyed felicity. Believe me, dear friend.

Yours very truly
Wm. Cooke [25]

About the same time, in 1822 I believe, I became a Member of the Society of Apothecaries having been previously a Licentiate, and a Member of the Council of the Associated Apothecaries and Surgeon-Apothecaries of England and Wales; and was admitted in 1823 and 1824 into the following learned societies of France: The *Société Linnéenne;* The *Société de Médecine;* The *Société de Pharmacie* of Paris; The *Académie Royale* of Marseilles; The *Société Académique de Médecine* of the same city; and the *Société Royale des Sciences* etc. of Nancy; and in 1824, not long before I left England, I was appointed a member of the Medical Committee of the *Royal Humane Society* of London.[26]

EARLY LITERARY CONTRIBUTIONS [26a]

It was in the year 1817, whilst a student, that I first began to write for the press; and one of the earliest articles was on the effect produced on vision by dilatation of the pupil which, however, was not published at that time. I had prepared some extract of Belladonna [27] from the juice of the fresh plant, and applied a small portion of it to one of my eyes; the effect was dilation of the pupil to such an extent that the iris was almost invisible. From the time the pupil attained to three times its natural dimensions, objects presented to this eye, with the other closed, were seen as through a cloud; and as it proceeded to the point of extreme dilatation this effect gradually increased, so that minute and near objects, as letter press etc. could not be at all distinguished. By means of a double convex lens, the focus of this eye was found to be at twice the distance of that of the sound eye; the iris, however, dilated upon the sudden admission of light; and although the pupil approached, by almost imperceptible degrees, for six days, to its natural size, at the end of that time, it was dilated to twice the size of the other, and in proportion as the contraction took place, the sight became more distinct, and the focus nearer the natural. In the open air, all objects, except those near, were distinctly seen but immediately on entering a room all was again enveloped in mist.

[25] William Cooke (1785–1873), orator (1839) and President (1841) of the Hunterian Society, edited an abridgment of Morgagni's *Seats and Causes of Diseases.*
[26] The Royal Humane Society for the "Recovery of Persons who are supposed Dead of Drowning," founded in 1744, was another organization Lettsom helped to establish.
[26a] The next few pages, which actually preceded the narrative part of the text, have been inserted here by the editor.
[27] Belladonna—so-called because the Spanish ladies made use of the plant to dilate the pupils of their brillant black eyes—was reported as a mydriatic in 1776 by Daries.

These experiments led me to infer, that the iris might be more concerned in the adaptation of the eye to vision at different distances within the limits of distinct vision than was generally admitted and they were cited by Dr. Fleming [28] as confirming this position (*The Philosophy of Zoology etc.* by John Fleming, 1:187, Edinb. 1822). They were published in *Thomson's Annals of Philosophy*, Vol. X, p. 432, for December 1820; [29] and I find, from Callisen, (*Medicinisches Schriftsteller-Lexicon*, 5er Band, S. 415) were copied into the *Journal Complémentaire du Dictionaire des Sciences Médicales*, T. 6, Cah. 23, Mai, 1820; [30] and into Gräfe und Walther, *Journ. der Chir.*, Bd. 2, 1821 and Bd. 3, 1822. [31] The work of Callisen, [32] by the way, is a monument of useful but unremunerative labour. It professes to give a bibliographical account of the then living medical writers (up to 1835); and I was surprised to find references in it, in my own case, to communications which I had forgotten. It is doubtless occasionally erroneous, as where it ascribes to me the authorship of the Annual Announcement of Jefferson Medical College for the Session of 1837-8, for which there was certainly no authority.

About the same time I sent a communication to the *Monthly Magazine*, then edited by Sir Richard Phillips, [33] on the subject of a floating island which makes its appearance from time to time in the lake of Derwentwater, and which—it appeared to me—was a loose stratum of soil buoyed up by the evolution of carburetted hydrogen gas beneath, and remaining at the surface until the gas had escaped, when it sank—to reappear subsequently. For a year or two after this, I was a regular correspondent of the *Monthly* and wrote several communications amongst which were one on the phenomena produced by the "wind of a ball" and several articles under the head "Collectanea Dietetica." All these were signed *Philos;* the real author was not known to Sir Richard; and it afforded me no little encouragement, when I was told by a friend, who was aware that I wrote them, that Sir Richard considered they were the production of a practiced observer and writer.

During my sojourn in London I was occupied in many literary undertakings, but none of them of any great moment. In the year 1822, I translated the Memoir of

Baron Larrey [34] on the use of the moxa, [35] which is contained in his *Recueil de Mémoires de Chirurgie*, Paris 1821, and prefixed to it a long dissertation on the history and properties of various forms of Moxibustion. The work was favorably received by the public, but it did not pay expenses. It was published altogether at my own cost and risk, the bookseller having no direct interest in it. This, according to my experience, is the least advisable plan for an author; and as a general rule it may be said, that an author, who is his own publisher, has a fool for a publisher. It is utterly impossible for him to distribute the book properly; and the bookseller, who has only a percentage on the sales, does not feel the same amount of interest as if he were wholly or in part owner. It is the only book on which I have lost money.

In January 1822, I became associated with my friend Dr. James Copland in the editorship of the *London Medical Repository*, at the time one of the most respectable medical journals of London. I had become intimately acquainted with Dr. Copland and his, to me, amiable wife, both of whom exhibited the most friendly feelings towards me. At all times, Dr. Copland was desirous of bringing me forward, and I owe much to his example and kind encouragement. His works, which are well known in both hemispheres, especially his *Dictionary of Practical Medicine*, [36] not yet completed, and which has been reprinted in this country under the editorship of Dr. C. A. Lee, of New York, is a monument of his skill and erudition in his profession, not exceeded by the work of any single author in modern periods.

I did not contribute to the *Medical Repository* more than one or two articles in the department of original communications. I attended more especially to that assigned to reviews and bibliographical notices. One of these original articles was a case of Arachnitis Cerebelli which appeared to me at the time "to be strongly corroborative of the opinion advanced by M. Serres, in the *Journal de Physiologie* of Magendie for April 1822,—that erection of the penis and irritation of the genital organs occurring during apoplectic affections, are diagnostic symptoms of derangement of the cerebellum," and it has been quoted by Dr. Copland in a recent work (*On the Causes, Nature and Treatment of Palsy and Apoplexy*, Amer. Edit. p. 240, Phila.,

---

[28] The Reverend Dr. John Fleming (1785-1857), F.R.S., was author of several work on physiology and natural history.

[29] Critically analyzed in the *London Medical and Physical Journal* 45: XXIII-XXVII, 1821.

[30] Pp. 285-286.

[31] *Journal der Chirurgie und Augenheilkunde* 2: 672-673; 3: 299-306.

[32] Adolph Carl Peter Callisen (1786-1866), professor of surgery in Copenhagen, is known particularly for his *Medizinisches Schriftsteller-Lexicon*, 33 v., Copenhagen, 1830-1845.

[33] Sir Richard Phillips (1767-1840), controversial author, bookseller and publisher whose *Monthly Magazine*, started in 1796, was carried on until 1843 (*Dict. Nat. Biog.* 45: 210, 1896).

[34] Jean-Dominique Larrey (1776-1842), Surgeon-General of the Grande Armée and closely attached to Napoleon, made his best contributions to military surgery.

[35] Moxa or Moxibustion is an ancient Chinese medical treatment in which combustible cones of artemisia moxa or common mugwort are applied to the skin and ignited, thus raising a blister.

[36] James Copland (1791-1870), L.R.C.P. of London, member of the London Medical Society, began the *Dictionary of Practical Medicine* in 1830 (Thomas Joseph Pettigrew, *Medical Portrait Gallery* 1, biographical sketch no. 11, London, 1838).

1850) to be corroborative of the views of Serres [37] and of Gall [38] and Spurzheim.[39]

Of the "Repository" I continued to be joint editor until October 1824, the month in which I left England; and in the number for that month appeared the following kind notice from the pen of my friend and colleague.

It is impossible for us to announce the departure of our colleague, Dr. Dunglison, to occupy a highly honorable professional appointment in a foreign country, without expressing great regret at the loss we shall experience in the privation of that assistance we have been accustomed to receive in our editorial labours. At the same time we gratify ourselves in stating our confidence, that his industry and talents will be more advantageously and beneficially exercised for the promotion of medical science, and that we shall have, from time to time, opportunities of seeing the result. We part with him, on every account, reluctantly; and it would be a commonplace unworthy of the occasion, did we confine our public adieu to a mere expression of regret, or a simple record of good wishes. He is now Professor of the Institutes and Practice of Medicine in the University of Virginia, a university recently endowed by the American Congress through the influence of Mr. Jefferson, the Ex-President of the United States:—and we are quite sure he will be well received among our Transatlantic brethren.

It need scarcely be said, that the mode of endowment of the University of Virginia stated by Dr. Copland was erroneous. It is a state institution.

Owing to my friend Mr. Charles Thomas Haden being compelled, from the state of his health, to leave London, I took charge, anonymously, of the London *Medical Intelligencer* and edited the fourth and last volume of it. The plan of the journal was a good one, and the preceding volumes had been edited by Dr. Armstrong,[40] Mr. Alcock [41] and Mr. Haden. It professed to give a brief analysis of the papers contained in other medical journals, published both in England and in foreign countries, with critical remarks, and it certainly contained a large amount of information. Dr. Copland was aware of the individual who had editorial charge of it; but none of the editors of the other medical journals were, for a time at least—a circumstance,

which placed me, occasionally, in an awkward position as it was the custom with the different editors of the leading medical journals to meet once a month at the White Horse Cellar, Piccadilly, for the purpose of good fellowship. I was there by virtue of my being one of the editors of the Medical Repository. The *Medical and Physical Journal*,[42] often called the "Yellow Journal" from the colour of its cover [42a] was represented by Dr. Mcleod [43] and Mr. Bacot [44]—the then editors. The *Medico-Chirurgical Review*, often called Johnson's Journal, was represented by its editor Dr. James Johnson,[45] and *The Quarterly Journal of Foreign Medicine and Surgery* by Mr. Rennie,[46] and last, and not least, my colleague, Dr. Copland, were generally present. At these social reunions the animadversions of the *Medical Intelligencer* were not unfrequently the subject of comment, and I had to bear the brunt of criticism of a varied character, without giving any outward visible signs of being more affected by it than the others. During a portion of the time, indeed, that I attended these meetings, a somewhat acrimonious controversy was carried on between the editor of the *Medico-Chirurgical Review*, and him of the *Intelligencer*, on the subject of removing calculi from the bladder, without cutting, as described by Prosper Alpinus.[47]

The *Medical Intelligencer* died from pecuniary inanition with the number for November 1823; the publishers not being able to make their payments to the editor, his labours terminated *ex necessitate rei;* and it was never resuscitated.

The infirm state of Mr. Haden's health also gave occasion to my editing the second edition of Magendie's *Formulary for the Preparation and Mode of Employing Several New Remedies.*[48] The first edition was trans-

---

[37] Antoine Étienne Renaud Augustin Serres (1786–1868), French physiologist for whom Serres' gland and Serres' angle are named.

[38] Franz Joseph Gall (1757–1828) originated the theory of organology, believing that brain centers, or "organs," presided over emotional and moral traits and could be measured by various protuberances on the surface of the skull.

[39] Spurzheim was not actually mentioned in this connection by Copland. Johann Caspar Spurzheim (1776–1832), disciple of Gall, introduced their joint researches in a treatise on phrenology which became an object of derision among scientific men.

[40] John Armstrong (1784–1829), famous for his published works on fevers, was a friend of Dr. Charles T. Haden and taught at the medical schools on Webb Street and on Little Dean Street in London (Francis Booth, *Memoir of the Life and Medical Opinions of John Armstrong, M.D.*, 1833).

[41] Thomas A. Alcock (1784–1833), biographer of Dunglison's friend Haden, was a surgeon of London trained at the Sunderland Dispensary and at Brooke's Anatomical School and a frequent contributor to the medical press.

[42] This journal, later renamed the *London Medical and Physical Journal,* ran from 1799 to 1833.

[42a] The *American Journal of the Medical Sciences* has also long been so dubbed for the same reason.

[43] Roderick Macleod (–1852), physician to the Westminster General Dispensary, lectured at the Medical School in Great Windmill Street, London.

[44] John Bacot (1781–1870), friend of Sir Benjamin Brodie, was appointed to St. George's and St. James' Dispensary and the Examination Commission of the Apothecaries Society.

[45] James Johnson (1777–1845) or, rather, Johnston, originated the *Medico-chirurgical Review* at his own risk and expense, and no quarterly medical journal up to that time ever attained a larger circulation or exerted wider influence (Necrology, *Medical Examiner*, n.s., 2: 76, 1846).

[46] James Rennie, editor of the *Quarterly Journal of Foreign and British Medicine*, was lecturer on Chemistry, Natural History and Philosophy and author of a number of pharmacopeal works (*London Medical and Physical Journal* 57: 78, 1827).

[47] Prospero Alpino (1553–1617), "the last of the Methodists," Professor of Botany at Padua, discusses removal of calculi from the bladder in *De Medicina Aegyptiorum*, Venice, 1591, Lib. 3, Cap. 14, "De lapidis è vesica extractione, etc."

[48] François Magendie (1783–1855), best known for his work in physiology, in 1821 published his *Formulaire pour l'emploi et la préparation de plusieurs nouveaux medicamens* which became very popular. The third edition, 1822, translated into English by others, both in England and America, was used by Dunglison

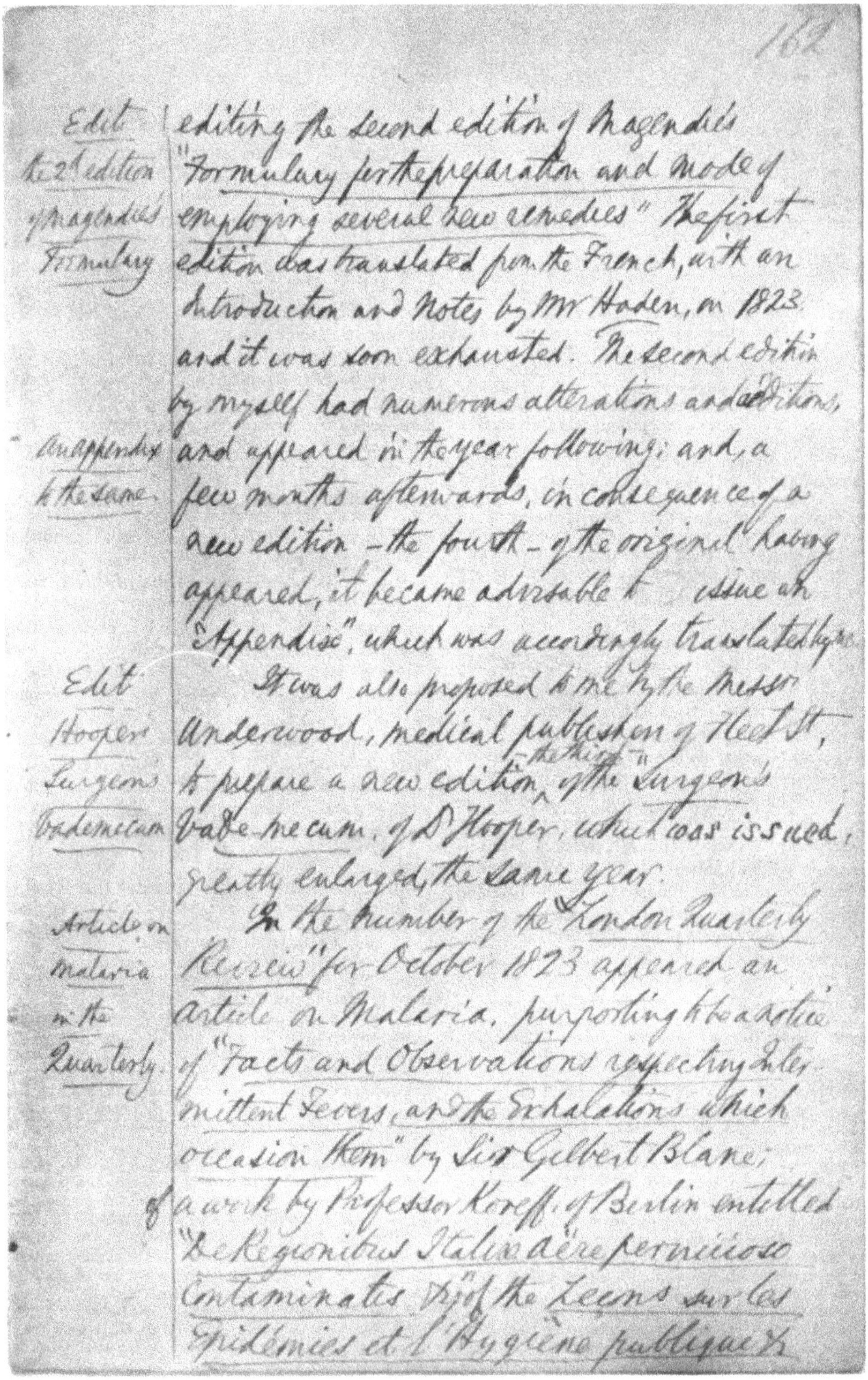

Edit the 2d edition of Magendie's Formulary

editing the Second edition of Magendie's "Formulary for the preparation and mode of employing several new remedies" The first edition was translated from the French, with an Introduction and Notes by Mr Haden, in 1823. and it was soon exhausted. The second edition by myself had numerous alterations and additions,

An appendix to the same

and appeared in the year following; and, a few months afterwards, in consequence of a new edition – the fourth – of the original having appeared, it became advisable to issue an "Appendix", which was accordingly translated by...

Edit Hooper's Surgeon's Vademecum

It was also proposed to me by the Messrs Underwood, medical publishers of Fleet St, to prepare a new edition "the third" of the Surgeon's Vade-mecum, of Dr Hooper, which was issued, greatly enlarged, the same year.

Article on malaria in the Quarterly

In the number of the London Quarterly Review" for October 1823 appeared an Article on Malaria, purporting to be a notice of "Facts and Observations respecting Intermittent Fevers, and the Exhalations which occasion them" by Sir Gilbert Blane; a work by Professor Koreff of Berlin entitled "De Regionibus Italiae aëre pernicioso Contaminatis &jc the Leçons sur les Epidémies et l'Hygiène publique &

FIG. 4. A page from the *Ana*.

14

lated from the French, with an introduction and notes by Mr. Haden, in 1823; and it was soon exhausted. The second edition by myself had numerous alterations and additions, and appeared in the year following; and, a few months afterwards, in consequence of a new edition—the fourth—of the original having appeared, it became advisable to issue an "Appendix," which was accordingly translated by me.

It was also proposed to me by the Messrs. Underwood, medical publishers of Fleet Street, to prepare a new edition—the third—of the *Surgeon's Vade-Mecum* of Dr. Hooper,[49] which was issued, greatly enlarged, the same year.

In the number of the *London Quarterly Review* for October 1823 appeared an article on Malaria, purporting to be a notice of "Facts and Observations respecting Intermittent Fevers, and the Exhalations which occasion them" by Sir Gilbert Blane;[50] of a work by Prof. Koreff,[51] of Berlin entitled *de Regionibus Italiae Aere Pernicioso Contaminatis etc.;* of the *Leçons sur les Epidémies et l'Hygiène Publique etc.* of Prof. Fodéré[52] of Strasbourg; and of the *Recherches Historiques, Chimiques et Médicales sur l'air Marécageux* of Professor Julia.[53]

The article was written by me for the purpose of replying to certain strange statements contained in the seventy-second number of the *Edinburgh Review*, which was ascribed to Dr. McCulloch,[54] as well as to discuss the subject of malaria in its varied relations, and the comparative prevalence of that fitful pest in Rome and its vicinity at different periods of history. The views, embraced in it, have been confirmed by the diversified and ample opportunities for observation I have since had.

Amongst the visionary assertions contained in that strange article in the "Edinburgh" we find the following:

It is commonly held, that it (the Malaria) cannot travel far from the place of its production; a fallacy often leading to very pernicious consequences. But the east wind has the power of transporting it to considerable distances; and we have little doubt ourselves, that whenever it occurs in

---

to prepare the second edition of Haden's translation; it was promptly reprinted in Philadelphia (November, 1824), New York, New Haven and elsewhere.

[49] Robert Hooper, M.D. (1772–1835), wrote also a *Physician's Vade Mecum* as well as a dictionary.

[50] Sir Gilbert Blane, M.D. (1747–1834), physician to the British fleet in North America during the Revolutionary War, introduced lemon juice in the sailors' diet to eradicate scurvy.

[51] Johann Ferdinand Koreff (1783–1851). This article appeared in Berlin in 1817 and in Rust's *Magazin* in 1821.

[52] François Emanuel Fodéré (1764–1835), Nestor of Public Hygiene in France, published these lectures in four volumes, 1822 to 1824.

[53] Jean Sébastien Eugène Julia de Fontanelle (1790–1842) published this prize-winning essay in 1823.

[54] John *MacCulloch*, M.D. (1773–1835), a good geologist, made malaria his special medical study. His book on *Malaria* (1827) was also reviewed in the American *Quarterly Review* 4: 286, 1828.

this city (Edinburgh) where it is now rare, the *poison is transported from Holland!*

It was farther affirmed that the malaria was generated in sufficient abundance in St. James Park, where there is a canal, to affect injuriously the health of those residing in the neighbourhood, and it was attempted, by means of an animated "miasmometer,"—an officer who had suffered at Walcheren,[55]—to mark with something like mathematical accuracy the distance from the canal to which this fancied malaria passed up the different streets of the neighbourhood. I was unknown to Mr. William Gifford, the then editor of the Review, and sent the article to him without any introduction. Not hearing from him, after some time I wrote to him, when he replied that he was afraid the Ms. of the article had been burnt up with other papers, which had taken fire in his study. Soon afterwards, however, he found it, and sent it to press. In a note he informed me, that the article in the *Edinburgh* had excited so much alarm, that property holders in the vicinity of the park, had found it difficult to rent their houses. Mr. Gifford is the "very old friend"—as he designated himself—alluded to in the following extract from the article in the *Quarterly*.

We should not have considered the preceding reveries worthy of notice, were it not, that the minute and categorical manner in which the progress of this 'airy phantom' has been described might—

—"draw on some better natures
To run in that vile line"—

and induce a belief, that such a focus of disease is really in existence. In spite, however, of all these mischievous assertions, principally founded on the information obtained from this animated *miasmometer,* who seems to have taken a pleasure in administering pretty largely to the credulity of the Reviewer, we can affirm, from an intimate acquaintance with the medical topography not only of the western but of the eastern districts referred to, that in the whole line of march, which has been ascribed to it, and even in situations most in proximity to the western focus of this fitful pest, there is no sensible evidence of the presence of such a deleterious agent,—that these very situations are as healthy as others more remote, and that some of them are even remarkable for their salubrity, and the longevity of the inhabitants; thus, in one street in the immediate neighbourhood of St. James' Park, situated to the westward of this very canal

"from whose humid soil and wat'ry reign
Eternal vapours rise"—

and, consequently, exposed to the pestiferous exhalations, were any such in existence, during the domination of an eastern blast, they seem to take a delight in falsifying the visionary assertions contained in that strange article. One gentleman, who had lived for more than half a century in the street, died there lately at the advanced age of 82. There are, at present, several septuagenarians in it, and a very old friend of our own has resided there for the last five and twenty years, labouring under a pulmonary complaint nearly coeval with his existence. To anyone ac-

---

[55] In 1809 an army of 40,000 was swept off by malaria in a few short weeks at Walcheren, an island in Holland.

quainted with these districts where malaria is most prevalent, it is needless to state, that the canal in St. James' Park is not the situation which gives rise to it in any "abundance": the water can never be said to be stagnant: and consequently even during the summer heats, except under circumstances of great neglect, no decomposition can, in our opinion, ever take place to a sufficient extent for the production of epidemic disease.

Experience, of no trifling extent, within the last thirty years—as I have remarked—has but tended to confirm the views I then entertained on the subject of malaria. In the *Virginia Literary Museum,* published at the University of Virginia, and in the number for July the 1st, 1829, I wrote an article on the causes of endemic disease, suggested by the appearance of typhoid fever there, in which I maintained the view that malaria, contrary to the then almost universally received opinion, does not originate in vegetable or animal putrefaction, singly or combined; and in my *Elements of Hygiene, Human Health,* and *Practice of Medicine,* the same views are inculcated. They are now embraced by some—many—of the best minds, and it is not easy to understand how the converse, and especially that malaria is the product of vegetable decomposition, could ever have been countenanced. Intelligent observers not of the medical profession have been dissatisfied with the evidence brought forward in its favour; and it is not many years since Mr. Nicholas Biddle [56]—a man of undoubted talent—in an agricultural address, which he delivered before the Agricultural Society of the County of Philadelphia, took strong ground against what he jocosely characterized as the last universally admitted medical truth; properly maintaining, that if malaria were owing to vegetable decomposition, every farm yard ought to present evidence of it, as both vegetable and animal decomposition are perpetually going on in periods of the year when malarious fevers prevail. As soon as I saw the published address, I wrote to Mr. Biddle informing him of the position I had taken on the subject in the article in the *London Quarterly* as long ago as the year 1823; and that I was at the time publishing and annually teaching the same doctrine to my class; to which he returned a courteous reply expressive of his satisfaction, that his views should accord with mine.

The same doubts had impressed Mr. Louis McLane,[57] from the results of his own observation on Bohemia River, a highly malarious locality in which he resided. We were then at Cape May, and as I was about to return to Philadelphia, I proposed to send him a copy of my *Elements of Hygiene* in which I had discussed the

subject at length. I subsequently sent him a copy of the work as a present, but, by some accident, it was forwarded to Boston, and Dr. Oliver Wendell Holmes sometime afterwards informed me that he had received it, and that it was evidently a presentation copy to some friend of mine as a memento of the pleasant time we had spent together at Cape May. I immediately wrote to Mr. McLane informing him of the mistake and that I had directed another copy to be forwarded to him, to which he sent the following reply.

Baltimore
May 18

Dear Sir:

On my return to the city after a short visit to the West, I had the pleasure to receive your kind note of the 10th inst.

The volume, to which it refers, has not yet come to hand; and I should regret if the former mistake at Boston had exposed me for a moment to the suspicion of being insensible to such a mark of your attention.

Soon after my return from Cape May, I procured a copy of your work, and read it with delight and instruction.

I have always regarded the pleasure of making your acquaintance as the chief advantage of my visit on that occasion, and will be proud and happy to receive the volume you have obligingly directed to be sent to me, as an evidence that our intercourse has not been forgotten by you.

I am,
Dear Sir,
Very faithfully yours,
Louis McLane

To R. Dunglison Esq. etc. etc.

It may be worthy, by the way, of mention, that in conversing with Mr. Thomas Biddle [58] of Philadelphia, on the nature of malarious emanations and soils, he spoke to me of certain opinions contained in an article in the *London Quarterly* on those subjects, published a quarter of a century before, but with no idea, that he was addressing the author of the article.[59] How little did I dream when it was written, that my lot would ever be cast on this side of the Atlantic!

In the year 1824, I wrote a review of *The Meteorological Essays and Observations* of Professor Daniel for the *Quarterly Journal of Science, Literature and the Arts,* edited at the Royal Institution of Great Britain; and Mr. Underwood, the bookseller, informed me, that Professor Daniel [60] had expressed a wish that the author of the article on "Malaria" in the *Quarterly* should prepare the review of his work. I also occasionally wrote for the *Eclectic Review,* and for the *Universal Review,* which was commenced in 1824, under the competent editorship of the Reverend Dr.

[56] Nicholas Biddle (1786–1844), Philadelphia antagonist of Andrew Jackson, after his retirement delivered two addresses before the Agricultural Society of Philadelphia (Thomas P. Govan, *Nicholas Biddle, Nationalist and Public Banker,* Phila., 1959).

[57] Louis McLane (1786–1857), President Jackson's first minister to England, was also Secretary of the Treasury and Secretary of State.

[58] Thomas Biddle, Philadelphia broker.

[59] Dunglison was an early advocate of the use of quinine instead of calomel in the treatment of malaria (Northern Schools and Southern Medicine, *The Medical Examiner,* n.s., 1: 140, 1845).

[60] John Frederick Daniell (1790–1845), F.R.S., inventor of the hygrometer, first published his *Meteorological Essays* in 1823.

Croly,[61] afterwards appointed, by Lord Brougham,[62] I believe, to the Rectory of St. Stephens, Walbrook. The Messrs. Whitaker, of Ave Maria Lane, were the publishers, and whilst Mr. Whitaker was Sheriff of London I had an opportunity of meeting Dr. Croly, and other literary persons, at his table. When Dr. Croly took leave of me on my setting out for America he facetiously recommended me to take care, that I was not eaten up by the bull frogs!

In the *Medical Portrait Gallery* of Mr. Pettigrew, in the biographical memoir of Dr. Copland, it is stated—

in 1825, Dr. Copland projected an *Encyclopaediac Dictionary of the Medical Sciences,* and drew up a prospectus of the undertaking. In this he was to have been assisted by his friend Dr. Dunglison, now of the United States, and by the late Dr. Gordon Smith; and the work was actually agreed upon by Messrs. Underwood, medical publishers, when the panic of this period caused them to relinquish it.

The work was in reality projected as early as 1823 or 1824; and in the first instance it was proposed, that Mr. Henry Earle,[63] of Bartholomews Hospital, should be associated as an Editor. It was afterwards determined, however, that the department of surgery should be omitted, and that the work should be a *Dictionary of Medicine,* essentially resembling the *Surgical Dictionary* of Mr. Cooper.[64] In October 1824, I left England, and heard no more of the Dictionary, until I saw it announced as forthcoming from the pen of Dr. Copland alone. The project, probably, also suggested the *Cyclopaedia of Practical Medicine* of Drs. Forbes, Tweedie and Conolly, the American edition of which was edited by me.[65]

In the *London Medical Repository* for November, 1823,[66] there was announced as preparing for publication *A Treatise on Organic Chemistry, containing the Analyses of Animal and Vegetable Substances,* founded on the work of Professor Gmelin[67] on the same subject, by Robley Dunglison, member of several learned Societies, Foreign and Domestic, and one of

the editors of the *Medical Repository,* but altho I made considerable progress in the work, I abandoned it.

The last work I wrote before leaving England was *Commentaries on the Diseases of the Stomach and Bowels of Children,*[68] which was published by Mr. Whitaker. It was intended as a commencement of a treatise or series of treatises on the diseases of children. In the preface I remarked: "at a future period, it is the intention of the author to resume the consideration of some other of those diseases, which are incident to children; not under the arrogant expectation of being able to communicate much important information from his own stores, but in accordance with the motto at the head of these prefatory observations: *Facem exiguam accendere, qua alii egregiis animi dotibus ornati opus imperfectum limato suo ingenio perpoliant.*[69]

The preface was written, or rather dated, on the 15th of September, a short time before I left England; and the work had to be hurried through the press, so that neither in its scope nor execution was it what I could have wished. The subjects treated of, besides the prefatory observations, which comprise an account of the "physiognomical system" of M. Jadelot[70] as applicable to the diagnosis of the diseases of childhood, were intestinal worms; constipation; acidity, flatulence and colic; diarrhoea; procidentia ani; vomiting; cholera; aphthae; inflammation of the stomach; inflammation of the intestines and intussusception.

In regard to the physiognomical system of Jadelot, of which but an imperfect account has been printed, there is, doubtless, much that rests on too slender a support of observation; but, as I remarked in the preface of the *Commentaries,*

an attentive examination of the anatomical expression of an infant's countenance will materially assist us in the diagnosis of the seat of a disease under which it may be labouring,—for example, whether it be in the head, or lower belly, no one who has paid any attention to the subject can doubt. There is a marked difference in the expression of the countenance which indicates the presence of violent pain in these two situations, even in the adult; lesser degrees of it are of course disregarded; and it is only in severe affections, that physiognomy can be inservient to diagnosis; but in the infant, which readily gives expression to any pain or uneasiness which it may experience, the countenance is an excellent medium of discrimination, and will frequently indicate, at the first glance, the seat of the derangement. The expression of the countenance, when suffering under pain, should, consequently, be always attended to; or, to use the words of an individual, whose

[61] George Croly (1780–1860), Irish clergyman and author.

[62] Henry Peter Brougham, Baron (1788–1868), literary and public figure, took a lively interest in Francis Gilmer's mission to England in search of professors for the University of Virginia.

[63] Henry E. Earle (1789–1838) became House Surgeon to St. Bartholomew's Hospital in 1808 and succeeded Abernethy there as Surgeon in 1827.

[64] Samuel Cooper (1780–1848), *A Dictionary of Practical Surgery,* first published in 1809, went through seven editions during his lifetime and was translated into French, German, and Italian.

[65] Sir John Forbes (1787–1861), John Conolly (1794–1866), and Alexander Tweedie (1794–1884) issued the "cyclopaedia" in four volumes, 1832 to 1835, a revised edition of which, by Dunglison, was published in Philadelphia in 1845 (William M. Gafafer, John Forbes's British and Foreign Medical Review, *Human Biology* 6: 639, 1934).

[66] Vol. 20, p. 243.

[67] Leopold Gmelin (1788–1853), one of a famous family of Gmelins, published this book in Paris in 1823: *Chimie Organique Appliqué à la Physiologie et à la Médecine,* etc.

[68] For a critical analysis see the *London Medical and Physical Journal* 52: 493, 1824.

[69] "Kindle a slender flame by which others blessed with eminent qualities of mind may brighten an imperfect work with their own elegant talent." Concerning the heroic measures Dunglison advocated in this youthful venture, see John Claxton Gittings, Pediatrics One Hundred Years Ago, *American Journal of Diseases of Children* 36: 5, 1928.

[70] Jean François Nicolas Jadelot, director of the Hospital for Sick Children of Paris, 1802 to 1818.

attempts to localize as much as possible the various diseases have gained him much renown—the celebrated Broussais—practitioners should learn to recognize *"le cri des organes qui souffrent."*

M. Jadelot promised a more detailed account of his *Semeiologie physiognomique;* but nothing of the kind has fallen under my observation. All that I knew of it was derived from the French edition of Underwood on the diseases of children translated into French by my friend M. Eusèbe de Salle of Paris, who had the most exalted opinion of it in detecting the diseases of childhood. "By it," he observes, in the language of hyperbole—

mediocre talent may hereafter observe and cure with as much promptitude and safety as medical genius itself. By it, our art may be hereafter cleansed of the reproach that has been long brought against it, and unfortunately with too much truth, that, as regards children, we only employ hippiatric [71] medicine.

These remarks were published by M. de Salle in 1823,[72] and it is melancholy to reflect, that they have not been corroborated by subsequent observation and experience.

My original intention of resuming the consideration of some other of those diseases which are incidental to children was never executed, excepting so far as they formed part of a general treatise on the *Practice of Medicine,* when they met with their due share of attention at an after period.

II. THE UNIVERSITY OF VIRGINIA AND THOMAS JEFFERSON

Voyage to America. Arrival at Norfolk. Journey to Charlottesville. University of Virginia. Jefferson's taste for architecture. First faculty. Sickness of Mr. Jefferson. Visitors. Lafayette. University rules. Visitors. Last illness of Jefferson. The Randolphs. Private life of Thomas Jefferson. Robert Patterson and John K. Kane. Dunglison at the University. On college degrees. Offered professorship elsewhere. Negotiations with University of Maryland. Typhoid fever at the University of Virginia. University of Virginia activities. Dunglison's *Dictionary.* Dunglison's *Physiology.*

VOYAGE TO AMERICA

On the 27th of October 1824, accompanied by my wife I left London to join the ship Competitor, Captain   which had dropped down the river to Gravesend. Being acquainted with the owners of this vessel, which was bound direct to Norfolk, in Virginia, and under the impression, that we should be able to obtain more roomy accommodations than in one of

the regular packets, we agreed to pay a larger price but made a serious mistake in adopting this course. All experience indeed, has demonstrated, that much greater advantages, in all respects, are to be obtained in vessels, which are regularly engaged in conveying passengers, and whose success must be greatly dependent upon the degree of satisfaction which they afford.

In an ordinary merchant vessel, the passengers are but incidents, and, too often, as in our case, only administer to the cupidity of the Captain, to whom every arrangement is generally left by the owners. At Gravesend, we found two of my colleagues,—Mr. Bonnycastle,[1] who was the newly appointed professor of Natural Philosophy, and Mr. Thomas H. Key,[2] the Professor of Mathematics, who was accompanied by his wife, an amiable and estimable lady.

In the evening of the 27th, the ship weighed anchor. but was not able to proceed farther than the Nose, where we remained until the wind and tide favoured. when we made another attempt, but we could proceed no farther than the Downs, where we were detained, in miserable weather for some days; and when, at last, we weighed anchor, we were compelled, after a rough and disagreeable passage, to run in between the Isle of Wight and Portsmouth, where we cast anchor at the Motherbank. Here we were detained by unfavourable winds for three or four weeks, during which time we lived on shore; first at the extravagant George Hotel; and, afterwards, at a private boarding house in the same street, where, if the landlady was illiterate, she at least made us comfortable. The bill she rendered us at our departure, was such an amusing example of the want of the schoolmaster, that I kept a copy of it, as a curiosity, for a long time, and published it in the Charlottesville newspaper. Only one or two specimens of it I recollect: "left too, on account, three shillings and three pence," was rendered "lef to in act. three shlines and three pines." At one charge for *shrimps,* as I conceived, I demurred, having had none of the article. It turned out, however, that the real word was *thrimps* or "three pence"!

It was whilst we were lying at the Motherbank, that we were exposed to the severe storm, which will be recollected by everyone that witnessed it, as one of the most violent, that was ever experienced on the southern coast of England more especially. The weather had been somewhat squally, and my colleagues, with Mrs. Key, had determined to pass the night in Portsmouth. Unluckily, we decided on staying on board. Previous to this, on the 18th and 19th of November, a hurricane had occurred, almost unprecedented in the physical history of Europe. It appeared to have originated on

---

[71] Hippiatric: pertaining to the diagnosis and treatment of the horse.

[72] *Traité des Maladies des Enfans de Michael Underwood* . . . par Eusèbe De Salle, avec des notes de M. Jadelot, et un Discours Préliminaire Contenant l'Exposition de la Nouvelle Séméiologie-Physiognomonique.

[1] Charles Bonnycastle (1792–1840) was educated at the Royal Military Academy of Woolwich, where his father was a noted mathematician.

[2] Thomas Hewitt Key (1799–1875), Master of Arts at Trinity College, Cambridge, studied medicine for several years in England.

the coasts of England and Holland, whence it swept along the North Sea, causing dreadful shipwrecks on the coast of Jutland. Thence it traversed Sweden, prostrating whole forests in its course. Gottenburg and Stockholm were in a state of the utmost terror, and suffered much. It is affirmed, indeed, that there is no instance recorded of such a storm.

It was not, however, so severely felt at Portsmouth as the one of the 22 and 23d of November. In the former, we were luckily on shore. In the latter, the wind increased during the after part of the day, until in the night and the next day, it blew a perfect hurricane and although we were lying at anchor, protected, to a great extent, by the island over which the wind came, it appeared as if it had not been shorn of a portion of its terrific power by that circumstance. The vessel pitched so as to render all locomotion impracticable to a landsman. She was constantly dragging her anchor, and at one time, the danger became so imminent, that she would be driven ashore, that the carpenter was present for a long period with axe in hand ready to slip the chain cable when the word of command should be given by the Captain; to give her a chance—a forlorn one, however, of running into the harbour of Portsmouth; and in addition to all the horrors of that night, a large Indiaman was beached under our lee, having dragged her anchors, and gone bodily ashore, and during the night was constantly firing signal guns of distress. No lives, however, were lost. Fortunately, our ship held on to her anchors; and, in the course of the following day, the weather abated; and we were rejoined by our companions who had been full of anxiety on our accounts.

It is indeed impossible to imagine a more dismal night than we passed. My wife, however, conducted herself with great fortitude as she did on other similar occasions during the voyage; and her example was not lost on me.

A few days after this, the wind became favourable; and we again set sail. It is curious, however, that on three successive Fridays we had made the attempt to proceed on our voyage but had as often failed—a much larger amount of coincidence by the way, than is needed with the feeble minded to convert an idle belief into "confirmation strong as proof of holy writ."

On this occasion our progress was not great. On the following morning we approached the Eddystone light house, when the wind became boisterous and contrary, and it was deemed by the Captain advisable to put into Plymouth, which was accordingly done. It was a most appalling spectacle to witness the immense amount of mischief and the great loss of life that had been the consequence of the great storm of the 22nd and 23rd inst. through which we had passed. Some of the shipping in the Sound had parted or cut their cables, and, being unmanageable, drove afoul of other vessels, carrying away their masts, bowsprits, etc. and altogether drifting on the rocks. Along the Devonshire coast, the

desolation was of the most melancholy description; wrecks were to be seen in every direction, and valuable property lay floating about without an owner. A large vessel, the Hibernia, was dashed to pieces under the platform of the Citadel. Her cargo, which consisted of hemp and tallow, was scattered about in every direction and five of the crew were drowned. In the Catwater, the havoc was astonishing and melancholy. When our ship entered it, there were, in a narrow space not more than 300 yards in length, the remains of 16 fine merchantmen, all crowded together in one vast ruin and destruction; and it was believed, that had not the breakwater presented a bulwark against the terrific sea and tide coming in from the southward, the lower part of Plymouth must have been almost demolished, and scarcely a ship in port have survived the hurricane.

During our stay of a couple of weeks at Plymouth, on the watch, at all times, for the signal of the blue Peter, when the wind became favourable, we had an opportunity of becoming acquainted with some of the scientific gentlemen of Plymouth and Davenport;—for example with Dr. Cookworthy,[3] a respectable physician there, and with Mr. Wm. Snow Harris,[4] since knighted. He was then busily engaged with experiments on electricity and especially with a plan for protecting vessels at sea from the effects of lightning by a proper adaptation of the lighting rod. I also renewed my acquaintance—as I remarked in an early part of these "Ana"—with Mr. Derwent Coleridge[5]—the son of the poet, who had studied for the ministry, and held a curacy in Plymouth.

At length, after having spent six weeks in the channel, we succeeded, under favouring auspices, in clearing it; but still were unable to pursue the course that was desired by the Captain. We were driven toward the Azores; and on a beautiful day—Sunday—we coasted so near one of the Islands, Terceira, as to be able to recognize distinctly with the perspective glass the people proceeding to and from Church. During the same night however, one of those sudden storms arose, which are so common amongst the western islands, and we were driven with intense velocity between two of them, having our mizzen topsail blown from its fastenings. The gust soon, however, subsided and the following morning we had the beautiful prospect of the sun illumining the lofty Pico—7000 feet high. The wind had entirely ceased; but the long swell continued, so that we were greatly incommoded by the heavy rolling of the vessel. In the course of a few hours, however, a breeze sprang up and it was determined by the Captain that advantage should be sought for in

---

[3] M.D., Edinburgh, 1812.

[4] William Snow Harris (1791–1867), electrical expert.

[5] Derwent Coleridge (1800–1883), second son of Samuel Taylor Coleridge, was born at Keswick and spent his boyhood there when Dunglison lived in the town. He was ordained and became an elementary schoolteacher and Latin linguist.

the tradewinds, and that the southern passage by the Bermudas should be attempted. Already did we experience the error we had committed in preferring a private to a public vessel. The ladies had fortunately laid in a private supply of certain articles, which were, nevertheless, to be furnished by the owners; but on reaching the Azores, the sugar gave out; and, for a long period, the tea and coffee had to be taken without it. Our skipper was an unpolished selfish person—a West Indian, he said, by birth—who had become practically well informed in his calling, employed the quadrant when opportunity occurred, and calculated his lunars,—for we had no chronometer on board. He was, however, conceited withall and had no great liking, whatever respect he might have felt, for "philosophers." Altercations consequently arose and before we reached Norfolk, I was the only one of the philosophers, as he called my colleagues and myself, with whom he held any communication; and with me it was restricted and constrained; for his manners were at times not a little offensive. His want of gentlemanly and proper consideration was sufficiently shown on our approach to Terceira, when he stated, in the presence of the ladies, that great danger was to be apprehended, in the night, in our passage through the narrow channels, which separated the western islands,—a communication which—it need be scarcely said—might have excited intense alarm in their minds; had we not, in some measure, appreciated his character, of which vanity formed a large part; and he felt desirous that great credit should be given him on the following morning for having navigated his Ship safely through so many suppositious intricacies and dangers.

But his want of delicacy and of that *savoir faire,* which is often so beautifully exhibited by the experienced and considerate Captain, was more strikingly shown in a terrific storm which we experienced off Cape Hatteras. The "vexed Bermoothes" maintained their character in our experience by a succession of squalls; and the proverbial dread of Cape Hatteras was not belied in our case.

> If the Bermudas let you pass
> Beware then of Cape Hatteras.

We had been sailing rapidly—at 12 knots the hour—with almost a gale from the S.E. On going on deck in the evening a "bank" of clouds appeared in the N.W. which I remarked to the helmsman, who replied that off the coast of America—as he knew by experience—the winds sometimes shifted round, on the instant, from S.E. to the N.W., so that ships were not unfrequently taken aback and exposed to imminent danger. Fortunately, this man remained at the helm. I had retired to rest when I was suddenly aroused by the roaring of the wind and waves, and by the skipper calling down the companionway—"all hands on deck, the ship's going down." It appeared that, as the man at the helm had described, the wind changed on the very instant

to the N.W. Being on the alert, however, he slightly altered the course, and prevented the vessel from being taken wholly aback. The danger was, therefore, of short duration. When the Captain's coarse announcement was heard below everyone of course started up. We had no deadlights up, and if the ship had made sternway would have been the first to suffer. Mr. Bonnycastle rushed on deck, almost in *puris naturalibus,* with his nightcap on, and his appearance was so droll as to excite the laughter of the sailors—all immediate danger having passed away. Mr. Key, who inhabited the apartment next to my own kindly but most tremulously called to me not to be alarmed. My wife exhibited great fortitude, scarcely uttering an exclamation; and I was surprised at my own freedom from agitation. Yet an anecdote occurred in her case, which is worthy of being recorded, and exhibits with what equanimity evils of magnitude may be tolerated, whilst those of a trifling character may occasion intense feeling. On the following night, the gale from the N.W. still continuing, I was awakened by her sobbing bitterly, and on inquiring into the cause, she said, in great alarm—she was sure "there was a mouse in the cabin."

In the evening of the storm we were under the belief that we should be able to reach Norfolk on the following day. In the morning, however, when I rose, everything had changed. The day before I had tried the temperature of the air and of the water, and found them to differ but little—being about 72°. We were then in the gulph stream. In the morning, the water was 72°; the air at 44°; and on going on deck the change of temperature was severely felt; and the gale well deserved the name given it by the sailors of "the barber." All around were mists, and water spouts of various shapes, often presenting the appearance of distant ships. In this condition we were kept for nearly a week, lying to under a storm tripsail, when at length the wind changed, and we directed our course to Norfolk, entering between the capes on the 10th of February 1825 and casting anchor in the river at Norfolk the same evening, after having eaten our last portion of fresh meat that day at dinner.

At some distance from the capes we took a pilot on board, with much more stolidity in him than in the Americans in general of that class, who are usually remarkable for the amount of their information especially on everything that relates to their own country. We were extremely anxious to learn from him whether Mr. Adams or General Jackson had been elected President of the U.S., but he was unable to inform us. We were soon, however, boarded by a loquacious custom house officer who gave us the necessary information,[6]

---

[6] There were four candidates running for the presidency in 1824, but none of them secured the necessary majority. The House of Representatives had, therefore, to decide, and their choice fell upon John Quincy Adams. Andrew Jackson was

and immediately made us acquainted with the notabilities of Norfolk and especially Senator Tazewell,[7] who appeared to him to be the lawyer and statesman *par excellence*.

ARRIVAL AT NORFOLK

Our arrival occasioned no little sensation. The length of our passage—fourteen weeks—had excited the greatest apprehensions for our safety in the mind of Mr. Jefferson; and in all those who felt concerned in the success of the new University.[8] It had been determined to open it on the 1st of February; but our tardy arrival had caused a postponement until the 1st of April.

"We are dreadfully nonplussed here"—says Mr. Jefferson to Joseph C. Cabell Esq.—one of the visitors of the University [9]

by the non arrival of our three professors. We apprehend, that the idea of our opening on the 1st of February prevails so much abroad (although we have always mentioned it doubtfully) as that the students will assemble on that day without awaiting the further notice which was promised. To send them away will be discouraging, and to open an University without Mathematics or Natural Philosophy, would bring on us ridicule and disgrace. We therefore published an advertisement, stating that on the arrival of these professors, notice will be given of the day of opening of the institution. (*Memoir, Correspondence and Miscellanies from the Papers of Thomas Jefferson, Edited by Thomas Jefferson Randolph*. Vol. IV, p. 411, Charlottesville, 1829.)

As soon as the vessel came alongside the wharf we were boarded by hotel runners, newsmen etc. Yet everything appeared to differ but little from the country we had left, if we except the number of the coloured population,—as when one of the editors of a newspaper asked me "Where we intended to locate?" A kind hearted Captain Cooper, of the Steamboat from Norfolk to Richmond, immediately sent on board a turkey and various fresh provisions, under the impression, in which he was not mistaken, that they would be grateful to us after the restriction to a different kind of food to which we had been so long subjected.

We took up our lodgings at the Steamboat Hotel; and I immediately, by desire of my colleagues, wrote to Mr.

Jefferson to announce our arrival, and informed him, that we would, at the earliest period, proceed to the University.[10]

JOURNEY TO CHARLOTTESVILLE

We were a little struck at the desire manifested, both here and at Richmond, to extol the salubrity of the town, of which the speaker was an inhabitant, over that of the other; and whilst at Norfolk were cautioned against the efforts that would be made to induce us to believe that Norfolk was less healthy.

At Richmond we remained for a short time to lay in a stock of articles for commencing housekeeping at the University, as we were warned that but little choice of material was to be found at Charlottesville. Much attention was paid us by some of the respectable citizens: and we were informed, that it was soon noised abroad by a Mrs. Camp—a kind of Mrs. Malaprop—that the "Universal confessors" had arrived.

We here first became acquainted with Governor Randolph,[11] who was at the time in office—an exceedingly well informed gentleman but eccentric even to aberration. Luckily, too, Mr. Thomas Jefferson Randolph [12] was in Richmond, and kindly made his arrangements to accompany us across the country, the roads of which—always of the worst kind, were, if possible, heavier than ever in consequence of the thaw that had been going on for some time. The accounts which we had of them from the Richmond ladies were sufficient to intimidate the stoutest hearts. But there was no help. The distance from Richmond to Charlottesville by the stage route was about 70 miles, which could not be mastered under two entire days. On the first day we travelled about 35 miles to the house of a Mrs. Tinsley, well known at that time as the resting place for the night. Not unfrequently in the course of the day, Mr. Randolph left his seat by the driver to prop up the stage when it was likely to be overset; and it may be imagined with what relief we sat down at 10 o'clock by a cheerful wood fire to Mrs. Tinsley's well supplied table, after having toiled through roads since 6 o'clock of the morning such as most assuredly we had never witnessed in all our previous experience in England or on the Continent of Europe.

On the following morning before break of day we were again enroute; and passed as on the preceding day, along the vilest roads and through a country gen-

one of the unsuccessful candidates at this time but defeated Adams at the next election in 1828.

[7] Littleton Waller Tazewell (1774–1860), of Norfolk, later Governor of the state.

[8] In a letter to Gilmer, January 25, 1825, Professor Long wrote, after his arrival in Virginia, that Mr. Jefferson was "like all of us, very uneasy about the delay of our friends. I do not yet, being acquainted with all the circumstances of the case, entertain any apprehensions about their safety, but I regret, both for the University and my own personal comfort, that they were so foolish as to embark in an old log." (Herbert B. Adams, *Thomas Jefferson and The University of Virginia*, p. 115, footnote, 1888.)

[9] Joseph Carrington Cabell (1778–1856), one of Jefferson's principal co-workers in establishing the University of Virginia (*ibid.*, p. 53).

[10] See letter, February 10, 1825, *The Jefferson-Dunglison Letters*, ed. by John M. Dorsey, p. 11, University of Virginia Press, 1960.

[11] Thomas Mann Randolph (1768–1828), Governor of Virginia 1819–1822, was in Richmond at this time as a state senator. He married Thomas Jefferson's daughter, Martha, and built Edgehill on his estate, which adjoined Jefferson's in Albemarle County, Virginia.

[12] Thomas Jefferson Randolph (1792–1875), son of Martha Jefferson and Thomas Mann Randolph, was Jefferson's oldest and favorite grandson.

erally by no means inviting; and it must be admitted, that the prospects were highly unfavourable to the beauty or cultivation of the region in which our lot was to be cast, were we to judge from the specimens we had seen. It had not, however, the least dispiriting influence upon us. I felt I was about to be transferred to a sphere in which I could be useful. Although in England, attached to the side embraced by Pitt, Percival, and Liverpool in succession, my political principles had never weighed heavily on me,[13] certainly not to such an extent as to make me believe they were the only ones that could be conscientiously embraced; and I may remark, *en passant,* that in all my observations since I have been in the U.S. I have found those who would be regarded as tories in their own country more tolerant of the institutions and political movements in this than the whigs and especially the radicals, who often form Utopian notions of the working of the political machinery in this country and not finding it in all respects corresponding with them, become dissatisfied and not unfrequently disgusted. A confirmation of this was witnessed in our own corps. In our journey from Richmond towards the University, Mr. Key, whose sentiments in politics were decidedly liberal, if not radical, emphatically declared, that this was the Country in which he would wish to live and die: yet, in the course of a year or two he has emphatically stated, that "he would rather live in England on six pence a day" than on the handsome income which he received from his labour at the University.

At a few miles from the University, Mr. Randolph [14] left us, stating that he could now be sure of our safe journey onwards, as he was the overseer of the road, and knew its condition. We had not proceeded far, however, before the driver of the stage cautioned us, that there was a bad place in the road, which he would endeavour to pilot us through securely. He failed, however. The wheels on one side sank into a deep hole, and we were fairly overturned. Fortunately, the horses stood firm; and we were extricated from the stage, which was by no means an easy matter; for at that date, the only means of entrance and exit was by the front. The accident occurred at a short distance from Moore's Creek, over which there was no bridge, and we were, consequently, compelled to ford it, bearing the ladies over; and such was our unpropitious introduction to the neighbourhood of the University. A Roman would have regarded it as a bad omen, and been disheartened. We were not Romans, however, and the affair only excited amusement.

After this we still had a short distance, about two miles, to walk before we reached Charlottesville, which we did late in the evening; and put up at the only hotel built of stone in the town, and therefore distinguished by the name of "The Stone Tavern."

Our arrival was made known to Mr. Jefferson by his grandson; and on the following morning we were visited by Mr. Nicholas P. Trist,[15] who had a short time before married one of Mr. Jefferson's granddaughters— Miss Virginia Randolph—and who was residing with Mr. Jefferson at Monticello. Soon afterwards, the venerable ex President presented himself and welcomed us with dignity and kindness for which he was celebrated. He was then eighty-two years old, with his intellectual powers unshaken by age; and the physical man so active, that he rode to and from Monticello and took exercise on foot with all the activity of one 20 or 30 years younger.[16] He sympathized with us on the discomforts of our long voyage; and on the disagreeable journey we must have passed over the Virginia roads; and depicted to us the great distress he had felt lest we had been lost at sea; for he had almost given us up when my letter arrived with the joyful intelligence that we were safe. Mr. Long,[17] the Professor of Ancient Languages, and Dr. Blaetterman,[18] the Professor of Modern Languages, both of whom had been chosen in London by Mr. Gilmer, had judiciously taken passage in regular packets to New York and had arrived a considerable time before us.

UNIVERSITY OF VIRGINIA

As it was not convenient for us to remain in Charlottesville, a distance of a mile and a half from the University by a road almost impassible for the pedestrian, and of the most wretched character for beasts of burden, we took board and lodging at one of the "hotels" of the University, as they were called, which were destined to furnish meals to the students; and remained there until our goods arrived from Richmond, which, along with the necessaries we were able to procure in Charlottesville, enabled us to occupy our

---

[13] Dunglison is here proclaiming himself a Whig. In England the Whig party and the Tory party were politically opposed groups. In America, during Revolutionary days, Whig was practically synonymous with patriot, while the loyalists were called Tories.

[14] Thomas Jefferson Randolph, who lived at Edgehill. The main road from the Western part of the state to Richmond ran between Monticello and Edgehill.

[15] Nicholas Philip Trist (1800–1874), who enjoyed the affection of President Jackson, was a favorite of Thomas Jefferson, was with him in his last illness and was named one of the executors in Jefferson's will.

[16] Mr. Jefferson was six feet two and one-half inches tall, lean and well proportioned and very strong in his prime (Hamilton W. Pierson, *Jefferson at Monticello,* p. 70, 1862). By 1824, however, the tall and powerful frame of Jefferson was bent and emaciated (Henry S. Randall, *Life of Thomas Jefferson* 3: 503, 1858).

[17] George Long (1800–1879), Master of Arts and Fellow of Trinity College, Cambridge, had a great influence upon the faculty and students, notwithstanding his short stay at the University of Virginia. He later helped found the Royal Geographical Society and the popular *Penny Cyclopedia* in England (Adams, *ibid.,* p. 117).

[18] George Blaetterman, born in Germany, was an Anglo-Saxon scholar and noted as a teacher of modern languages and comparative philology (*ibid.*).

own pavilion. It was not long before we were able to sleep in it; but somewhat longer before our bedsteads arrived, until which time we slept on the floor. This was the most trying period of our residence at the University of Virginia. We had not yet succeeded in procuring valuable domestics, and had to put up with those of the most inferior kind; for the better class were in employment. The market people had scarcely begun to present themselves, and it was therefore no easy matter to procure the material for subsistence; but still, everything promised well.[19] The houses were much better finished than we had expected to find them, and would have been far more commodious, had Mr. Jefferson consulted his excellent and competent daughter—Mrs. Randolph—in regard to the interior arrangements, instead of planning the architectural exterior first, and leaving the interior to shift for itself. Closets would have interfered with the symmetry of the rooms or passages, and hence there were none in most of the houses; and in the one which was furnished with a closet, it was told as an anecdote of Mr. Jefferson, that not suspecting it according to his general arrangements, he opened the door and walked into it on his way out of the pavilion.

JEFFERSON'S TASTE FOR ARCHITECTURE

He was fond of architecture and anxious that the Rotunda and the different pavilions should present specimens of the various orders; and although, from the necessity of building them of brick and wood, the effect was greatly diminished, it was, on the whole, agreeable. The heavy cornices in the interior of the rooms—of the Palladian style—were, however, anything but pleasing. Undoubtedly, too, the desire for having everything architecturally correct according to

[19] Soon after this, the house of Dr. Dunglison at Charlottesville was described as in every respect a most charming one, graced by all the elegancies that can adorn private life (Lonsdale, *ibid.*, 268).

FIG. 5. The University of Virginia when Dunglison was there. From a print in possession of Dr. Radbill.

FIG. 6. Model for Pavilion X, originally occupied by Dunglison. Sketched by Jefferson's granddaughter, Cornelia. Reproduced from Adams, *Jefferson and the University of Virginia*, p. 14, 1888.

his taste, induced him in more cases than the one I have mentioned to sacrifice convenience. He could not but admit the anomaly of having windows arranged as in modern habitations, but farther than this it was difficult to induce him to go, and when I consulted him in regard to a distinct building for anatomical purposes, which he agreed to,[20] he at the same time told me, that he must choose the position and the architectural arrangement externally, whilst all the interior arrangements should be left to me.[21]

"This plan" (of the University) says Professor Tucker (*Life of Jefferson* ii, p. 431, Philad. 1837)

has indeed been the subject of frequent criticism; yet it is probable, that any other plan, which could have been devised, would have incurred as much censure, since architecture is a branch of art in which all—*docti indoctique*[22]—think themselves equally competent to judge. It is certainly remarkably shewy to the eye, and the view of its exterior is always very imposing to him who beholds it for the first time. Though expensive for the accommodation it affords, in consequence of its spreading over so large a surface, it is on that account more favorable to order and quiet than if the students had been congregated into one or two large

[20] In this anatomical "theatre" dissection was provided for from the first, in spite of the theoretical nature of Jefferson's pedagogical plan. No doubt the bodies of negro slaves furnished ample materials for such purposes. In 1826 a dispensary was also established in this building (Wyndham D. Blanton, *Medicine in Virginia in the Nineteenth Century*, p. 21, 1933). For further details about this outpatient teaching clinic see John M. Dorsey, *Jefferson-Dunglison Letters*, pp. 47–48, 1960.

[21] Concerning some of the details of these "interior arrangements," see John M. Dorsey, *ibid.*, pp. 14–16, 17 and 29. In the quest for anatomical teaching materials, Jefferson also wrote to John Patterson on November 22, 1825, in reference to the purchase of an anatomical museum (A.L.S., *Collection of Autographs of Ferdinand J. Dreer* 1: 329, Phila. 1890).

[22] Docti-indoctique: schooled and unschooled.

FIG. 7. The anatomical "theatre." Reproduced through the
courtesy of the University of Virginia.

buildings. It is also more secure from destruction by fire,
by reason of the ease with which every part can be approached and subjected to the action of the fire engine, and
because but a small part can be consumed at one time. . . .
There was no employment whatever in which he could have
found such agreeable occupation, as in thus carrying into
execution the long cherished schemes of his patriotism—in
providing for the education of the youth of the country,
and, at the same time, gratifying his taste, or rather his
passion for architecture; especially for Grecian architecture. The pavilions, provided for the professors, were each
adorned with a portico, where he exhibited to his admiring
countrymen models of all the orders, rigidly copied to the
smallest minutiae; and to furnish these models probably
more money was spent in the ornamental parts of the edifice
than in those which were indispensible.[23]

FIRST FACULTY

As I before remarked, the opening of the University,
which had been fixed for the first of February, was
postponed on account of our late arrival in the country,
until the 1st of April, when it took place. All the professors, except the incumbent of the law chair, were
on the spot: and the faculty consisted of Mr. Long,
Professor of Ancient Languages; Mr. Key, Professor of
Mathematics; Mr. Bonnycastle, Professor of Natural
Philosophy; Dr. Blaetterman, Professor of Modern
Languages; Dr. Emmet,[24] Professor of Chemistry; Mr.

[23] Jefferson planned the college buildings as examples of the
principle of architecture for the students. By 1821, ten pavilions
had been completed, each with a lecture room, four apartments
for the professor and his family, and a garden (Herbert B.
Adams, ibid., p. 101). Jefferson wanted the ground floor of
each pavilion for lecture rooms, but the professors' wives soon
changed that idea, and the classes were driven out (ibid., p.
116).

[24] John Patten Emmet (1796–1842), son of Thomas Addis
Emmet, Irish patriot, M.D., 1822, College of Physicians and
Surgeons of New York, was first Professor of Natural History

Tucker,[25] Professor of Moral Philosophy; and Dr.
Dunglison, Professor of Medicine. All the Professors
were foreigners; for Dr. Emmet was born in Ireland;
and Mr. Tucker in the Island of Bermuda. In the organization of the faculty, it was deemed advisable to
choose for our first chairman, who was charged with
certain points of discipline, Professor Tucker—the oldest amongst us; and who had been recently in Congress.
It was presumable that he was better acquainted with
the manners, customs, and feelings of the young men
of the country than we could be supposed to be; and
would be more acceptable to them, and their parents,
than entire strangers, as all, except Dr. Emmet, were.
The post of Secretary to the faculty, one of some responsibility, as he was the organ through which all of
the acts of the body were promulgated, was conferred
upon me.

The fact of all the professors being foreigners, it
might be imagined, would be unfavourable to discipline;
and might lead the disorderly to rebel against the
authorities of the University. It is but justice, however,
to a highly numerous body of generous young gentlemen to say, that during the whole period of my residence
at the University, which amounted to nine years, no
single act came to my knowledge of insubordination
from that cause; whilst ample evidence was afforded of
their great respect for those who had left their homes,
and were zealously engaged in instructing them. Mr.
Jefferson was, however, severely criticized for having
gone abroad for professors; but he vindicated himself
from the few, who objected to this course, by similar
arguments to those which have been brought forward
by this biographer. "Mr. Jefferson"—observes Professor Tucker—

has been censured for this course, as reflecting on the
science and literature of his own country. But surely
nothing can be more defensible. The institution, which had
been reared by his efforts, and for whose success he had
every motive, personal and patriotic, aspired to give a
course of education equal to any other in the U.S. As the
most capable professors were presumed to be already occupied, that description of talent being not yet redundant
in the country, and scarcely equaling the demand, to have
confined himself to such professors as could have been obtained here would have subjected the visitors to the alternative of either taking inferior men, such as had not found
employment elsewhere, or of enticing them from some other
institution. The first course would have been unfaithful to
themselves, their own promises, and the public expectations.
The last would have been invidious, would have subjected

at the University of Virginia. This was changed to chemistry
and materia medica in 1827. Dr. Dunglison assisted at the birth
of his son, Thomas Addis Emmet (1828–1919), who became a
well known gynecologist in New York City (Thomas Addis
Emmet, Incidents of My Life, p. 108, 1911).

[25] George Tucker (1775–1861), Professor of Moral Philosophy and Political Economy, was a prolific writer, a member of
the Virginia legislature and the Congress of the United States
(Robert Colin McLean, George Tucker, Moral Philosopher
and Man of Letters, University of North Carolina Press, 1961).

them to a still severer censure, and their own injustice might have been retorted on them. Vol. 2, p. 475.

My own selection perhaps excited the most feeling amongst my professional brethren. The announcement in the *London Medical Repository* of my appointment gave occasion to a long and excited article in the ninth volume of *The Philadelphia Journal of the Medical and Physical Sciences* edited by Dr. N. Chapman under the caption of "American Medicine," in which, whilst the writer disclaimed having any reference in his observations to the individual selected in this or any similar instance, he unequivocally suggested that such violation of propriety on the part of the appointing power should be visited on him. "We do not"—he remarks—

hesitate to pronounce the policy of such appointments as most injudicious and subversive of the general good of our citizens, and especially of the dignity and interests of our professional brethren. We fully agree with those who have felt and expressed their indignation on this subject, without the slightest reference to individuals who may have been chosen, though we regret *the unpleasantness of the situation in which they will be placed, especially as it is probable the American Medical Societies throughout the country will protest against this act of injustice.*

The article is upward of five pages in length, and is signed/.

I felt that it was natural for such an impression to exist to a greater or less extent as that expressed by the writer, and determined so to act as to neutralize it as far as practicable; and it has been a source of no little satisfaction to me, that whatever may have been the feeling adverse to me for my not having been born in the country in the minds of some, the circumstance has been so little regarded by the many, that almost every appointment and compliment that had been conferred upon me has been unsolicited and hence the more highly appreciated. Upon principle, I have been opposed to the system of bestowing scientific and literary distinctions on those only who apply for them; and I think it will be generally admitted by the reflecting, that where a vacancy occurs in a professorship in a scientific or literary institution, the most desirable course would be for the appointing power to look abroad as well as at home and select him who, it would be admitted by the wisest in such matters, is the best qualified to fill it. Such has been the policy of the University of Virginia, and such the mode of conduct and belief with many of the trustees of the Jefferson Medical College; and I do not know that they have had to regret the liberal course they have pursued. Local influences are, indeed, apt to be all powerful; and where qualifications of candidates are equal should generally perhaps be decisive; but farther than this they ought not to be permitted to weigh.

Under the depressing influence produced in my mind by the article in the so called *Chapman's Journal,* which was at the time the most prominent medical periodical published in America, it was, I confess, not a little

Fig. 8. Robley Dunglison as he appeared when he was first Professor of Medicine at the University of Virginia. From a portrait in possession of the University of Virginia.

gratifying to me to find the suggestion contained in it disregarded, in a few months afterwards, by the authorities of one of the most celebrated colleges of the country, who, altogether unknown to me, conferred on me the honorary degree of *Doctor of Medicine.* The first intimation I had of this was in a newspaper paragraph, in which I saw, that such a degree had been conferred on Robley Dunglison amongst others. Under the impression that there was no other of the name in the country, I wrote to the College authorities; and sometime afterwards, my diploma arrived addressed to me at "Hampden Sidney College" instead of the University of Virginia; yet the official letter informing me of the honor miscarried. In a review of my *Human Physiology,* in Professor Silliman's Journal (Vol. 64, p. 165, New Haven, 1833) written, I was informed, by Professor Knight[26] of Yale College, it is stated, the compliment

---

[26] Jonathan Knight (1794–1864) was on the teaching staff of Yale Medical School from its beginning in 1813 until his death. He was President of the American Medical Association in 1853.

was paid me on the "proposition" of N. Smith, the professor of Surgery in that learned institution. The diploma is dated the 13th day of September 1825—about six months after my arrival in the country.[27]

SICKNESS OF MR. JEFFERSON

Not long after my arrival at the University, Mr. Jefferson found it necessary to consult me in regard to a condition of great irritability of the bladder under which he had suffered for some time, and which inconvenienced him greatly by the frequent calls to discharge his urine.[28] Few, perhaps, attain that advanced age without suffering more or less from disease of the urinary organs. On examining the urethra I found that the prostatic portion was affected with stricture, accompanied and apparently produced by enlargement of the prostate gland. This required the use of the bougie,[29] which he soon learned to pass himself; one of the smallest size being inserted at first with difficulty; but as the urethra became gradually enlarged his intervals became prolonged; and his inconvenience greatly diminished. This condition interfered, however, materially with his horseback exercise to which he had been accustomed on his excellent and gentle horse Eagle—long a favourite with his illustrious master.[30] Mr. Jefferson was considered to have little faith in physic; and has often told me he would rather trust to the unaided, or rather uninterfered with, efforts of nature than to physicians in general. "It is not," he was want to observe, "to physic that I object so much as physicians."[31] Occasionally, too, he would speak jocularly especially to the unprofessional of medical practice; and on one occasion gave offense where most assuredly if the same thing had been said to me no offense would have been taken. In the presence of Dr. Everett,[32]

afterwards private secretary to Mr. Monroe, who was sensitive and somewhat cynical, and moreover not particularly partial to Mr. Jefferson, he remarked, that whenever he saw three physicians together he looked up to discover whether there was not a turkey buzzard in the neighborhood. The annoyance of the Doctor I am told was manifest. To me when it was recounted it seemed a harmless jest.

But whatever may have been Mr. Jefferson's notions of physic and physicians, it is but justice to say, that he was one of the most attentive and respectful of patients. He bore suffering inflicted upon him for remedial purposes with fortitude; and in my visits shewed me by memoranda the regularity with which he had taken the prescribed remedies at the appointed times. From the very first, indeed, he kindly gave me his entire confidence and at no time wished to have anyone associated with me. It was about this time that Mr. Short[33] wrote to him urging that he should consult Dr. Physick in Philadelphia.[34] His reply to Mr. Short—who communicated it to my venerable friend Mr. Duponceau[35]—who mentioned it to me with great satisfaction upwards of fifteen years afterwards—was, that he had his Dr. Physick near him in whom he reposed his trust.

VISITORS

I generally visited him at Monticello two or three times a week and always had my seat at table on his left hand. His daughter Mrs. Randolph or one of the grandaughters[36] took the head of the table; he himself sat near the other end, and almost always some visitor was present. The pilgrimage to Monticello[37] was a favourite one with him who aspired to the rank of the patriot and philanthropist; but it was too often undertaken for idle curiosity; and could not under such cir-

[27] Nathan Smith (1762–1829), after completing his medical studies abroad, was elected a corresponding member of the London Medical Society and was so informed, in a letter dated November 6, 1823, by Dr. Dunglison, who was then Corresponding Secretary for the Society. In a P.S. Dunglison informed Smith: "I should feel highly honored by being proposed a member of any of your Scientific Societies in America" (Oliver P. Hubbard, *Introductory Lecture, Dartmouth College, July 31, 1879*, p. 36, 1880).

[28] This was on May 17, 1825; during the next two weeks it was necessary for Dunglison to make eight professional visits to the ailing Jefferson (Dorsey, *ibid.*, p. 13).

[29] Dunglison discusses the use of the bougie in his *Dictionary* (1: 124, 1833).

[30] For an account of Jefferson's love for his favorite horse, see Pierson, *ibid.*, p. 58.

[31] In a letter to Dunglison (U. Va. Collection) he said ". . . time and experience as well as science are necessary to make a skillful physician, and Nature is preferable to an unskilful one." Jefferson had expressed his views concerning doctors in a letter to Caspar Wistar written June 21, 1807 (Adrienne Koch and William Peden, *Life and Selected Writings of Thomas Jefferson*, pp. 583–585, New York, 1944). For a further exposition of these views, see John M. Dorsey, *ibid.*, pp. 93–106.

[32] Charles D. Everett, M.D., 1795, University of Pennsylvania, was also Monroe's family physician after the latter retired from the presidency.

[33] William Short, American diplomat, private secretary to Jefferson in Paris, was his protégé and intimate friend.

[34] Philip Syng Physick (1768–1837) had acquired particular skill and renown in the treatment of urethral obstructions, developing catheters and bougies that were probably better than any that could be produced at that time in Europe. He was well known to Jefferson, Madison, and Monroe (George F. Sheldon, Rush and Physick, An Important Medical Friendship, *Transactions & Studies of the College of Physicians of Philadelphia*, 4th ser., 29: 28–38, 1961).

[35] Peter Stephen Du Ponceau (1760–1844), one of the French officers who served through the Revolution and as Assistant Secretary of State for many years, was President of the American Philosophical Society.

[36] Martha Jefferson Randolph lived with her father and kept house for him after he returned from Washington. She had six daughters and five sons. Mr. Jefferson was devoted to his grandchildren, and they delighted to follow him about (Pierson, *ibid.*, p. 85).

[37] Monticello, having fallen into decay not long after Jefferson's death, was restored years later through the efforts of Commodore Uriah P. Levy and a descendant, Jefferson Levy; now maintained by the Thomas Jefferson Memorial Foundation it is a tourist attraction once more (Rabbi Malcolm H. Stern, Monticello and the Levy Family, *Journal of the Southern Jewish Historical Society* 1: 19–23, 1952).

cumstances have afforded pleasure to, whilst it entailed unrequited expense on, its distinguished proprietor. More than once, indeed, the annoyance has been a subject of regretful animadversion. Monticello, like Montpellier, the seat of Mr. Madison, was some miles distant from any tavern and hence without sufficient consideration the traveler not only availed himself of the hospitality of the Ex Presidents but inflicted upon them the expense of his quadrupeds; on one occasion at Montpellier, when my wife and myself were paying a visit to Mr. and Mrs. Madison, no fewer than nine horses were entertained during the night; and in reply to some observations, which the circumstances engendered, Mr. Madison remarked, that whilst he was delighted with the society of the owners he confessed he had not as much feeling for their horses.

Sitting one evening with Mr. Jefferson on the porch at Monticello, two gigs drove up each containing a gentleman and a lady. It appeared to me to be evidently the desire of the party to be invited to stay the night. One of the gentlemen came up to the porch and saluted Mr. Jefferson, stating, that they claimed the privilege of American citizens in paying their respects to the Ex-President and inspecting Monticello. Mr. Jefferson received them with marked politeness, and told them they were at liberty to look at everything around, but as they did not receive an invitation to spend the night, they left in the dusk and returned to Charlottesville. Mr. Jefferson, on that occasion, could scarcely avoid an expression of impatience at the repeated, though complimentary, intrusions to which he was exposed.

At all times dignified and by no means easy of approach to all, he was generally communicative to those on whom he could rely; and in his own house was occasionally free in his speech even to imprudence to those of whom he did not know enough to be satisfied that an improper use might not be made of his candour. As an early example of this, I recollect a person from Rhode Island visiting the University, and being introduced to Mr. Jefferson by one of my colleagues. This person did not impress me favorably; and when I rode up to Monticello I found no better impression had been made by him on Mr. Jefferson and Mrs. Randolph. His adhesiveness was such, that he had occupied the valuable time of Mr. Jefferson the whole morning, and stayed to dinner; and during their conversations Mr. Jefferson was apprehensive, that he had said something which might have been misunderstood and be incorrectly repeated. He therefore asked of me to find the gentleman, if he had not left Charlottesville, and request him to pay another visit to Monticello. He had left, however, when I returned, but I never discovered that he had abused the frankness of Mr. Jefferson. Mr. Jefferson took the opportunity of saying to me how cautious his friends ought to be in regard to the persons they introduce to him.

It would have been singular if in the numerous visitors some had not been found to narrate the private conversations held with such men as Jefferson and Madison; yet they were few; and one of them, Mr. Charles J. Ingersoll,[38] who, I am sure, had none but the kindliest motive as regarded Mr. Madison, on one or more occasions, published in the newspapers of the day, sentiments said to have been expressed to him by Mr. Madison on public events, which were the cause of much regret, in as much as it is difficult—as Mr. Madison stated in remarking to me on one of these publications—always to convey accurately the views of the individual when orally communicated under such circumstances; and hence he is subjected to animadversion and criticism, without having always the power or desire of replying.

It has always appeared to me a gross breach of propriety either for political or any purposes, to parade before the public opinions freely and unreservedly communicated in unsuspecting private intercourse, without the consent of the parties concerned.

In Mr. Jefferson's embarrassed circumstances in the evening of life, the immense influx of visitors could not fail to be attended with much inconvenience. I had the curiosity to ask Mrs. Randolph what was the largest number of persons for whom she had been called upon unexpectedly to prepare accommodations for the night, and she replied fifty. In a country like our own, there is a curiosity to know personally those who have been called to fill the highest office in the republic; and he who has attained this eminence must have formed a number of acquaintances, who are eager to visit him in his retirement, so that when his salary as first officer of the state ceases, the duties belonging to it do not cease simultaneously; and I confess I have no sympathy with the feeling of economy—political or social—which denies to the Ex-President a retiring allowance, which may enable him to pass the remainder of his days in that useful and dignified hospitality which seems to be demanded by the citizens of one who has presided over them.

On one of those who visited Mr. Jefferson I desire to make a few comments, partly to correct some errors of fact into which he fell many years afterwards in referring to this subject. On paying a visit to Monticello Mr. Jefferson asked me if I had seen a strange personage from the West to whom he had given a letter of introduction to me that morning. He had amused both him and Mrs. Randolph with his egregious vanity, and both were desirous of hearing the impression he had made upon me. We appeared to have crossed each other on the way; and on my return to the University I found Mr. Jefferson's note of introduction, couched as he explained to me, in exaggerated

---

[38] Charles Jared Ingersoll (1782–1862), democratic member of Congress, lawyer and author, was a member of the American Philosophical Society.

terms, as the note was opened, and would doubtless be scanned by the bearer. That note is now before me:

Th. Jefferson presents his compliments to Dr. Dunglison and begs leave to introduce to him Dr. Caldwell [39] one of the professors of Transylvania College. He is highly considered in that institution, and very justly so. He is anxious to see our establishment, and especially to become acquainted with its professors, which good office, Th. J. requests Dr. Dunglison to render him.

Th. J. is going on steadily well. His intervals have improved less than any other symptoms. He will pray a supply of pills by his grandson, either today or tomorrow.
June 16, 1825.

When I returned to the University I found my colleagues full of amusement at the singular remarks and opinions of Dr. Caldwell, who had visited the University in my absence. He was then, as I believe he has since ever been, enthusiastic in the belief of cranioscopy, and of the information on individual aptitudes, which he conceives it is capable of affording. In his appreciation of the individual aptitudes of the professors, however, they considered he had made palpable blunders. My friends, Professors Bonnycastle, Key, Long, and Blaetterman, informed me that they had defended me against his assertion, that it was impossible for me to treat American disease properly, and for a flourishing medical school to be established at the University; although *he* could found a successful school on the Rocky Mountains. I certainly was not present at any such conversation; and was, therefore, not a little astonished to peruse the following observations from the pen of Dr. Caldwell, written seventeen years afterwards in *The Western Journal of Medicine and Surgery* for June 1842:

About twenty years ago (a few years perhaps more or less), Dr. Dunglison arrived by invitation in the United States, as one of the professors in the University of Virginia. Soon after that period, the writer of this article had the pleasure of an introduction to him, and found him confident in the belief that though he had scarcely, perhaps, at the time of the interview, felt an American pulse, or made a single visit to an American sick room [see the letter of introduction of Mr. Jefferson, whom I was then and had been for some time attending], he had notwithstanding a complete knowledge of American diseases, and was perfectly competent to their treatment and care, and also to become an instructor in them to others. Nor is this statement made in a form of an inference from a casual conversation. It is the substance of an opinion, that was frankly avowed. And although, during the conversation, the professor himself was perfectly calm, courteous and self-possessed [I certainly was not present] one or two of his colleagues deported themselves differently, when the correctness of the belief was delicately questioned. An insinuation quite civilly made, that an English physician *might* be surpassed and even instructed in his profession by an American, appeared to be regarded in the light of an indignity. And thus was the conversation suddenly interrupted, or the subject of it changed.

[39] Charles Caldwell (1772–1853), M.D. 1796, University of Pennsylvania, founded two medical schools west of the Alleghenies and was a very busy man, writing, talking and manoeuvering in medical politics all his life.

Yet in regard to my action in the matter, all this was undoubtedly fabulous; and it may perhaps admit of a question, whether the part taken in the conversation by Dr. Caldwell could have been, from its very nature, and under the circumstances, either "delicate" or "civil," especially as the subject—my colleagues informed me—was introduced by Dr. Caldwell himself.

It may be well to observe that the article from which I have made the preceding extract is a bitter review of my *Practice of Medicine,* as well as of its author, immediately suggested perhaps by one that appeared in a previous number of the *same* journal, and which was as eulogical of the work and the man as the former was disparaging. The earlier article was written, I believe, by Professor Gross.[40] To both, I may have occasion to refer hereafter. Professor Gross's article was contained in the number of the *Western Journal of Medicine* etc. for April, 1842, p. 282.

LAFAYETTE

In the summer of 1825, the monotonous life of the college was broken in upon by the arrival of General Lafayette to take leave of his distinguished friend, Mr. Jefferson, preparatory to his return to France. A dinner was given to him in the Rotunda by the Professors and students, at which Mr. Madison and Mr. Monroe were present, but Mr. Jefferson's indisposition prevented him from attending. "The meeting at Monticello" says M. Levasseur, the secretary to General Lafayette during his journey, in his *Lafayette in America in 1824 and 1825* Vol. 2, p. 245 [41]—

of three men, who, by their successive elevation to the supreme magistracy of the state, have given to their country twenty-four years of prosperity and glory, and who still offered it the example of private virtues, was a sufficiently strong inducement to make us wish to stay there a longer time, but indispensable duties recalled General Lafayette to Washington, and he was obliged to take leave of his friends. I shall not attempt to depict the sadness, which prevailed at this cruel separation, which had none of the alleviation that is usually left by youth. For in this instance the individuals who bade farewell, had all passed through a long career, and the immensity of the ocean would still add to the difficulties of a reunion.

Mr. Levasseur has evidently confounded this banquet with that given by the inhabitants of Charlottesville the year preceding during the first visit of Lafayette to Mr. Jefferson. At that period, there were neither professors, nor students, as the institution was not opened until six months afterward. "Everything" —says M. Levasseur (Vol. 1, p. 220)

[40] Samuel David Gross (1805–1884), M.D., 1828, Jefferson Medical College; taught at the Medical College of Ohio (1833–1835), Cincinnati College (1835–1839), University of Louisville (1840–1849 and 1851–1856) and became Professor of Surgery at Jefferson Medical College in 1856. A leading surgeon and medical educator of his time, he was a member of the American Philosophical Society, the College of Physicians of Philadelphia, and many other organizations.

[41] Originally issued in Paris, 1829.

had been prepared at Charlottesville by the citizens and students, to give a worthy reception to Lafayette. The sight of the nation's guests seated at the patriotic banquet, between Jefferson and Madison, excited in those present an enthusiasm which expressed itself in enlivening sallies of wit and humor. Mr. Madison, who had arrived that day at Charlottesville to attend this meeting, was especially remarkable for the originality of his expressions, and the delicacy of his allusions. Before leaving the table, he gave a toast "To liberty, with virtue for her guests, and gratitude for the feast," which was received with transports of applause.

The same enthusiasm prevailed at the dinner given in the rotunda. One of the toasts proposed by an officer of the institution I believe was an example of forcing a metaphor to the full extent of its capability: "The Apple of our Heart's eye—Lafayette!"

In referring to Mr. Jefferson's bodily condition at this period, Mr. Tucker has the following remarks (*Life of Jefferson*, Vol. 2, p. 478)

His health began now to be seriously impaired. An infection of the bladder had for some time given him uneasiness, and of late the symptoms were so aggravated as to call for medical aid. On these occasions Dr. Dunglison of the University was always his physician. He bore his sufferings with exemplary patience, and except in the paroxysms of his pain, had his wanted equanimity and good humour. The disease had returned with more than usual severity in August, when he received another visit from General Lafayette, then about to leave the United States for France. The general found him on a couch in the drawing room, evidently altered since he saw him the year before, and then suffering acute pain. He manifested a good deal of solicitude for his friend, had conferences with Dr. Dunglison, and having learned that certain preparations, useful in his disease, could be obtained better in Paris than elsewhere, he remembered the fact, and as soon as he returned, sent a supply which would have been sufficient for twenty patients. Lafayette again received a public dinner at the University, given on this occasion, by the professors and students, and although Mr. Jefferson could not partake of the entertainment, it afforded him no small gratification.

The preparations referred to by Mr. Tucker were elastic gum catheters. Soon after Lafayette's return to France he wrote to Mrs. Randolph stating that he had sent me a small *caisse* as a present. What this could be puzzled us not a little. When it arrived, which was some time after the death of Mr. Jefferson, it proved to be a case of nearly 100 elastic gum catheters!

The announcement of Mr. Jefferson's decease gave occasion to another letter in which, after expressing his grief at the loss of his old friend, he asked Mrs. Randolph to inform me how much he sympathized with me at the sad, although, by him, not unexpected, event.

From the Life of General Lafayette by M. Jules Cloquet, M.D., it appears that those instruments were selected by him. "Some years ago"—he remarks—

Lafayette instructed me to choose for him some surgical instruments, which he wished to present to President Jefferson at the period of his last illness. When I handed to him the box containing them, he thanked me with his usual kindness, and added: "What think you of my friend's health? The situation causes me the greatest anxiety. Why can I not send him in this box not only the instruments which he requires but your experience and your guiding hand?" At that period he little foresaw that one day he himself would be attacked with a similar malady and that all my care would be ineffectual to preserve his life! (*Recollections of the Private Life of General Lafayette*, by M. Jules Cloquet, M.D., p. 41).[42]

UNIVERSITY RULES

In the framing of a code of laws for the government of the University, Mr. Jefferson—for he was their chief author—was under the erroneous impression, that more might be done with the students by an appeal to their patriotism and honor, than by positive punishment; but whilst it may be admitted, as a general rule, that certainty of punishment is more effective than severity; and that the best spirits amongst a body of highminded young gentlemen may be ruled and governed by such feelings as those invoked by Mr. Jefferson, the result proved, that all were not thus influenced; and that, in many cases, separation from the University was indispensable. It was fancifully believed by that distinguished personage, that the students themselves might be induced to form a part of the government, to constitute a court for the trial of minor offenses, and to inflict punishment on a delinquent colleague; and farther, that their cooperation might react beneficially in the prevention of transgressions. The scheme had a republican appearance, and was favorably thought of by the rector and board of visitors. In the first printed copy of the enactments of the Institution (1825) is the following:

The major punishments of expulsion from the University, temporary suspension of attendance there, or interdiction of residence or appearance within its precincts, shall be decreed by the professors themselves. Minor cases may be referred to a Board of six censors, to be named by the Faculty from among the most discreet of the students, whose duty it shall be, sitting as a board, to inquire into the facts, propose the minor punishment which they think proportioned to the offense, and to make report thereof to the professors for their approbation or their commutation of the penalty, if it be beyond the grade of the offense. These censors shall hold their offices until the end of the session of their appointment, if not sooner revoked by the faculty.

During the very first session of the University, events occurred to exhibit the insufficiency of any enactment of the kind; and, accordingly in the very next edition of the "Enactments" published in 1827, it was stricken out. So long, indeed, as I have elsewhere remarked (See an article on "College Instruction and Discipline" in the *American Quarterly Review* for June 1831, p. 294) as the *esprit du corps* or Burschenschaft, prevails amongst students, which inculcates that it is a

---

[42] First English edition issued in London, 1835. Germain Jules Cloquet (1790–1883), Baron, was Physician to the Hospital St. Louis and Professor of Surgery.

stigma of the deepest hue to give testimony against a fellow student, it is vain to expect any cooperation in the discipline of the institution from them. This "loose principle in the ethics of school boy combinations"—as it was termed by Mr. Jefferson—has indeed, led to numerous and serious evils. It has been a great cause of the combinations formed in resistance of the lawful authorities, of intemperate addresses, at the instigation of some unworthy member, and of repeated scenes of commotion and violence. It is rare for a youth to hesitate to depose in a court of justice touching an offense against the municipal laws of his country committed by a brother student. The youth, and the people at large are, indeed, distinguished for their ready attention to the calls of justice. Yet it is esteemed the depth of dishonor to testify when called upon by the college authorities, against the grossest violation, not only of collegiate, but of municipal law; as if it could be less honorable to give the same testimony before one tribunal than another; or as if the morality of the act differed in the two cases.

The fallacy of placing any reliance on appeals to reason and to sense of propriety on the part of the students; and the evils of this *Burschenschaft* were apparent before the termination of the first session. Offenses of a disturbing character were committed; and when the offenders were detected they were first admonished, and then mildly punished; until, at length, riot and disorder occurred, which could no longer be tolerated. "Nightly disorders"—says Professor Tucker in his *Life of Jefferson,* Vol. 2, p. 480—

were habitual with the students, until passing from step to step, they reached a point of riot and excess, to which the forbearance of the professors could no longer extend, when the students considered their rights violated, and openly resisted the authority of the faculty. This happened in October, immediately before the annual meeting of the visitors. The subject was laid before them by the faculty. More deep mortification, more poignant distress could not be felt than was experienced by Mr. Jefferson. The following day he came down with the other visitors from Monticello, which was their headquarters, summoned the students into their presence, and they were addressed in short speeches by himself, Mr. Madison, and Mr. Chapman Johnson.[43] The object of these addresses was not merely to produce in the young men a disposition to obey the laws, and return to their studies, but to induce the principal rioters to give up their names. The address of these men—the two first, venerable by their years, their services, and their authority—could not be resisted. The offenders came forward, one by one, and confessed their agency. Among those who thus almost redeemed their past error, by this manly course, was one of his own nephews (great-nephews). The shock which Mr. Jefferson felt when he for the first time discovered that the efforts of the last ten years of his life had been foiled and put in jeopardy by one of his family, was more than his own patience could endure, and he could not forebear from using, for the first time, the language of indignation and reproach. Some of the

offenders, among whom was his nephew, were expelled by the faculty; and others were more lightly punished. Their offensive memorial was withdrawn; the exercises of the University were resumed; and under a system liberal without being lax, a degree of order and regularity has been progressively increasing, and is supposed to be now nowhere exceeded.[44]

But not the least embarrassing circumstance connected with this riot was the resignation or proffered resignation of two of the professors—Professors Long and Key. When the faculty were deliberating on the punishments to be inflicted on the participators in the riot, they declared they were no longer members of the faculty. Yet these gentlemen had entered, like myself, into a covenant for five years; both were liberal in their political creed at home; and one—as I before remarked —had exclaimed a few months before that this was "the country in which he would wish to live and die." Subsequent reflection induced them to withdraw the plea of resignation, but the affair was, doubtless, the commencement of dissatisfactions, which induced them to seek employment elsewhere, and quit the University. Both these gentlemen had been students of Cambridge University: Mr. Long was, indeed, a fellow; both were dogmatical, as the students of Cambridge have often been considered to be; but Mr. Long was much the more amiable of the two; and having married Mrs. Selden, a sister to Mrs. Brockenbrough, who was the wife of the proctor of the University,[45] we presumed he would continue at the institution. Having obtained, however, a professorship at the then commencing University of London, he accepted it. Since then, his career has been varied. He studied law; but never practiced it, I believe, extensively. He was an excellent classical scholar; gave great satisfaction to his pupils whilst at the University; and, since he returned to London, has enriched the domain of literature by many valuable contributions, particularly in the department of philology. Before he left the university, he put into my hands the manuscript and printed copy of *An Introduction to the Study of Grecian and Roman Geog-*

---

43 Chapman Johnson (1779–1849), lawyer and member of the Virginia legislature, was a member of the Board of Visitors of the University of Virginia.

44 This riot was partly student reaction to "the European professors." Henry Tutwiler described the meeting at which Jefferson addressed the students as follows: "At a long table near the center of the room sat the most august body of men I have ever seen—Jefferson, Madison and Monroe. . . . Chapman Johnson . . . Cabell . . . Cocke . . . Jefferson arose to address the students. He began by declaring it was one of the most painful events of his life, but he had not gone far before his feelings overcame him, and he sat down, saying he would leave to abler hands the task of saying what he wished to say. Johnson then spoke, and when he called on the rioters to come forward and give their names, nearly all did so (Irving Brant, *James Madison, Commander in Chief 1812–1836,* p. 455, 1961). See John M. Dorsey, *ibid.,* pp. 39–41, and Robert Colin McLean, *George Tucker,* pp. 31–32, for further accounts of these student riots.

45 When Mr. Brockenbrough, the proctor, died on April 27, 1832, Dunglison wrote to Nicholas Trist requesting his support in obtaining the post of postmaster for Mr. Brockenbrough's son or wife, who were left in straightened circumstances (Library of Congress, Trist Papers, A.L.S. by Robley Dunglison).

raphy asking of me to see it through the press; and as he had only prepared the part which related to the Greeks, begging of me to do the same in regard to the Romans. This I endeavored to do; and in the year 1829, it was published at Charlottesville. Soon after his return to London, he became the editor of the *Journal of Education;* was an active Member of the Society for the Diffusion of Useful Knowledge; and edited the excellent *Penny Cyclopaedia.* Still, he would have been happy, at an after period, to return to this country; and, as late as 1850, would have accepted a professorship of ancient languages in one of our best universities, could such a situation have been offered him; and with this view, I was informed by Professor Tucker, had written more than one letter to him. He was, in many respects, a superior man to Mr. Key, but, not unfrequently, conveyed the impression of being unduly swayed by him in his opinions and decisions.

Mr. Key likewise received an appointment in the London University—as professor of Latin, I believe—Mr. Long being professor of Greek—and left the University of Virginia before his colleague. He was by no means amiable; was fond of controversies; and not always courteous, so that he got into numerous personal altercations. He was, indeed, one of the most impracticable men with whom I have ever been thrown in contact. He was, however, a competent and faithful teacher, and was highly thought of by his class.

VISITORS

Situated, as the university was, near the residence of Monticello, and on the high road from Richmond and Fredericksburg by Staunton to the Virginia Springs and to Guyandotte, it could not fail to be visited by numerous strangers, many of whom brought me letters of introduction. In 1825—the year of the commencement of the exercises—I noticed a stranger on the lawn in front of the pavilions, who was evidently at a loss; and who, on my approaching him, addressed me in regard to the institution. It proved to be the Duke of Saxe-Weimar Eisenach, who was on his way from the western part of Virginia, and intended to visit Monticello. In his published travels,[46] of which a translation appears in Philadelphia in 1828, under the title *Travels Through North America, During the years 1825 and 1826. By His Highness Bernhard Duke of Saxe-Weimar Eisenach,* he thus speaks of the University, as he saw it in November, 1825. "This establishment has been opened since March 1824." (It was really opened as I have already remarked, in April 1825)

and it is said to have already one hundred and thirty students; but a spirit of insubordination has caused many of the pupils to be sent away. The buildings are all new, and yet some of them seem to threaten to fall in, which may be

the case with several others also, being chiefly built of wood. The interior of the library was not yet finished, but according to its plan, it will be a beautiful one. The dome is made after the model of the pantheon in Rome, reduced one-half. This place is intended for public meetings of the academy; but it is said, that an echo is heard in case of loud speaking, which renders the voice of the speaker unintelligible. Under the rotunda are three elliptical halls, the destination of which is not yet entirely determined. The set of columns on the outside of this building, I was told, is to be a very fine one; the capitals were made in Italy. As for the rest, the ten buildings on the right and left are not all regularly built; but each of them in a different manner, so that there is no harmony in the whole, which prevents it from having a beautiful and majestic appearance. The garden walls of the lateral buildings are also crooked lines, which gives them a singular but handsome appearance. The buildings have been executed according to Mr. Jefferson's plan, and are his hobby; he is rector of the university, in the construction of which the State of Virginia is said to have laid out considerable sums of money. We addressed a gentleman whom we met by chance, in order to get some information, and we had every reason to be satisfied with his politeness. It was Dr. Dunglison, Professor of Medicine. He is an Englishman, and came last year with three other professors from Europe. He showed us the library, which was still inconsiderable and had been provisionally arranged in a lecture room; it contained some German belles-lettres works, amongst others a series of Kotzebue's "Calendar of Dramatic Works" (Vol. 1, p. 197).

Duke Bernhard was a frank, kind-hearted, liberal gentleman; not distinguished, however, for much stretch of intellect. He visited Monticello the same day, and made himself quite agreeable. I was to have dined at Monticello, and, had I done so, should have had a more extended opportunity for appreciating him, but was unable to be there. I was informed by the family, that he was prevented from reaching Monticello until the dinner had been cleared away; a circumstance, which he alludes to in his *Travels;* but he evidently enjoyed the remnants which were again brought to table; to all appearance, indeed, as much as if he had sat down at the commencement of the repast.

Immediately after the university went into operation, it was visited by Mr. Stanley [47]—now Lord Derby, Mr. Dennison,[48] Mr. Stuart Wortley [49] and others; and soon afterwards, by a gentleman who had been previously, and was afterwards, an extensive traveler; and lost his life, some time subsequently, by falling into a crater in one of the South Sea Islands. This was Count Vidua.[50] On introducing him to the Proctor, for he brought me a letter of introduction, I had the

---

[46] *Reise . . . durch Nord-Amerika in den Jahren 1825 und 1826,* Weimar, 1828. Karl Bernhard, Duke of Saxe-Weimar (1792–1862), a German general. See also Appendix I, p. 186.

[47] Edward Geoffrey Smith Stanley, Twelfth Earl of Derby (1752–1834), English politician and sportsman. He; Denison and Stuart-Wortley were traveling together.

[48] John Evelyn Denison (1800–1873), member of the British Parliament, later Viscount Ossington.

[49] Either James Archibald Stuart-Wortley-Mackenzie (1776–1845), first Baron Wharncliffe, grandson of Lady Mary Wortley-Montagu (of smallpox inoculation fame); or his eldest son, John Stuart-Wortley (1801–1855), both of whom were in Parliament at this time.

[50] Count Charles Vidua (1765–1832), famous Italian traveler.

mortification to forget, for the moment, the name of Mr. Brockenbrough, and exhibited more honesty than tact in not attempting to conceal my forgetfulness. It is a misfortune, that has probably befallen everyone in the course of his existence; but I was distressed to find, that in narrating the circumstance, the Count should have ascribed it to affectation on my part.

LAST ILLNESS OF JEFFERSON

In the Spring of 1826, the health of Mr. Jefferson became more impaired; his nutrition fell off; and, at the approach of summer, he was troubled with diarrhea, to which he had been liable for some years; ever since— as he believed—he had resorted to the Virginia Springs —especially the White Sulphur and had freely used the waters internally and externally for a psoric[51] eruption, which he had acquired in his journeying between Monticello and Washington, and which did not readily yield to the ordinary remedies. I had prescribed for this affection early in June, and he had improved somewhat; but on the 24th of that month, he wrote me the last note I received from him begging of me to visit him as he was not so well. This note was perhaps the last he penned. On the same day, however, he wrote an excellent letter to General Weightman, in reply to an invitation to celebrate, in Washington, the 50th anniversary of the Declaration of Independence which he declined on the ground of indisposition. This, Professor Tucker says, was probably his last letter. It has all the striking characteristics of his vigorous and unfaded intellect.

The tone of the note I received from him satisfied me of the propriety of visiting him immediately, and having mentioned the circumstance to Mr. Tucker, he proposed to accompany me. I immediately saw, that the affection was making a decided impression upon his bodily powers, and, as Mr. Tucker has properly remarked in his life of the distinguished individual, was apprehensive that the attack would prove fatal. Nor did Mr. Jefferson himself indulge any other opinion. From this time his strength gradually diminished and he had to remain in bed. The evacuations became less numerous but it was manifest that his powers were failing.

At this time he was visited by Major Lee,[52] a well

known political writer, whose feelings towards Mr. Jefferson had never been—I believe—very favorable; and whose moral character was certainly far from being without reproach. He had an interview with Mr. Jefferson, and found him so clear in intellect; and vigorous in discourse, that he expressed to me his feelings, that he would recover. I told him my anticipations were of the most gloomy kind, as I had already stated to the family; how much then ought I to have been surprised, when I saw in the *Richmond Enquirer* the particulars of Major Lee's visit, with the assertion, that I told him I thought he would recover. He did not, however, add what was nevertheless a fact—his having suggested to the family that they ought not to trust the life of such a distinguished individual wholly to the hands of a foreigner!—a sentiment which I have no doubt was more offensive to Mr. Jefferson's excellent daughter than it was to me.

Until the 2nd and 3rd of July he spoke freely of his approaching death; made all his arrangements with his grandson, Mr. Randolph, in regard to his private affairs, and expressed his anxiety for the prosperity of the University; and his confidence in the exertions in its behalf of Mr. Madison and the other visitors. He repeatedly, too, mentioned his obligations to me for my attention to him.[53] During the last week of his existence, I remained at Monticello; and one of the last remarks he made was to me. In the course of the day and night of the 2nd of July, he was affected with stupor; with intervals of wakefulness and consciousness; but on the 3rd, the stupor became almost permanent.[54] About seven o'clock in the evening of that day, he awoke, and seeing me standing at his bedside, exclaimed "Ah! Doctor are you still there?" in a voice, however, that was husky and indistinct. He then asked "Is it the 4th?" to which I replied "It soon will be." These were the last words I heard him utter. In Mr. Wirt's eulogy of him, it is said that he clasped his hands and said *"Nunc dimittis."*[55] No such expression was heard

---

[51] Scabies, for which Dunglison considered sulfur ointment the best application. Jefferson's diarrhea first appeared in 1801 and may have been due to self-medication with calomel, to which Jefferson was prone, or to amebiasis; but malignancy of the bowel would be the most likely diagnosis (Karl C. Wold, *Mr. President—How Is Your Health?* p. 31, Bruce Publishing Co., Saint Paul, 1948). Dunglison was most apprehensive about the dire consequences of this longstanding diarrhea (John M. Dorsey, *ibid.*, pp. 66–67).

[52] Henry Lee (1787–1837) corresponded with Jefferson while re-editing the Memoirs of his father, General Henry Lee, who was a bitter critic and political foe of Jefferson. William Cabell Rives, in 1831, tried to dissuade the younger Lee (irked by Randolph's account of Jefferson and the elder Lee) from an

attack upon Jefferson's character, joining the name of Jefferson, Madison, and Monroe with that of Washington, and saying "I would prefer to see these our demigods, occupying their respective niches in our national Pantheon . . . undisturbed by . . . the echo of contemporary passions . . ." (A.L.S. Nov. 4, 1831, Library of Congress, Personal Papers, Misc. Collection).

[53] For letters between Jefferson and Dunglison on this point, see John M. Dorsey, *ibid.*, pp. 34–45.

[54] This was not diabetic coma, as someone has apparently concluded. A medical student visiting Monticello stated that Jefferson was suffering from diabetes, but by this he meant excessive urination. Jefferson himself, in a letter to James Monroe, explained that Dr. Dunglison was familiar with "diabetes or incontinence of urine," but that his (Jefferson's) complaint was "difficulty of making water"; to William Short he wrote that it was "an affection of the bladder and prostate gland" (James A. Bear, Jr., Curator of the Thomas Jefferson Memorial Foundation, *Medical Chronology of Thomas Jefferson,* unpublished manuscript, Charlottesville, Va.). Dunglison knew how to test for sugar in the urine.

[55] Now lettest thou thy servant depart—Luke 2: 29.

by me; and if any other person had heard it, it would certainly have been communicated to me. Until toward the middle of the day—the 4th—he remained in the same state, or nearly so; wholly unconscious to everything that was passing around him. His circulation was gradually, however, becoming more languid; and for some hours prior to dissolution, the pulse at the wrist was imperceptible. About one o'clock he ceased to exist.

THE RANDOLPHS

Soon after we had closed the eyes of the venerable patriot, I was called to visit Mrs. Randolph, by desire of her husband; when a singular scene presented itself. That excellent lady and her equally excellent daughters were in the deepest distress, whilst Governor Randolph was taunting her for not shedding a tear, and ascribing it to a morbid condition, which I might—he urged—rectify. Nothing as a matter of course could be done, but to enjoin quiet which she was not likely to obtain. I have before alluded to the eccentricity, if not mental aberration, of that gentleman. On two successive summers my family and myself spent the vacation of the university at Monticello; and although Governor Randolph generally returned after dark from his plantations, I never saw him there. His entrance and his exit were equally unknown to me. He took a violent dislike to those who were most intimate with his family; and was on the worst possible terms with his own son, Thomas Jefferson Randolph, whom he had attempted to stone in a lonely place, but was prevented by the latter putting spurs to his horse.

I had the pleasure, however, of witnessing an entire reconciliation on his deathbed with all the family. In his last illness, evidently produced by narrowness of the bowels, and adhesion of them to each other, I was called in consultation by Dr. Bramham [56] of Charlottesville; and attended him until he died. Nothing could be more correct than his deathbed conduct. All animosity was forgotten; and he died at peace with those whom he had treated with so much cruelty. "Originally" says Mr. Tucker (*Life of Jefferson* Vol. 2, p. 301)

one of the most generous, disinterested, and high minded men on earth, he was gradually transformed into a gloomy, unsocial misanthrope—his proud spirit suffering intensely, but suffering in silence, seeking solace of no one, but showing too plainly the discontent, which secretly preyed on his mind, by the harshness or coldness with which he treated all around him.

He was exceedingly fond of natural history and, it was said, had composed a treatise on that of the hog; but, since his death, I have not heard of it. He was

said, too, to have introduced horizontal plowing—an important improvement in hilly lands in preventing the formation of gullies.

PRIVATE LIFE OF THOMAS JEFFERSON

The opportunities I had of witnessing the private life of Mr. Jefferson were numerous. It was impossible for anyone to be more amiable in his domestic relations; and it was delightful to observe the devoted and respectful attention that was paid him by all the family. In the neighborhood too he was greatly revered. Perhaps, however, according to the all-wise remark, that no one is a prophet in his own country, he had more personal detractors there—partly owing to the differences in political sentiments, which are apt to engender so much unworthy acrimony of feeling; but still more perhaps owing to the views, which he was supposed to possess, on the subject of religion; yet it was well known, that he did not withhold his aid when a church had to be established in the neighborhood, and that he subscribed largely to the Episcopal Church erected in Charlottesville. After his death, much sectarian intolerance was exhibited, owing to the publication of certain of his letters, in which he animadverted on the Presbyterians more especially; yet there could not have been a more unfounded assertion than that of a Philadelphia episcopal divine, never celebrated for his liberality or tolerance, that Mr. Jefferson's memory was detested in Charlottesville and the vicinity. It is due, also, to that illustrious individual to say, that in all my intercourse with him I never heard an observation that savoured, in the slightest degree, of impiety. His religious belief harmonized more closely with that of the Unitarians than of any other denomination, but it was liberal, and untrammeled by sectarian feelings and prejudices. It is not easy to find more sound advice, more appropriately expressed, than in the letter, which he wrote to Thomas Jefferson Smith,[57] dated February 21st, 1825:

This letter will, to you, be as one from the dead. The writer will be in the grave before you can weigh its counsels. Your affectionate and excellent father has requested that I would address to you something which might possibly have a favorable influence on the course of life you have to run; and I, too, as a namesake, feel an interest in that course. Few words will be necessary, with good dispositions on your part. Adore God. Reverence and cherish your parents. Love your neighbor as yourself, and your country more than yourself. Be just. Be true. Murmur not at the ways of Providence. So shall the life into which you have entered be the portal to one of eternal and ineffable bliss. And if to the dead it is permitted to care for the things of this world, every action of your life will be under my regard. Farewell.

---

[56] Horace Bramham (1798–1834), M.D., son of Peggy Marshall and Colonel Nimrod Bramham, practiced in Charlottesville, Va., presumably at his home, "Oak Lawn," where he was a man of some affairs in the community.

[57] Son of Jefferson's old friend, Samuel Harrison Smith, one-time editor of the *National Intelligencer* (Adrienne Koch, and William Peden, *Life and Selected Writings of Thomas Jefferson*, p. 717, 1944).

The portrait of a good man by the most sublime of poets, for your imitation.

Lord! Who's the happy man that may to Thy blessed
  courts repair
Not stranger like to visit them, but to inhabit there?
Tis he whose every thought and deed by rules of virtue
  moves;
Whose generous tongue disdains to speak the thing his
  heart disproves.
Who never did a slander forge, his neighbor's fame to
  wound;
Nor harken to a false report, by malice whispered round.
Who vice and all its pomp and power can treat with just
  neglect;
And piety, though clothed in rags, religiously respect.
Who to his plighted vows and trust has ever firmly stood,
And though he promise to his loss, he makes his promise
  good.
Whose soul in usury disdains his treasure to employ;
Whom no rewards can ever bribe the guiltless to destroy.
The man, who by his steady course has happiness insured,
When earth's foundations shake, shall stand by Providence
  secured.
(*Memoir, Correspondence and Miscellanies*, Vol, IV p. 413).

On the last day of the fatal illness of his granddaughter, who had married a most unworthy person of the name of Bankhead,[58] a man of the most intemperate habits, and, so far as I know, possessed of no redeeming virtues, Mr. Jefferson was present in the adjoining apartment, and when the announcement was made by me, that but little hope remained,—that she was, indeed, moribund, it is impossible to imagine more poignant distress than was exhibited by him. He shed tears; and abandoned himself to every evidence of intense grief.

It was beautiful too to witness the deference that was paid by Mr. Jefferson and Mr. Madison to each other's opinions. When, as secretary, and as chairman of the faculty, I had to consult one of them, it was a common interrogatory—what did the other say of the matter? If possible, Mr. Madison gave indications of a greater intensity of this feeling; and seemed to think, that everything emanating from his ancient associate must be correct. In a letter, which Mr. Jefferson wrote to Mr. Madison a few months only before he died (Febr. 17, 1826) he thus charmingly expresses himself:

The friendship which has subsisted between us, now half a century, and the harmony of our political principles and pursuits, have been sources of constant happiness to me through that long period. And if I remove beyond the reach of attentions to the university, or beyond the bourne of life itself, as I soon must, it is a comfort to leave that institution under your care, and an assurance, that it will not be wanting. It has also been a great solace to me to believe, that you are engaged in vindicating to posterity the course we have pursued for preserving to them, in all their purity, the blessings of self-government, which we had assisted, too, in acquiring for them. If ever the earth has beheld a system of administration conducted with a single and steadfast eye to the general interest and happiness of those committed to it;—one which protected by truth, can never know reproach, it is that to which our

[58] Anne Randolph married Charles S. Bankhead, an inebriate.

lives have been devoted. To myself you have been a pillar of support through life. Take care of me when dead, and be assured, that I shall leave with you my last affections. (*Memoir* etc. Vol. IV. p. 428).

It is somewhat singular however that about the very time this letter must have been penned, Mr. Jefferson should have declared at table, in my presence, that he had no desire for posthumous reputation, nor could he well understand how anyone could be anxious for it. I was surprised at the time to hear the sentiment expressed. The prospect of future rewards and punishments is confessedly one of the greatest incentives to correctness of conduct, and the transmitting of a good name to posterity must enter largely into the consideration of the good as one of those future rewards; and such could scarcely fail to have been the feeling with Mr. Jefferson when he asked Mr. Madison to take care of him when dead. Some paradox may have been involved in the remark, which it is not easy to unravel.

As a relic of Mr. Jefferson, I possess the thermometer,[59] made by Jones of Holborn, with which, for forty years of his life, he regularly registered the varying condition of the instrument, and he startled the professor of Natural Philosophy—Mr. Bonnycastle—not a little, when he informed him, that he considered one of the best times for taking the observation to be three o'clock in the morning. This relic was presented to me by the family; as well as the clock, which stood in his bedroom. By this he rose, whenever he was able to see the hands. It is made by Voigt of Philadelphia; is a specimen of excellent workmanship, and had, when at Monticello, a mercurial pendulum, but this was broken in moving it down to the university. I hope it will be taken care of by my family, for I prize it greatly from its associations rather than from its extrinsic excellence. Its rate of going is marked on the inside of the case in the handwriting of Mr. Jefferson; and the days of the week, as reached by the weight of the clock, day after day, are indicated, the clock going eight days, and being wound up on the Sundays. Isolated, as it were, on Monticello, and often in the recess of his own study, the precise day of the week might readily escape him, and this was a convenient method of reminding him thereof.

I had no knowledge of the intention of Mrs. Randolph to bestow this clock on me; but had determined to possess it, if it went at a reasonable rate, at the sale. When put up, it soon reached *one hundred dollars.* General Cocke, of Fluvanna,[60] bade, I think, *145 dollars.* I bade 150; and it was knocked down to Mr.

[59] This and two goblets from a set that Dunglison received from Jefferson were presented by his great-grandson, Mr. Tifft, to the Historical Society of Pennsylvania. "Fever heat" on this thermometer had an indicator marked at 112°.
[60] General John H. Cocke (1780–1866), from the beginning a great friend to the University of Virginia, was often at Monticello in Jefferson's later days (Pierson, *ibid.*, p. 57). Bremo was the name of his country seat in Fluvanna County, Virginia.

Trist for one hundred and fifty-five dollars. I immediately went up to Mr. Trist, apologizing for having opposed unwittingly the desire of the family to possess the clock, when he told me I might make my mind easy, as he had been commissioned by them to buy it, in order that they might present it to me.[61]

ROBERT PATTERSON AND JOHN K. KANE

On the withdrawal of Mr. Key from the university, the chair of Mathematics became vacant; for which, as being more lucrative, Mr. Bonnycastle was an applicant and was appointed. The professorship of Natural Philosophy thus became vacant, and was conferred on Dr. Robert M. Patterson,[62] at the time, or previously, Professor of Natural Philosophy and Vice-Provost of the University of Pennsylvania. Previous to accepting the appointment, Dr. Patterson, in company of Judge Kane[63]—then Mr. Kane—visited the university, to inspect the whole ground. This was the commencement of an acquaintance, that has been the source of no little pleasure to me. The day—a broiling one in June 1828 —was one of the hottest of the season. The gentlemen brought me a letter of introduction—if I recollect rightly —from Mr. Walsh,[64] and although my family was by no means in a fitting state, from indisposition, for receiving company, I insisted, that they should spend the day with me and return to Charlottesville in the cool of the evening. To this they consented; and I cannot help flattering myself, that the reception they received had something to do with the determination of Dr. Patterson to accept the chair. It is now (1852) twenty-four years that we have been intimately acquainted with each other, and my regard and esteem for both the gentlemen has continued without any diminution. For a portion of the time—as I shall mention more particularly hereafter—we formed a part of the same social club "The Five," and had been thrown sufficiently into each other's society to enable a correct judgment to be formed; and I believe, that judgment has been mutually favorable. In the autumn of that year (1828) with the commencement of the session, Dr. Patterson entered upon his duties at the university; and took his meals with me until his pavilion was ready.[65]

DUNGLISON AT THE UNIVERSITY

It was during this time, that I had the misfortune to lose my second child—a fine boy eleven months old. In October 1825, my wife was confined of a daughter; and, in November 1827, of a son. He was a noble, hearty child; full of intelligence, when he was attacked with bronchitis, which proved fatal in the course of ten days. We had passed a part of the summer at Brown's Tavern at the foot of Blue Ridge; and on our return to the university, an old miller, of the name of Maupin, struck with his good looks, rudely remarked, that he was a fine boy, but might be dead in a week. The observation would have been forgotten, had it not grated harshly on the mind at the time, and had not the child died so soon afterwards. He died at the time of the annual meeting of the visitors of the university; and his remains were followed to the grave by Ex-presidents Madison and Monroe, the visitors, the professors and the students of the institution. Rarely perhaps has so distinguished an assembly attended to the grave the body of one so young.

At the university I undertook, according to the arrangements of the board of visitors, the whole domain of medicine:—it being at first intended to afford to the general student a knowledge of medical science preparatory to his farther cultivation of it elsewhere. It was not deemed probable, that many of those, who were intended for the active exercise of the profession, would be satisfied with the amount of knowledge they could obtain, in the absence of clinical advantages.[66] It soon became apparent however, that opportunities for practical information on anatomy and on medicine and

[61] Henry S. Randall in his *Life of Thomas Jefferson* (3: pp. 512–519, 547–549, New York, 1858) presents long excerpts from Dunglison's "Ana" that relate to Jefferson (see Appendix I), and succeeding biographers have often quoted from this source (William Eleroy Curtis, *The True Thomas Jefferson*, p. 369, Lippincott & Co., Phila. & London, 1901; Saul K. Padover, *Jefferson*, pp. 402, 420, Harcourt, Brace & Co., 1942; Gilbert Chinard, *Thomas Jefferson: The Apostle of Americanism*, pp. 402, 423, 425, 516, 520, 531, 2nd ed., Univ. Mich. Press, 1957).

[62] Robert Maskell Patterson (1786–1854), A.B. 1804, M.A. 1807, M.D. 1808, University of Pennsylvania, Professor of Natural Philosophy, Chemistry and Mathematics at the University of Virginia from 1828 until 1835, was elected a member of the American Philosophical Society at the age of twenty-two, the youngest ever to be so honored up to that time, and in 1843 wrote a history of the society. In 1814 he married Helen Hamilton Leiper, daughter of a friend of Thomas Jefferson, and sister-in-law of Judge John K. Kane, close friend of Dunglison (Simpson, *Lives of Eminent Philadelphians*, p. 761, 1859).

[63] John Kintzing Kane (1795–1858), Judge of the United States District Court, associated with the Erie Railroad, Chesapeake and Delaware Canal, Girard College, Girard Bank, the Second Presbyterian Church of Philadelphia, The Academy of Fine Arts, The Musical Fund Society and the American Philosophical Society, was a supporter of Andrew Jackson and wielded great political power in the Democratic party (Simpson, *ibid.*, p. 617). He married Jane Duval Leiper, and two of their sons (not without being influenced by Dr. Dunglison, no doubt) became physicians. His *Autobiography*, written in 1849 and published in 1949, seems to have set the example for Dunglison's own *Ana*, and a college club of five, "The Highlanders," seems to have been the inspiration for a similar club of five which included Dunglison and is described by him in these *Ana*.

[64] Robert Walsh (1784–1859), lawyer, writer and publisher, was for twelve years editor of the *National Gazette*, in its time a leading afternoon paper of Philadelphia (Simpson, *ibid.*, p. 938).

[65] For further remarks on Dr. Patterson, see pp. 62–66, 170–171 of these *Ana*.

[66] When the University was opened on March 7, 1825, with 68 students, 20 were enrolled in the School of Anatomy and Medicine under Dunglison (Andrew De Jarnette Hart, Jr., Thomas Jefferson's Influence on the Foundation of Medical Instruction at the University of Virginia, *Annals of Medical History*, n.s., 10: 58, 1938).

surgery might be afforded; an anatomical hall was constructed; and in the course of a year or two, at my suggestion and with his consent, the department of Materia Medica was added to that of Chemistry and taught by Dr. J. P. Emmet; and a demonstratorship of anatomy and surgery was appointed,[67] the first incumbent of the office being a nephew of Mr. Chapman Johnson, one of the visitors.[68] He did not, however, give entire satisfaction, and left the university the year after I did.[69] The duties, under this arrangement, which devolved upon my chair were lectures on the theory and practice of medicine, Physiology, Obstetrics and Medical Jurisprudence. The last branch was made the subject of a special ticket, with a fee of $15.00 and could be attended not only by the medical student, but by any one desirous of obtaining such information; and, accordingly, it was followed by many gentlemen, who have since become prominent as lawyers and legislators.

To facilitate them in this, I published, in the year 1827, a syllabus of my lectures on that subject as well as on the treatment of poisoning and suspended animation.

ON COLLEGE DEGREES

It was a novel feature, as regarded the institutions of this country, to permit, in the University of Virginia, graduation in the separate schools of ancient languages, modern languages, mathematics, chemistry, natural philosophy and moral philosophy as well as in those of law and medicine; but this liberal provision was, at first, rendered nugatory in every school except that of Ancient Languages by another provision in the enactments, that no diploma should be given to anyone, who had not passed such an examination in the Latin language, as proved him able to read the highest classics in that language with ease, thorough understanding and just quantity. As regarded the department of medicine, I considered it to be equivalent to a prohibition, in as much as it would require a greater amount of time to be spent in the acquisition of such a preliminary education than was practicable; or would be deemed advisable; seeing that, in the most prominent medical schools of the country, no knowledge of the ancient languages was required as a preliminary to graduation. Moreover, it did not appear consistent to declare that a youth might obtain a degree for chemistry, for example, which is in no wise connected with the ancient languages; and then to demand such an acquaintance with the latter as would render the declaration null and void. Accordingly this subject was brought by me before the faculty, during the very first session; and they agreed to suggest a modification of the enactment relating to graduation,[70] which was adopted by the board of visitors, with—I was informed—but one dissenting voice—that of Mr. Jefferson. It was then determined, that any candidate for graduation in any of the schools shall give satisfactory proof of his ability to write the English language correctly; and the enactment thus exists at the present day (1852) (See an article by the author, *American Quarterly Review* for June 1831, p. 308 [71] and his *Medical Student* p. 17).

So far as it went, this plan of permitting graduation in the separate schools of the university was judicious; but an evil existed, in no special course of study being recommended, no goal of more elevated attainment, which might excite the attention and emulation of those whose opportunities admitted of their being well educated.[72] This deficiency led me to suggest a higher university degree in the article in the *American Quarterly* above referred to; which should require an attention to language, to the sciences relating to magnitude and numbers and to those that embrace the phenomena of mind and of matter. It should comprise, in other words, an acquaintance with ancient languages, and with mathematics; and when the student has attained this more elementary instruction, he is capable of undertaking satisfactorily the study of physics, and of becoming acquainted with the bodies that surround him, and the laws that govern them, as well as of entering upon the science of moral philosophy, and of comprehending the interesting subject of his own psychology. These it seemed to me, were the only departments of knowledge that needed to be acquired for a university degree. They comprise an acquaintance with the ancient classics, and the philosophy of language, as well as with mathematical, physical and metaphysical facts and reasonings; and their acquisition will enable the student to enter upon professional or political life with every advantage. I did not consider it necessary to include a knowledge of the modern languages, in as much as the valuable stores to be drawn from them, especially from the French and German, are, of them-

---

[67] This was in 1827, when the school was enlarged to a department and organized as follows: Robley Dunglison, M.D., Professor of Physiology, Theory and Practice of Medicine, Obstetrics, and Medical Jurisprudence; John P. Emmet, M.D., Professor of Chemistry and Materia Medica; Thomas Johnson, M.D., Demonstrator of Anatomy and Surgery (Herbert B. Adams, *ibid.*, p. 178).

[68] The first catalogue of the University of Virginia lists the Board of Visitors as follows: "Thomas Jefferson, Rector. Thomas Jefferson, James Madison, Chapman Johnson, Joseph C. Cabell, James Breckenridge, John H. Cocke, George Loyall and Secretary of the Board, Peter Minor."

[69] Thomas Johnson was advanced to the professorship in 1832 but he resigned in 1834 to become Professor at the short-lived Richmond Medical School. In 1838 Hampden-Sidney College established a medical department, on the faculty of which he served until 1844 (Wyndham B. Blanton, *ibid.*, p. 46).

[70] Dunglison then directed a long letter, April 3, 1826, to the Rector and Visitors of the University of Virginia (John M. Dorsey, *ibid.*, pp. 54–61), from which it would seem that the Latin requirements for graduation retarded student enrollment and, consequently, the emolument of the Professor of Medicine.

[71] On College Instruction and Discipline. Dunglison and Jefferson did not see eye to eye on the "honor system" for students.

[72] Jefferson desired to foster the elective system of study, allowing the students uncontrolled choice in the lectures they chose to attend.

selves, attractions, which render unnecessary collegiate restraint or recommendation. No one indeed can be esteemed well educated who is thoroughly ignorant of them.

These views were laid by me before the Faculty and a higher degree was recommended by them to the Board of Trustees at their next meeting which was adopted in accordance with the views stated above, excepting that they required in addition a knowledge of any two of the modern languages taught at the University; and this more perhaps with the view of attracting the attention of the students to the modern languages, which were not as well followed from some cause or other as they ought to have been; and from the representations of the professor of that department, that the exclusion of these languages from the requirements for a higher degree would cause them to be still farther neglected, than from their conviction, that a *Bachelor* or *Master of Arts*, or University Graduate no matter under what appellation, should, as a general rule, be required to possess an acquaintance with the modern languages.[73]

This degree was difficult of attainment, and hence, whilst the graduates of other institutions are annually numerous, the University of Virginia did not confer its highest degree, at any time perhaps on more than half a dozen young gentlemen, and in some years, I believe, not a single candidate presented himself. It is indeed by no means true, that the literary or scientific institution which "raises the standard," as it has been termed, and makes its highest honors the most difficult of attainment, will have the greatest number of candidates, and of students pressing forward for distinction. It may be said however, that they who succeeded would be more respected, and successful, than the alumni of other institutions where graduation was more easy; but I have not had any evidence of this. On the contrary, I do not know of a single case in which marked advantage has accrued to the graduate, for the toil which he had to undergo for this more elevated collegiate distinction. A certain, or rather uncertain, amount of literary and scientific knowledge is admitted by all to be requisite; but the general impression in this country—erroneous as it may be—is, that high literary and scientific distinction is unnecessary and, many even believe, injurious in the transaction of the business concerns of life. Hence, in my own profession, we not unfrequently hear of the learned and accomplished physician being dismissed, and the lowest empiric consulted in his place. It is strange, but not less true than strange, that an individual ordinarily—and even more than ordinarily—intelligent will consent to confide in the professional judg-

ment of one whose opinions on other matters he would contemn.

OFFERED PROFESSORSHIP ELSEWHERE

Whilst I resided at the university I had many applications to leave and accept a professorship elsewhere. The first of these was from Dr. Wm. P. C. Barton, a professor in the Jefferson Medical College, asking me if I would accept a chair in the Jefferson Medical College, and desirous also that my friend Dr. Robert M. Patterson should accept another.[74] His letter is dated June 15, 1829. This offer I declined.

In December 1830, I received a letter from Dr. Drake, dated Philadelphia, Decr. 20, 1830, to the following purport,—

Sir:
The Miami University in Ohio has resolved to establish a medical department in the city of Cincinnati, and has authorized me to select suitable professors for the different chairs, subject, of course, to the ratification of the board of trustees. The object of this note is to tender to you the chair of anatomy, which I hope you will feel inclined to accept.

As it is the wish of the board to complete the organization of the faculty without delay, I must request the favor of an early reply to my proposition.

I have the honor to be, respectfully,
Your obedient servant
Danl. Drake M.D. Dean of the
Medical Faculty
Prof. Dunglison
Univ. of Virginia

In this case too it was desired that Dr. Patterson should be professor of chemistry, if he would consent to accompany me. After much correspondence however we declined the proposition, although it was accompanied by the offer of a guarantee of three thousand dollars for the session.[75]

Soon after this, in Jan. 1831, I received letters from Drs. S. McClellan, Dean;[76] and Dr. Jacob Green,[77]—

---

[73] Except for the M.D., the first graduates of the University of Virginia did not receive degrees, but certificates which declared them eminent in the school or schools attended. In 1828 the faculty reported it favored the title of Master of Arts upon graduation, and this was adopted constitutionally in 1831, when the visitors also desired the faculty to consider and report whether higher or other degrees should be provided for.

[74] In a reshuffling of the Jefferson Medical College faculty during the summer of 1828, Dr. Robert M. Patterson had been appointed to the Chair of Anatomy, but in consequence of having transferred to the University of Virginia, he declined. Dr. W. P. C. Barton took the Chair of Materia Medica and Botany at that time (James F. Gayley, *History of the Jefferson Medical College of Philadelphia*, p. 19, Phila., 1858).

[75] Daniel Drake (1785–1852) was Professor of Medicine for the term of 1830–1831 at the Jefferson Medical College of Philadelphia, where he went for the avowed purpose of recruiting teachers for the projected Miami Medical College in Cincinnati. Even George McClellan, the founder of Jefferson Medical College, received an appointment but changed his mind about going to Cincinnati (Emmet Field Horine, *Daniel Drake*, pp. 248–258, University of Pennsylvania Press, Phila., 1961).

[76] Samuel McClellan (1800–1854), M.D. 1823, Yale Medical School, was on the Jefferson Medical College faculty from 1828 until 1839, when he left with his brother to form the Pennsylvania Medical College. He was like the gentle Zephyrus compared to his brother, George, who was more like Jupiter Tonans on the whirlwind (Wm. H. Pancoast, *Introductory delivered October 5, 1874*, p. 22, Phila., 1874).

[77] Jacob Green (1790–1841), M.D. 1827, Yale (honorary?),

Professors of the Jefferson Medical College, informing me of vacancies in that institution [78] and asking of me if I would accept a chair—the former gentleman kindly offering to give up his chair of anatomy if I preferred it. I replied to the former at some length asking for information on certain topics, which he communicated to me in a letter, dated Philadelphia, Feb. 3, 1831. The inducements held out were not however by any means strong enough to induce me to leave my position at the University of Virginia; and I therefore declined the proposition.

NEGOTIATIONS WITH UNIVERSITY OF MARYLAND

These offers had become known; as well as the fact, that repeated attacks of acute rheumatism accompanied with endocarditis had made me disposed to think that the locality was not favorable to the health of my wife, and had led me to think it might become proper for me to change my rural for a civic residence. A report, that I designed to relinquish my chair at the University of Virginia had reached Baltimore, and I received a letter from Professor N. R. Smith,[79] dated Mar. 15, 1831, stating this circumstance, and likewise, that my name had been mentioned in relation to the vacant chair of anatomy in the University of Maryland, but whether with my approbation he did not know. He stated, at the same time, that Dr. Wright [80] of Baltimore and Dr. Geddings [81] of Charleston, had been spoken of; and offered his services to aid me in any manner in making my application, should I desire to become a candidate. About the same time, the pro-

fessor of chemistry, Dr. DeButts died; [82] and Dr. Smith, with other of the Faculty, were anxious that Dr. Patterson should be appointed to the vacant chair. Neither of us was desirous of being regarded as an applicant; but both expressed their willingness to accept the chairs should they be tendered us. Local interest, however, independently of every other consideration, was overwhelming; and Drs. Wright and Ducatel,[83] both of whom were highly respectable gentlemen, received the appointments. Dr. Wright was an exceedingly nervous man, and his apprehensions and sensitiveness became ultimately so great, that he resigned the office before the commencement of the session, and at so late a period that I could not, consistently with my engagements to the University of Virginia, have availed myself of the situation if it had been offered to me. I was regarded, according to Dr. Smith, as unattainable, and the faculty determined to recommend Dr. Geddings to the board of trustees, who appointed him accordingly.

The following letter was written to me by Mr. Somerville of Baltimore, one of the trustees of the University of Maryland, with whom I had become acquainted at the Virginia Springs.

Bloomsbury, September 13th, 1831

Dear Sir:

I was disappointed on my arrival in Baltimore to learn, that you had not visited our friends there agreeably to your intention when I parted with you at the Virginia Springs. I have not seen since my return any one of our Board of Trustees, but from an interview with several distinguished gentlemen of our Faculty I am seriously apprehensive that we shall be obliged to go into a new election of professor of anatomy in order to fill the chair for the approaching course. The ill health of our professor elect, Dr. Wright, will prevent him from attending to the arduous and responsible duties to which he had so recently been appointed. I have not yet had an interview with the Doctor, but I have such thorough evidence of the correctness of the fact of his determination to withdraw, that I consider the chair of anatomy as now vacant. While I deeply lament this misforutne to our school, it becomes our duty to remedy the evil with as much promptness as the nature of the circumstance will possibly permit. At the late election we had all the talents of the country before us, and the question now is, whether the field of selection has since become more limited, in consequence of the private arrangements or subsequent views of the distinguished gentlemen whose services were then at our command. From our conversation at the White Sulphur I fear it would be impossible for you to remove to Baltimore immediately, even should our Board be disposed to give you its unanimous support. I write *privately* to you on this subject; but with a view of communicating immediately with some of the members of the appointing power I would like you to

held the Chair of Chemistry at Princeton from 1818–1822 and assumed the same chair at Jefferson Medical College when it opened in 1825. Refraining from faculty controversy, he succeeded in holding his chair until his death.

[78] The vacancies, of course, were due to the defection of Drake and Eberle to Ohio.

[79] Nathan Ryno Smith (1797–1877), son of Nathan Smith, who had been instrumental in Yale's awarding the honorary M.D. to Dunglison, taught for fifty years (1827–1877) at the University of Maryland. He was President of the Medico-Chirurgical Faculty of Maryland and Honorary Fellow of the College of Physicians of Philadelphia.

[80] Thomas H. Wright (died 1856), honorary M.D. 1819, University of Maryland, was physician to the Baltimore Almshouse and a frequent contributor to the medical press (Eugene Fauntleroy Cordell, *Historical Sketch of the University of Maryland School of Medicine*, p. 72, 1891).

[81] Eli Geddings (1799–1878), M.D. 1825, Medical College of South Carolina, served in the Confederate Army and after the war revived the Medical College of South Carolina. When John Beale Davidge died in 1829, he was first succeeded by Wells, of Boston, for one term, then by Lincoln, of Vermont, for another term, after which it was announced that competition would be open to all comers. Six applied for the position, Dr. Dunglison among them, but Wright was chosen because he was a Baltimorean. He soon resigned, however, and for the fourth time in two years an election was held, in 1831, and this time Geddings was chosen (Cordell, *ibid.*, pp. 67–74).

[82] Elisha De Butts (1773–1831), M.D. 1805, University of Pennsylvania, had been Professor of Chemistry since 1809.

[83] Jules Timoleon Ducatel (1796–1849), son of a Baltimore pharmacist, transferred from the Chair of Chemistry and Geology in the Arts & Science Department to fill the Chair of Chemistry in the Medical Department left vacant by De Butts. He was a member of the American Philosophical Society and other scientific bodies (Cordell, *ibid.*, p. 78).

inform me if it would be impracticable for you to accept the appointment of professor of anatomy in our school, should it meet the approbation of the Board of Trustees. I feel very confident the choice will rest between yourself and Dr. Geddings of South Carolina, or to the latter gentleman exclusively should it be totally impossible for you to serve us. At this late season, I fear we can procure no professor who will be able to do as ample justice to our Institution or to himself as if he had received the appointment at the proper period. But so it is, as Dean Swift says—"it is impossible to make a silk purse of a sow's ear." But if we can obtain an able man of prompt and decided character, firm in purpose, and determined to unite or identify his reputation with that of our University, we shall ultimately, in spite of all rivalship, succeed in giving to our School that celebrity which is so anxiously desired by its warmest patrons.

I remain dear Sir, yours very respectfully
H. V. Somerville
To Dr. Dunglison U. Va.

To this letter I immediately returned the following answer.

University of Virginia
Sept. 18, 1831.
Dear Sir,

I need hardly express to you my acknowledgments for your favour of the 13th inst. which I received yesterday, or my regret that you did not stay at the University on your way to Baltimore. With regard to the immediate subject of your communication it is due to you to express my sentiments fully and fairly. When the chairs of anatomy and chemistry were both vacant, I did think, that if they were filled by zealous and competent individuals success was almost certain, Baltimore offering so favorable a situation for an extensive Medical School. I think so still, and see no reason why, if properly organized and governed, the University of Maryland should not equal—nay, in process of time (and that time not distant) take precedence of that of Philadelphia. It is on the road to that town from the South, and is possessed or might be possessed of every facility which Philadelphia affords. In my brief conversation with you at the Springs I communicated in sincerity my fears regarding the prosperity of the Baltimore Medical School under present circumstances. The system of appointing to the chairs is one which cannot, I think, admit of justification, and must, sooner or later, inevitably prove fatal to any Institution where it prevails. Unquestionably, if two persons are candidates for any vacant appointment of equal talents and experience, and equally successful as lecturers, the one a resident and the other not,—the former should be preferred; but in the case of the late appointment to the chair of anatomy, the incumbent, estimable and excellent as he is described to be by yourself and others, has never—I am told—delivered a course of public lectures; whilst fears, at that very time, were entertained by his most intimate acquaintances, that he would fail in decision and promptness of character. It is right, likewise, that I should say in confidence, that some of the most distinguished members of the profession, and others, friends to your institution, have informed me both prior to and since, my name was before you, that if Dr. Patterson and myself had been appointed, the burden of the University would have fallen chiefly upon us, that the only able colleague was Dr. Smith, and of him I have never heard but one opinion. Notwithstanding this discouraging intelligence, which I find is circulating everywhere, and the want of harmony in the Faculty, we were willing to serve had we been elected. At the least we should have been zealous

and untiring in our exertions for the prosperity of the School, and with Dr. P. so satisfied am I of his value, could an opening be made for him in the department of chemistry or elsewhere, I would even yet join you in a future session. My emolument as Professor here I consider to have fluctuated from 2500 to 3000 dollars.[84] As Professor in Baltimore, neither Dr. Patterson nor I calculated on receiving more than 1800 dollars the first year and perhaps not so much; yet, in spite of the advice of many of our friends, we were disposed to try the field that presented itself, convinced then, and *still convinced, that it is but necessary for the College to be properly organized, to do well.* At present were I to join you, it would be, I conceive, connecting myself with a declining establishment. For Dr. Wright's fame it is better, I am satisfied, that he should now resign, for I am informed by one well capable of judging, that there certainly will be a considerable declension in the numbers resorting to the school under the Faculty as now advertised. In addition to this intelligence, a young gentleman, who spent his last winter in Baltimore and has joined us here, asserts, that the impression of future decrement prevails in the community, and that but few of those who are not compelled to return to graduate, and of whom he is one, will join your University the ensuing session. Were I, therefore, at liberty to accept the office if now tendered unanimously by the Board of Trustees, it would be under most disadvantageous circumstances. By this time of the year the Student has chalked out his course. We have, indeed, already commenced, and with unusually favorable prospects; and I trust it may not in any manner be considered as egotistical or assuming undue importance to myself, when I affirm, that had Dr. P. and myself joined you, we should have taken with us to Baltimore 50 or more additional students. The objections I have urged in this letter, and to which—with an openness, perhaps at the time, scarcely authorized, but for which, I trust, I was pardoned, I drew your attention when I had the pleasure of becoming acquainted with you at the Virginia Springs, would, under my present feelings, be fatal, even had I not the disqualification, which, at this period of the year, must apply to every teacher, that his engagements are already made for the session, and it would not be honorable—to me it is impossible—to break them. I fear, consequently, that much as it may injure your Institution, and your faculty in particular, you will be compelled to make temporary arrangements, which are ever unfortunate.

My visit to Baltimore has not been given up. It has only been postponed; but when it will be effected I cannot, at present, say. I am not as sanguine as when I saw you, that I shall remove to Baltimore at all. When I returned from the Springs I found a letter from a Professor of the Medical College of Cincinnati, telling me, that the Medical department of the Miami University had become amalgamated with the Medical College of Ohio:— that there was a Professorship in it vacant, and begging to know, whether I would accept it, if appointed. One of my greatest objections to joining the Miami University last winter was the fact of its seeming to be an opposition establishment. This has been done away with by the amalgamation; and if all the arrangements in the College can be made such as to be satisfactory, and the next session be

[84] From this it would appear that Dunglison never had over fifty students in any one class in Virginia. Originally, Gilmer had optimistically estimated fees from students for each professor at a probable $6,000 a year, but the average the first year was only about $1,000 (Davis, Richard Beale, *Correspondence Between Thomas Jefferson and Francis W. Gilmer*, p. 19, note 5, 1946).

successful (they expect 200 pay students—between 3000 and 4000 dollars) I may be induced to unite my destinies with those of that establishment, if my exertions are considered by its Governors important to its farther prosperity. It is not my choice, however, to go to the "far west." Pecuniary advantages, and an augmented field for exertion will alone take me. I prefer Baltimore from what I know of it, to any of the towns, and an Atlantic town to one that is inland. Under all circumstances, I have increased regret for the decision at which the Board of Trustees arrived at the late election in the University of Maryland,—regret not merely of a personal but public character, as another resignation in the anatomical department cannot but be injurious to the prospects and usefulness of the Institution.

May I beg of you to pardon the candid manner in which I have stated my sentiments. The impression left upon me by our short intercourse at the Springs has induced me to be open and explicit; to regret the few opportunities I had of enjoying your Society, and to assure you, that I am, with the most lively respect and esteem,

Truly yours,
Robley Dunglison

PS. Professor Smith of Baltimore, has written to me to say he intends to pay me a visit in the course of a few days. From him I shall be enabled to learn all particulars concerning your University.

H. V. Somerville Esq.

So little was I personally known to the professors—I had not indeed seen anyone of them—that a ludicrous mistake was made by Professor McDowell,[85] an aged individual of excellent moral character but limited intellectual powers, who called upon Dr. Patterson and his sister-in-law, Miss Leiper, as they passed through Baltimore prior to the election. He remarked, that the only objection he had heard against my eligibility was —that I was at an age when I was not likely to improve! Miss Leiper replied, that I was eleven years younger than Dr. Patterson whose services they were desirous of obtaining!

As soon as the result of the election was known, I received from Mr. Madison—to whom, as Rector of the University, and one who, moreover, had exhibited much interest in the movements of myself and family— I had communicated the possibility of my separation from the University,[86] a kind letter of which the following is a copy.

Montpellier, June 15, 1831
Dear Sir,

I have received your letter of the 11th, and will not disguise the fact, that notwithstanding my sympathy with the considerations which might have deprived the University of your valuable services, I learn with satisfaction, that the danger has not been realized, and I hope experi-

ence will prove that the mountain climate is less charged with rheumatic tendencies than occurred to you in your anxiety for the health of Mrs. Dunglison.

I am very desirous of giving on the approaching occasion [of the meeting of the visitors at the University] an attendance, which cannot be often, if at all, repeated. But the effort, I fear, will not be permitted by the decrepit state of my health. In case of its sufficient improvement, my inclination will insure a fulfilment of my duty. Be the event as it may, I beg you to be assured of my great esteem, and of the interest which Mrs. Madison jointly with myself feel in your and Mrs. Dunglison's health and happiness.

James Madison
Professor Dunglison

Certain letters, which I received from Dr. N. R. Smith, and which are still in my possession, give the secret history of the election not very favourable to the consistency of one of the Professors, who, I regret to say, has never had the reputation of unwavering fidelity to truthfulness. No warmer letter could certainly have been penned than the one of which the following is a copy. It was written—it will be observed— after the results of the election were known.

Baltimore June 14, 1831
Dear Sir,

Your favour was very grateful to me. I feel much obliged by your kind and courteous expression of goodwill. Eer this you have heard of our defeat; the causes of which you will learn from my letter to Dr. Patterson. To make assurance doubly sure, we made your recommendation unanimous with that of your Colleague. Nevertheless, several days before the election, we saw a determination of the city trustees to elect the Baltimore candidates. Everything was sacrificed to the election of Ducatel. Our recommendation was totally disregarded by the city electors, who constitute the majority. Unfortunately, four of our country members, all good men and true were absent. In any way I can serve you pray command,

Dear Sir,
Yours with great respect,
Nath Potter [87]

This letter does not tally well with one written to me, nearly about the same time, by Professor N. R. Smith, from which the following is an extract.

You have observed, that I have restrained myself with great difficulty when speaking of Professor Potter. *He has been the chief cause of our losing both Professor Patterson and yourself.* At the meeting of the Trustees, Mr. Gwynn, an influential and veracious member, rose and said, that Professor Potter had that day called upon him to say, that although he had signed the memorial urging the claims of Professor Patterson, yet he was favorable to the election of Ducatel, and that the only reason, why he had not, in the first instance, nominated him, was, that he had not known his wish to be a candidate sooner. Thus, said Mr. Gwynn, the asserted unanimity of the Faculty in nomi-

---

[85] Maxwell McDowell (1771–1848), Professor of the Institutes of Medicine (1814–1833) and several times Dean, was President of the Medical and Chirurgical Faculty of Maryland (Cordell, *ibid.*, p. 76).

[86] Dunglison had written (A.L.S. April 18, 1831, Library of Congress, Madison Papers) to prepare the Board of Visitors against the possibility of his leaving, informing them of various offers from other teaching centers and the need of a change of locality for the sake of his wife's health.

[87] Nathaniel Potter (1770–1843), M.D. 1796, University of Pennsylvania, pupil of Benjamin Rush, Professor of Medicine at the University of Maryland School of Medicine from its inception until his death, was Dean on two occasions (1812 and 1814), President of several of the Baltimore medical societies, and edited various medical books and journals (John R. Quinan, *Medical Annals of Baltimore*, p. 147, 1903).

nating Dr. Patterson is disproved, for Ducatel is virtually nominated by Dr. Potter. Dr. Semmes[88] then rose and expressed great indignation at the duplicity of P.; said, that he had been urged by the Faculty to be present on this occasion, and assured by them, that they unanimously nominated Dr. Patterson. It now appeared, that he had come here to be operated upon by an *upper* and an *under* current. That he had ever been disposed to favour the views of the Faculty, but did not know what to make of such conduct as this. Prof. Potter also agreed, in case Drs. Hall[89] and Baker[90] would nominate Geddings as *their* second choice, that he would nominate you as his *second* choice. He did sign our memorial, but what next did he do? He wrote a letter to the Board in which he strenuously laboured not only to shew, that Geddings was superior to any man living, but also denying your qualifications *in toto* for the anatomical chair, and reiterated the assertion, that you had never taught anatomy. P. was sincere in relation to Professor Patterson till he saw, that should he be elected you would also be placed in the anatomical chair. He resolved, therefore, secretly to throw the weight of his influence into the scale of the town interests, and to defeat you both. Some of Ducatel's friends (also D. himself) were favorable to Geddings, and endeavouring to make a common cause. Potter was resolved at all events to defeat you, and to do this he was willing to sacrifice Professor Patterson. Indeed when—a few days before the election I told him that if we continued divided, we should entirely lose our influence, and that the town candidates nominated by no one of us, would certainly be elected, he declared with all the effrontery in the world, that he did not care a *damn* who they elected, provided they did not elect you. This made me suspicious of him in relation to Professor Patterson, but still he managed to trick us in the end, affecting to be better pleased that you should be elected. His hostility to you was nothing personal, but arose solely from the consciousness that in medical science he would be entirely eclipsed by you, and from the fear that his consulting business (all the practice which he has) would take wing.

I have but little doubt, that Dr. Smith narrated the occurrences as they happened. My subsequent experience of Dr. Potter—as I shall mention hereafter—made it difficult for me to know at all times, where to find, or to place him; and in that I was by no means singular. His course, doubtless, was calculated to aid those who were desirous that the appointments should be made from Baltimore. "I understood" says Dr. Patrick Macaulay[91]—a highly respectable gentleman, in no way connected with the school—

that during my absence, a considerable excitement was gotten up on the part of the Faculty here about the introduction of gentlemen from a distance to fill seats in the medical department, when there were so many expectants at home. This, I believe, decided a majority of the Trustees to make the appointments exclusively from residents here. To me the disappointment was the more, since I had not only made large calculations upon the accession of strength, which the appointments I advocated would have brought to the school, but also upon the pleasure of adding you to our society (Letter to me dated Baltimore, 30th June 1831).

TYPHOID FEVER AT THE UNIVERSITY OF VIRGINIA

Perhaps a locality could not exist, which was more healthy than that of the University of Virginia. During the nine years I resided there, I never saw a case of intermittent, which could be presumed to have originated there. Yet on the banks of the Rivanna, not many miles distant, intermittents[92] prevail annually in summer and autumn. In the winter, however, of 1828–9, a malignant typhoid fever[93] made its appearance at the University, which proved fatal to several of the students; and prevailed so extensively, that it was deemed advisable to suspend the exercises, and dismiss the students for a time, until all the signs of disease had disappeared. It was an occasion of great anxiety to me, not only in regard to the cause of the malady; but as to the steps to be taken to prevent its extension, and it need scarcely be said, that the question of continuing or suspending the exercises would have to depend greatly upon the opinion of the Professor of Medicine;—and, intimately connected with this, and involving perhaps still more responsibility was the determination of the time at which the exercises should be resumed by the return of the students to the Institution.[94] In a letter to me from Dr. Patterson, who had left the University with his family—dated Mar. 13, 1829—he thus expresses himself on this matter.

I suppose we shall hardly get back to the University before hot weather. It is better, indeed, that the whole session should be lost, than that the students should be recalled

---

[88] Benedict J. Semmes (born 1789), one of the trustees of the University of Maryland School of Medicine, M.D. 1811(?), College of Medicine, Maryland, served in Congress 1829 and 1851 (Eugene Fauntleroy Cordell, *Medical Annals of Maryland*, p. 564, Baltimore, 1903).

[89] Richard Wilmot Hall (1785–1847), M.D. 1806, University of Pennsylvania, was Professor of Obstetrics from 1812 to his death, Secretary of the Board of Regents and twice Dean (Cordell, *Historical Sketch of the University of Maryland School of Medicine*, p. 111, 1891).

[90] Samuel Baker (1785–1835), M.D. 1808, University of Pennsylvania, Professor of Materia Medica, 1809–1833, at his death was extolled by Dunglison (Cordell, *ibid.*, p. 75).

[91] Patrick Macaulay (1792–1849), M.D. 1815, University of Pennsylvania, served as intermediary for Dr. Granville Sharp Pattison in the famous feud with Dr. Nathaniel Chapman (Cor-

dell, *ibid.*, p. 50). It is small wonder that Chapman never warmed up to Dunglison, who was friendly at the time to both of Chapman's adversaries. See Chapter IV, fn. 36, and Chapter V, pp. 84 and 87.

[92] Intermittents were mostly malaria, or ague, as Dunglison called it in his *Dictionary* (1: 539, 1833); he did not even list malaria, considering it just a descriptive word equivalent to miasm or poisonous marsh-exhalation that in some unexplained way became infectious when mixed with the atmosphere.

[93] Note that he calls this "a malignant typhoid fever"; to him, typhoid was a descriptive term, not the name of a specific disease as it is to us now. Dunglison discussed this outbreak in his *Practice of Medicine* (2: p. 544, 1842).

[94] The University issued a circular, February 21, 1828, suspending the resumption of classes "beyond the previous time mentioned." Dunglison was anxious to have the students return as soon as it was safe, but had to revise the reopening date several times.

prematurely. You will give your vote for reopening the schools under a heavy responsibility.

It was deemed safe and advisable to reopen the Institution on the first of April, with proper precautions, however. It was in an answer to the summons, that Dr. Pat erson, in a letter to me, dated Philadelphia, Mar. 22, 1829, thus wrote

I have duly received your call for Allfools day, and shall duly obey it. I wish you could in conscience have made it without reserves and doubts and conditions. If I had a son concerned, I would not send him to the University under the summons which you have sent. You acknowledge that the disease has not entirely disappeared, and you doubt the safety of living within the precincts, and you almost recommend, that the students be scattered around the neighbourhood, in taverns etc. removed from our immediate control lest they breathe the infected air of the University. You had doubtless good reasons for your resolution and could, I dare say, remove the strong impression, which I now feel, that it was premature, and that the interests of the Institution would have been better consulted by waiting until you could say, in your circular, that the disease had disappeared, and that you believed it to be perfectly safe for the students to resume their *places*. A week or two might have enabled you to take this course,—if not, then the classes ought not *now* to have been called together.

The students obeyed the summons pretty generally, and soon after the first of April, the affairs of the University went on as usual. No fresh case of fever occurred. During the recess I remained with my family at the Institution; a thorough system of cleansing the dormitories and grounds was practised, but, as in every similar case of endemic disease, no light whatever was thrown on the cause. In the *Virginia Literary Museum* for June 17, 1829—the first number of the work, I inserted the following paragraph.

Salubrity of the University—In the statement published by the Faculty, regarding the causes of the fever which afflicted the University a few months ago, it was remarked, that similar complaints are known to attack the most salubrious situations, commit their ravages for a while, and then disappear, without any possibility of accounting either for their origin or disappearance. Such has been the case with the epidemic of this University, although the physical causes on which it was dependent, have not been discovered, and consequently could not be directly combated, the disease has totally vanished and left us enjoying our usual and pre-eminent degree of salubrity. This result has evidently been anticipated by the Parents and Guardians of the students, who have exhibited the most praiseworthy feeling in the promptness with which they have sanctioned the call to reassemble, as well as by the body of students themselves, who calmly resumed their places in the Institution, undismayed by the melancholy occurrences of the few preceding months. The number of students, at present attached to the Institution, is ninety one,—fifteen have not returned since, and in consequence of the epidemic, five of whom had, themselves, been afflicted with the disease.

I deemed it advisable, too, to write an article "on the causes of endemic disease" in the third number of the same periodical, in which I attempted to show—as I did subsequently at greater length in my *Elements of*

Hygiene, that we are totally unacquainted with the cause of every endemic and epidemic disease of every kind.

The course pursued by the Faculty through this distressing and harassing period was generally approved; and at the next meeting of the Board of Visitors, held in July, it was *resolved*—

That on enquiry into the late endemic that has affected the University, the Board has been unable to discover anything in its local situation, the construction of the buildings, or police of the Institution, which can furnish any plausible solution of its cause, but in reviewing the measures adopted to arrest its progress and guard against its return, they feel it their duty to declare, that the conduct of the chairman and Faculty deserves much commendation, and receives their entire approval.

The ascribing of the malady to "physical causes" did not, however, satisfy the "righteous over much." They regarded it as an infliction on the University on account of there being no religious instruction given within its walls. This doctrine was, indeed, encouraged by an episcopal minister of eminence, who has since become Bishop of Virginia—Meade. At a meeting of the Faculty, it was proposed by Judge Lomax, and acceded to, that the Reverend Mr. Meade should be invited to improve the occasion by a sermon to be delivered at the University before the Professors and Students. He accepted, and the sermon he gave on the occasion advocated the idea of a special visitation and was more distinguished for ability and zeal than for good taste.

UNIVERSITY OF VIRGINIA ACTIVITIES

In the year 1830, my excellent friend Professor Lomax, who occupied the Law chair,[95] resigned, on being appointed a Judge of the General Court of Virginia. In 1825, at the commencement of the University, Mr. F. W. Gilmer was made Professor of Law but he did not live to enter upon his duties, and after his death the choice of the visitors fell upon Mr. John Tayloe Lomax, a distinguished lawyer of Fredericksburg, Virginia. With him, my associations were of the most agreeable character; and I greatly deplored his removal from the scene of his useful labours at the University. He was succeeded in his chair by another valued friend, Mr. John A. G. Davis, a young but highly promising lawyer of Charlottesville who became distinguished as a teacher; and whose life was prematurely terminated by a worthless student, about ten years afterwards, whilst he was engaged in the suppression of a disturbance as chairman of the Faculty.[96]

This office—always a disagreeable one—I was appointed to fill during the second session. The day after the election, which was then by the Faculty, I

[95] John Tayloe Lomax (1781–1862) was appointed in 1826.
[96] Davis, Professor of Law 1830–1840, published a treatise on criminal law in 1838.

visited Monticello, and, in the course of conversation, Mr. Jefferson suggested to me, that to conciliate the feelings of the community, it might be policy to choose Mr. Gilmer—whose death was not then anticipated by him, although he was very ill—as he was a Virginian and it would be presumed—better acquainted with the habits of the students, and more able perhaps to guide or repress them. I then informed him of the action of the Faculty with which he appeared to be satisfied. Soon afterwards indeed, Mr. Gilmer's health became rapidly deteriorated, and he was never able, so far as I can recollect, to set his foot in the University. In the years 1829 and 1830, the appointment of chairman having been assumed by the visitors, and a salary attached to the office of *Five hundred dollars per annum,* I again filled it and experienced, I believe, as little difficulty or annoyance as was usually met with by the incumbent of an office with which the discipline of the Institution was so intimately associated.

During the whole period of my residence at the University of Virginia I practised my profession within the walls; but was not permitted to exercise it out of the University except in consultation—a restriction, which I greatly preferred.[97]

I was much occupied, also, in various literary labours. When the *American Quarterly Review* was commenced in Philadelphia in 1827, Mr. Walsh wrote me to furnish him with occasional articles, which I consented to do; and, in the progress of the work, both whilst I was at the University and in Baltimore, furnished him several essays. The first of these was on the "Gastronomy of the Romans" in which I entered, at much detail, into the consideration of the eatables and the culinary preparations not only of the Romans but of the Greeks, interlarded with remarks on the gastronomic customs of our ancestors of Great Britain. The article gave much trouble in the preparation, on account of the vast research it required, and as I was unable to see a proof, it **was**, of course, inaccurately printed, but less so than I had anticipated under all the circumstances of the case and Mr. Walsh made the excuse for the errors that they were "periodical." In the next volume, I noticed the "Voyage to the Moon" of my friend Professor Tucker of the University. To the third volume of the work I contributed an article of much detail on the various popular superstitions, attempting to explain them on rational grounds, and refer them to their proper origin. In the eighth volume, there is by me an Essay on "Longevity," in which the total abstinence theory is examined, and sundry arguments on the subject advanced in the *Journal of Health,* published in Philadelphia, are combated. This article led to a very severe reclamation on the part of Dr. John Bell,[98] the Editor of that periodical, who

was by no means complimentary to the "young physician," as he characterized the author of the Essay in the *Quarterly,* whom Mr. Walsh had, he professed to conceive, employed to attack the temperance views maintained by the Editor of the *Journal of Health.*[99] This collision of opinion had, however, no effect whatever on the personal relations of Dr. Bell and myself. Both of us, indeed, subsequently referred to the matter as a circumstance that had passed away, and left no unpleasant feeling remaining.

An article in the 9th volume on the subject of "College Instruction and Discipline" was suggested by the meeting of a Convention of literary and scientific gentlemen to establish a University in New York—the present University of the City of New York—but a main object with me, in writing it, was to canvass the great questions of instruction and discipline in our higher schools, and to bring the plan adopted in the University of Virginia more prominently into notice.[100]

It was to this article that I referred, when speaking of the suggestion for a higher degree being established at the University of Virginia than was contemplated in the earliest enactments.

The last article which I wrote in this journal was on Madden's "Infirmities of Genius," in which I endeavoured to show, that his inferences as to the deleterious influence of the play of the imagination on the frame were not borne out even by the examples he adduced; and that the exercise of the great organ of the intellect in its normal acts, provided due attention be paid to regularity and propriety in diet; and to exercise, sleep etc., is not injurious, even when carried to the extent observed with many of the German scholars, who were equally distinguished for their application and longevity.

Early in the year 1829, it was deemed desirable by the Faculty, that a weekly periodical devoted to Belles Lettres, Arts etc. should be commenced at the University, to which the different members of the faculty should be contributors, with the expectation, however, of receiving communications from others. Such a journal was determined upon, the title of which should be *The Virginia Literary Museum and Journal of Belles-Lettres, Arts, Sciences* etc. Edited at the University of Virginia; and the editorship was assigned to

[97] Dunglison was thus the first full-time Professor of Medicine in this country.
[98] John Bell (1796–1872), M.D. 1817, University of Pennsylvania, medical author and editor, member of the American

Philosophical Society, Fellow of the College of Physicians of Philadelphia, President of the Philadelphia County Medical Society, was a protégé of Dr. Chapman (*Transactions of the Medical Society of the State of Pennsylvania* 10(2) : 746, 1875).
[99] In the last issue of the *Journal of Health,* Drs. John Bell and Francis Condie are named for the first time as editors. An active campaign against the use of ardent spirits was going on at this time, to the evident annoyance of the wine-loving Dunglison.
[100] For an interesting comment on this article, see Adams, *ibid.,* p. 208; also Dunglison, A.L.S. February 14, 1830, Madison Papers, Library of Congress, where he promises "rigid" midterm examinations to impress upon the students the necessity to study from the commencement of the session.

Professor Tucker and myself. The objects of the journal—as stated in the prospectus—were to communicate the truths and discoveries of science to the miscellaneous reader, and to encourage a taste for polite literature.

It will rely, chiefly, for its support on the professors of the University, whose minds, kept in a state of active inquiry by the lectures required of them, may be expected to afford original and interesting contributions on all the important branches of learning or science. The scientific portion of the work will generally be of a popular character, but should it occasionally contain discussions, which on account of their novelty or importance, may also interest the adept, it will be the aim of the editors to make such articles, so far as may be practicable, intelligible and instructive to the general reader; whilst the journal will be principally devoted to general topics of Moral or Physical Science, Philology and Polite Literature, the editors will not be unmindful of our local and peculiar concerns. They will endeavour to collect and diffuse what information they can concerning the history of Virginia, and the other states—their first settlement;—their progress as Colonies and as independent States,—their peculiarities in Laws, Manners or Dialect,—their statistical details and Natural Phenomena. Such a Repository is much wanted. The information, which now lies scattered among individuals, if collected, would shed great light on the past history and present state of our Country. On these and other subjects they solicit contributions. A part of the journal will communicate information concerning the University—the course of instruction pursued by the several professors—meetings of the visitors—public examinations—statutes and regulations of the university—lists of professors and students—honorary distinctions, and, occasionally, such productions of the students as may possess unusual merit. This information, peculiarly interesting to the parents and guardians of the students, will not be unacceptable to the public. The journal may, also, by receiving and transmitting hints on the difficult subjects of college government and instruction, render an important service to the cause of education. Party politics and controversial theology will be excluded; but such exclusion will not extend to religious or political topics of a general character, discussed with temperance and ability.

Such were the objects contemplated to be fulfilled in the *Museum*, as stated in the Prospectus. The first number appeared on the 17th of June 1829, and was favorably received. We soon found, however, as Editors—what, indeed, has been the experience, I believe, of everyone who has occupied the same post, that but little reliance could be placed on promised contributions; and that, in consequence, the labors of the editors became exceedingly onerous. I have elsewhere indicated the various articles furnished by myself,[101] which were necessarily numerous, and of varied character, to compensate for the disappointments the editors were constantly doomed to experience. On the 9th of June 1830—after one year's duration—they were compelled to bring it to a close. The reasons for this were stated in an address "to the public."

The most weighty of these are, that the editors, not having received the aid from contributors that they expected, find that to furnish the requisite material from their own resources demands more of their time than is consistent with their other duties and engagements. . . . A periodical, devoted exclusively to literature and science must rely, it seems, for its materials on the editors alone, and not on occasional contributions; and although such a paper may obtain subscribers enough to defray its expenses, and even something more, their number will scarcely be sufficient to reward the undertakers for giving their whole time and attention to it.

DUNGLISON'S DICTIONARY

From the earliest period of my professional labors I had been impressed with the great want of a Medical Dictionary containing a concise account of the subjects and terms. The Dictionary of Hooper was the only one available to the medical student. It had been stereotyped in this country, and had experienced but little modification for fifteen or twenty years.[102] Having been engaged in correspondence with Messrs. Gray and Bowen, of Boston, on matters connected with the library of the University, I made proposals to them to prepare such a work, and, in March 1833, it was agreed, that they should be at the expense of bringing out the work; and should pay me, for an edition of 1,000 copies the sum of *Twelve Hundred and Fifty Dollars*, and allow me twenty-five copies of the work, free of charge. One of the important features of the work was its extensive synonymy, not only in the Greek and Latin, but also in the French and German languages.[103] I endeavored, indeed, to make it a complete Dictionary of French medical terms, so that the medical students engaged in the perusal of a French medical author need not seek in a French Dictionary for the explanation of any medical term he might meet with. It contained also short biographical and bibliographical notices. In this form, although the work was most favorably received, it did not prove profitable to the bookseller. A mistake was made in publishing it in two volumes, which interfered most materially with its general reception; and, perhaps, the place of publication, Boston, situated at a distance from the center of populations, and with fewer facilities than at present for extensive distribution, interfered unfavor-

---

[101] See also: John R. Quinan, *Medical Annals of Baltimore*, 1894 (pp. 95–96) for detailed bibliography of Dunglison; and McLean (*ibid.*, pp. 95–104) for a full account of the *Virginia Literary Museum*.

[102] The dictionary of Robert Hooper first appeared in England in 1798. In 1822 an American edition appeared in' New York City and a second American edition appeared in Philadelphia in 1824, both from the fourth London edition. From 1824 to 1832 five subsequent editions were published in the United States, and it continued to appear at intervals until 1854.

[103] In a letter to James Madison dated March 10, 1830, Dunglison calls it a "Polyglot Dictionary of Medical Literature"; and on March 19, 1830, he wrote Madison that he intended to publish it in England as well as here (Madison Papers, Library of Congress). It seems not to have been published in England, however, until 1857.

ably with it. The preface was dated "University of Virginia, Oct. 1832"; and the Dictionary was published early in 1833, dedicated to my excellent friend and colleague, Dr. R. M. Patterson, as follows:

To:
Robert M. Patterson, M. D.
Professor of Natural Philosophy
in the
University of Virginia
the Enlightened Professor and Accomplished Gentleman
this work is inscribed
In testimony of the affectionate regard and esteem
of the Author.

Feeling satisfied that the work ought to have a large sale, if brought out under favorable circumstances, I applied to Messrs. Lea and Blanchard, in the year 1838, to undertake a second edition of it, provided Messrs. Gray and Bowen would be willing. A correspondence took place with the Boston publishers, and they declined having anything more to do with it. It was urged by Messrs. Lea and Blanchard, that the bibliographical and biographical parts, and the German synonyms should be omitted: for, strange to say, it was objected to the work, that it was too learned.[104] To the omission of the German synonyms I readily consented; but I reluctantly agreed to leave out the biographical and bibliographical notices, which I thought, and still think, formed a valuable feature in the undertaking.[105] It appeared, however, indispensable that this should be done; that the work should be brought out in one volume; and that, if possible, it should not exceed 600 pages of small type and double column. Owing, too—it was urged by the publishers—to the bad name the *Dictionary* had acquired, as regarded its slow sale, and the greater risk to them of disposing of the edition, they only offered me *Six Hundred Dollars*, at six months from publication, for an edition of 1500 copies; stipulating, at the same time, that they should have the right to publish all future editions; for which they should pay me not over *Sixty Cents* per copy at six months' credit from publication: allowing me, free of cost, for distribution among my friends, *ten copies* of each edition. This proposition I accepted; and the result has shown, that it was certainly most favorable to the publishers. This edition of 1500 copies appeared in 1839; a *third edition* of the like number, was issued in 1842; a fourth edition in 1844; and a fifth, in 1845. The sixth edition, published in 1846, was increased to *Thirty-seven hundred and fifty copies;* and the seventh edition, published in 1848, to *five thousand copies*. Before the appearance of this edition I wrote the following letter to Messrs Lea and Blanchard:

---

[104] The dictionary won Dunglison the nickname of "The Walking Dictionary" among his friends and students in Virginia.
[105] Dunglison wanted this part of his *Dictionary* published separately, but was so discouraged that, when John M. Toner contemplated a biographical dictionary, he advised him to shorten his biographical intentions (A.L.S. Aug. 23, 1866, Library of Congress, Toner Collection).

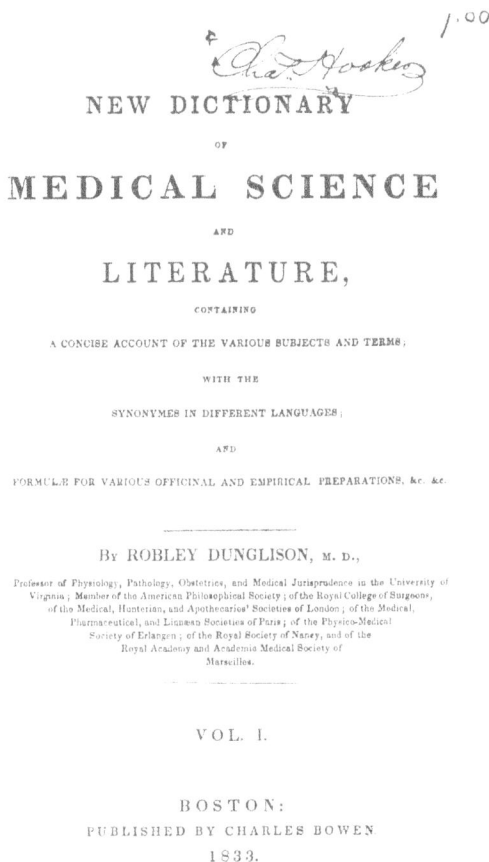

1.00

Cha. Hooker

NEW DICTIONARY

OF

MEDICAL SCIENCE

AND

LITERATURE,

CONTAINING

A CONCISE ACCOUNT OF THE VARIOUS SUBJECTS AND TERMS;

WITH THE

SYNONYMES IN DIFFERENT LANGUAGES;

AND

FORMULÆ FOR VARIOUS OFFICINAL AND EMPIRICAL PREPARATIONS, &c. &c.

By ROBLEY DUNGLISON, M. D.,

Professor of Physiology, Pathology, Obstetrics, and Medical Jurisprudence in the University of Virginia; Member of the American Philosophical Society; of the Royal College of Surgeons, of the Medical, Hunterian, and Apothecaries' Societies of London; of the Medical, Pharmaceutical, and Linnean Societies of Paris; of the Physico-Medical Society of Erlangen; of the Royal Society of Nancy, and of the Royal Academy and Academia Medical Society of Marseilles.

VOL. I.

BOSTON:
PUBLISHED BY CHARLES BOWEN
1833.

FIG. 9. The *Dictionary*, 1833.

Philadelphia Feb. 1848

Gentlemen,

The time is now approaching when it will be necessary to go to press with another edition of my *Medical Dictionary*. I have frequently stated to Mr. Blanchard, that I regarded the copy money as altogether inadequate, when consideration is had to the shape and character which the dictionary has assumed after successive editions, and that it is even an insufficient compensation for the time occupied in passing the work through the press, independently of the intellectual and manual labor required to render the work what it now is. In the forthcoming edition, the labor will be still greater, and the work rendered still more useful by the numerous old and new terms, which I have deemed it advisable to add; and I have, therefore, to suggest to you, whether, in equity, another understanding should not be substituted for the one now in existence between us. In the original contract, I agreed "to revise and improve all future editions"; but it was never, I presume, contemplated by either party, that the labor on my part would be as great as it has been to enable it to assume the position it has now attained. Additions have been made, I believe, by you, from time to time, to the price at which it was first sold, whilst *my compenstaion*—if I may use the inappropriate term—has remained stationary. Moreover, you have caused to be reprinted, under the sanction of an American editor, the English Medical Dictionary of Hoblyn,[106] which you sell at a much less price than mine is sold for, and from which you have doubtless derived emolument; whilst, to a certain extent, mine must inevitably have been diminished thereby.

Under these considerations, and under the idea, that the work has essentially changed its character, both as regards size and usefulness, and that each successive edition has not been merely modified, so as to make it keep pace with the improvements in medical science, but to render it more extensively useful to the profession, and more lucrative to you, I submit in that frankness which has ever characterized my varied dealings with you for a period of *sixteen years*, whether the arrangement, thus far existing between us, in regard to the *Medical Dictionary* may not be susceptible of modification, which may render it more equitable and agreeable to both parties. I am, gentlemen,

very truly your friend and servant,
Robley Dunglison.

P.S. The original stipulation was, that the work should make 600 pages; whereas it is, and has been, upwards of 800; and I presume, as a matter of course, it sold at a much higher price than if it had been of a smaller size.

Messrs Lea and Blanchard.

I ought to have stated, however, that a few years earlier, I addressed the following letter to my publishers.

Philadelphia, Jan. 25, 1845.
Gentlemen,

You are aware, that in every speculation in regard to medical works I have frankly and honestly stated to you, that I considered it to be your duty, as dealers, to look to your own interests, whilst I thought, that your attention to mine should be but incidental. Still I knew, that the proper course—entirely proper in my estimation—which you adopted must necessarily interfere with my emoluments from my labors. Numerous works on the same subject could not but affect the sale of anyone; and this must be the case whether published by others or by you. On one point, however, I do not feel disposed to be passive. My dictionary has been the offspring of immense labor and intellectual effort. It has been gradually attaining its position, and this partly—perhaps in a great measure, because, owing to *Hooper* being stereotyped, it has had no active rival. You have now advertised a cheaper dictionary, which, from my knowledge of the capacities of the pockets of medical students, must greatly interfere with mine. It will do so, I know. The object then, of my present communication is, to state to you, that my interests require, that this must be checkmated; and, with my present sentiments, I shall set actively to work to prepare a dictionary not larger than Hoblyn's. This I can do without violating my contract with you. Perhaps, indeed, you may yourself be disposed to favor my undertaking.

I throw these suggestions before you in the best possible feeling; for I can have no other towards you, and under the impulse, that the time has come when I must be up and

active for fear that years of intellectual and corporeal labor may be rendered nugatory.

Believe me, gentlemen,
Most truly yours,
Robley Dunglison.

Messrs Lea and Blanchard

This letter gave occasion to a reply in which the publishers expressed the opinion, that the publication of Hoblyn's small Dictionary might, in other hands, interfere more with our mutual interests than it would do in their's. "In our hands," they say, "in our views of its use, we think it will not impair the use of your dictionary, as it is intended to meet an opening for a small dictionary, to satisfy the wishes of those who may prefer two works, or one by a new author." In regard to the publication by me of an abridgement they add,—

While we think, that *Hoblyn* will not in our hands, interfere with your work, we suspect, that an abridgement of your large work, whether published by us or by others would do so. But if you differ from us, and feel the necessity of protecting your interests, by the preparation and publication of an abridgement, we should not object, but on the contrary shall be very happy to know that you have done well, and certainly shall not persuade you to withhold your plan. That we could not publish it to your advantage you can readily see, after what we have written, and you must be aware, that its immediate announcement and publication would impair the value of the large quantity of the last impression now in our hands. But, in these matters do not for a moment allow our interests to affect your decision. All we ask is, that your exertions to keep up the value of your large work be not abated, and that what you prepare shall be an abridgement *in fact*, as well as by announcement in the title and advertisements.

I did, however, allow the considerations set forth by Messrs. Lea and Blanchard to interfere, and abandoned the idea of preparing an abridgement of the *Dictionary*.

My letter of February 1848 gave occasion to a reply, in which the publishers disclaimed having increased the price of the dictionary, and any desire or intention to add to it in the future, whilst they hoped to improve its appearance, and increase the quantity of the impression over the preceding edition,—thus again placing more to my credit for the labour of one revision, whilst they would spend more money to produce it. "The extended sale of your work"—they add—

ought to satsfy you, as it has done us, that in printing *Hoblyn*, we really have protected your interests as our own, by parrying off the blows that would have been leveled at it by other dictionary makers, as we informed you in our note of January 1845. Such as have been published since are neither the one thing nor the other.

In a subsequent letter, dated March 10th, 1848, they state.

We have given farther consideration to the matter of the *Dictionary*, and the conviction still remains with us, that the increased compensation per copy as suggested by you, will so far affect your interest in the receipts from the work, that we cannot concur with you as to the expediency of

---

[106] The dictionary of Richard D. Hoblyn first appeared in 1835; as late as 1865 a New American Edition, revised by Isaac Hays, was published by Henry C. Lea, Dunglison's publisher.

making the change, seeing that it must carry up the price. Any such increase in price we are now desirous of avoiding, as other "Dictionaries"—as well those published as promised—must be met by an improved edition of yours at the old price. The above is said in entire frankness, and in the conviction, that it is the interest of both parties to make no change in the price. We desire to print, of the coming edition, five thousand copies, provided no abridgement shall be prepared by you until they are sold . . . trusting that all this will be perfectly satisfactory to you, and that we may all live to see many, many more such editions issued from the press, we are

Very respectfully,
Very truly,
Yours etc.
Lea and Blanchard.

Robley Dunglison, M. D.
S. 10th Street.

The following reply terminated our correspondence on this subject.

Philadelphia March 11th, 1848

Gentlemen,

I have to acknowledge the receipt of your letter of the 10th of March in which you remark that you deem it expedient, for reasons therein stated, to decline any modification of the contract existing between us, so far as regards the price per copy of the Medical Dictionary allowed me in our former understandings. As the power resides wholly in yourselves, I must submit; and will, if not cheerfully, honestly, endeavour to carry into effect the contract. I must decline, however, your proposition, that I should bind myself not to "prepare" an abridgement until the edition of five thousand copies, which you desire to print shall have been sold. I wish, in other words, now that you have determined, that the price to me per copy shall remain as heretofore, that the original contract between us shall continue in full force without any modification whatever.

I am, gentlemen,
Very truly yours,
Robley Dunglison.

Messrs. Lea and Blanchard

In the year 1851, it became necessary to prepare an eighth edition of the Dictionary, and propositions were made to me by my publishers to have it stereotyped. To this I objected for some time, but ultimately consented, provided that I should have the privilege of having the plates destroyed after a short period, so that the work should not suffer materially by its recomposition being delayed a little longer than had, thus far, been the case. The following arrangement was therefore entered into.

Philadelphia April 12th, 1851

Professor Dunglison
Dear Sir,

It is understood, that we are to stereotype your revised copy of the Medical Dictionary and that you are not to require us to destroy or modify the plates, nor are we to require you to revise the work for a new edition, for a period of five years from the publication of the first fifteen hundred copies, (unless the said steel plates are accidentally injured or destroyed). Should we not have printed and sold ten thousand copies from the plates, at the expiration of the said five years, then, the understanding exists until the quantity of ten thousand copies are sold. The existing contract for this work, of accounting for each fifteen hun-

dred copies, and other matters, are not affected by stereotyping it. Yours very truly,

Blanchard and Lea

P.S. It is understood, that our selling price of the edition is not to be increased.

It was the desire of my publishers, as well as my own, to introduce the Dictionary into the English market; but this was rendered impracticable by that of Hooper, which had been revised by Dr. Klein Grant, being owned by not less than twenty prominent firms in London, Edinburgh and Dublin. How much the bookselling interest was concerned in excluding it, was shown by notices of the eighth edition contained in the Lancet and in the London Journal of Medicine, early in 1852. The British and Foreign Medical Review for April 1852; and the Dublin Quarterly Journal of Medical Science for May 1852, were, however, uninfluenced by such considerations, and made no invidious comparisons between the two works, which are, in fact, very dissimilar. The London Lancet insinuated—indeed, stated—that mine appeared to be an abridgement of Hooper's—an abridgement containing, perhaps, 30,000 terms and their definitions more than the fancied prototypes!!

The following is an account of the number of copies and of the copy money paid to me for the Dictionary up to this date (June 1852):

Edition	Number of Copies	Copy Money
1833 First—	1000	$ 1250
1839 Second—	1500	$ 600
1842 Third—	1500	$ 900
1844 Fourth—	1500	$ 900
1845 Fifth—	1500	$ 900
1846 Sixth—	3750	$ 2250
1848 Seventh—	5000	$ 3000
1851 Eighth—	1500	$ 900
	17250	$10700

In the early period of my residence at the University I set to work vigorously to acquire a knowledge of the German language. In this I was assisted occasionally by the Professor of modern languages—Dr. Blaetterman,—a man of great philological knowledge, but by no means refined. He was kind hearted; and a greater enemy to himself than to any other person. Whilst engaged in this, I was impressed with the value of some of the German works on ancient history; and absolutely translated from the German the whole of Eichhorn's Ancient History which gave me the idea of writing a work on the subject of which Eichhorn should be the basis. I had, indeed, corresponded with Gray and Bowen, the publishers in Boston, regarding it, who offered me, if I would consent to its being printed in Boston, to publish it at their own risk and expense and share the profits equally with me. (Letter to me dated Jan. 12, 1830). I abandoned the idea, however, under the conviction, that it would be desirable for me

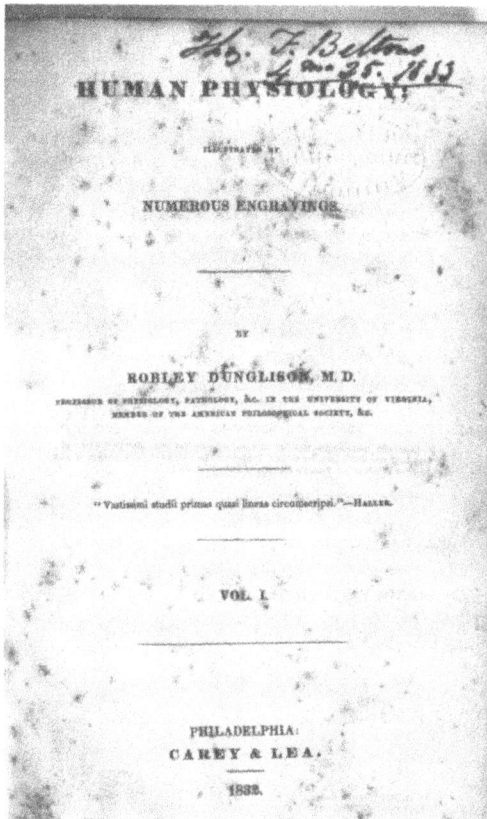

FIG. 10. *Human Physiology*, 1832.

the latter part of the year 1831; who entered into an arrangement with me on the subject. The treatises of Bostock, Richerand and Magendie [107] had been reprinted in this country; but it appeared to me, that there was required a more comprehensive work, which should be elucidated by wood cuts and copper plates; and which might ·prove attractive, on that account, to the student more especially. The arangement, entered into with the publishers, is shown in the following letter from Messrs. Carey and Lea.

Dear Sir,

The terms, agreed on for the publication of your work on Physiology, we understand to be as follows.—It is to be in two octavo volumes, and is to be printed at our expense and risk; and we are to pay you for the use of the copyright, for the first edition of 1500 copies, *One thousand dollars,* and should farther editions be required, we are to pay you at the same rate per copy, for each and every edition, settlement to be made by note at six months from the publication of each edition. You are also to receive *twelve* copies out of each edition that may be published. We believe you will find the above correct. If so, be so good to write us to that effect and oblige

Dear Sir,
Yours truly, and respectfully,
Carey and Lea

Dr. R. Dunglison   Philad. Nov. 19, 1831.

The first edition appeared in the year 1832—the dedication to my venerable friend Mr. Madison being dated July 3, 1832,[108] which was as follows:

To
James Madison, Esqr.
Expresident of the United States, Rector of the University of Virginia etc, etc.
alike distinguished as an illustrious
Benefactor to his Country
a zealous promotor of Science and Literature and
The friend of mankind;
this work,
intended to illustrate the functions executed by
that being, whose moral and political condition
has been with him an object of ardent and
successful study,
is, with his permission, inscribed,
in testimony of the unfeigned
respect
entertained for his talents and philanthropy, and of
gratitude
for numerous evidences of friendship
by his obedient and obliged servant
The Author.

University of Virginia
July 3, 1832.

to adhere to subjects connected with my own profession, and, more especially, as the work of Heeren on the same subject had already been translated by, or under the auspices of, Mr. Bancroft.

After I had made some progress in German, the Revd. E. Smith, who was at the time officiating at the University, agreed with me, that if I would aid him in reading German, he would do the same to me in regard to Hebrew; and, for some time, we were agreeably engaged in those studies, and I hope were of advantage to each other. I know he was to me; but I have since forgotten, from disuse, the little Hebrew that I acquired by his assistance. He was a man of considerable learning and ability; and has since been usefully engaged as a Christian pastor in Washington and New York.

DUNGLISON'S PHYSIOLOGY

Desirous of undertaking a work on physiology to serve as a textbook to my students, I commenced a correspondence with Messrs. Carey, Lea and Co., in

---

[107] John Bostock (1773–1846), *Elementary System of Physiology,* 1824–1827 (American edition 1825–1828); Anthelme Balthasar Richerand (1779–1840), *Nouveaux élémens de physiologie* (frequently reprinted and translated; American editions 1808 to at least 1825); François Magendi (1783–1855), *Précis élémentaire de physiologie,* 1816–1817 (4th ed. 1836; American editions 1822 to at least 1844).

[108] Dunglison's Letters to Mr. Madison dated March 10, 1830, and March 19, 1830, requesting permission to publish this dedication, are in the collection of Madison Papers of the Library of Congress.

The work under the title *Human Physiology, Illustrated by Numerous Engravings* was certainly very successful.[109] A *second edition*, of the same number of copies, was called for in 1836, and a third edition in 1838. To the first two editions no special references to authorities, or rather to books, were appended. It appeared to me, however, that the value of the work would be greatly augmented by such references; and, at great labor I added them to the third edition; thus anticipating the suggestion of a friendly reviewer of the second edition of the *Physiology* in the *British and Foreign Medical Review* for January 1838, who remarked—

In closing our notice of Dr. Dunglison's work, we would suggest to its learned author one addition, which would much enhance its value; viz., that of references to the authorities from which the facts are derived, especially those which are not usually consulted by the student. We deem it the duty of everyone, who has traversed such an extensive labyrinth as that which our author has so diligently explored, to leave a sufficient number of direction posts for the guidance of those that may come after, especially where those hidden and devious paths are to be indicated, which might escape the notice of an ordinary treatise.

In this third edition, I was careful to give exact references both for facts and opinions; yet, strange to say! in the year 1848, *sixteen years after* the publication of the first edition, and *ten years* after that of the third edition modified in the manner, which I have mentioned, there was permitted to appear in the *Southern Medical and Surgical Journal*, published at Augusta, Georgia, and of which Dr. Paul F. Eve was editor,[110] an article, in which the reviewer charged me with copying from the *Physiologie de l'Homme* of Adelon[111] certain physiological statements; without a word being stated of my having absolutely referred the reader to Adelon for them in all the editions from the third to the seventh inclusive; or making the slightest allusion to the fact, that in the very first edition, in the Preface, I remarked.

In preparing the present work, the author has availed himself freely of the labors of his predecessors. His object has been to offer a view of the existing state of the science, rather than to strike out into new, and perhaps devious, paths. To the labors of Adelon and Chaussier[112]—especially of

the former—of Blumenbach,[113] Richerand, Magendie, Rudolphi,[114] Broussais,[115] Sir Charles Bell,[116] and others, who have had the chief agency in raising physiology to its present elevated conditon, he has been indebted for essential aid; and many of the illustrations have been taken from the admirable graphic delineations of the last mentioned distinguished physiologist.

The *animus* of the author of the article was so transparent, that I took no notice of it. On the topic of original works of science, I shall have something to say hereafter, when referring to the *Practice of Medicine*. It may be sufficient, at present, to observe, that a treatise on the existing condition of any branch of science, which is not founded on the accumulated observation and reflection of an author's predecessors and contemporaries would be an absurdity, and therefore of but little or no value.

The *fourth edition* of the *Human Physiology* published in 1841; and the *fifth edition*, in 1844, consisted of 1500 copies each. The *sixth edition*, published in 1846, was increased to 1750 copies; and likewise the *seventh edition*, published in 1850.

The following is an account of the number of copies and of the copy money paid to me for the *Human Physiology* up to this date (June 1852).

A.D.	Edition	Number of Copies	Copy Money
1832	First—	1500	$1000
1836	Second—	1500	1000
1838	Third—	1500	1000
1841	Fourth—	1500	1000
1844	Fifth—	1500	1000
1846	Sixth—	1750	1166.67
1850	Seventh—	1750	1166.67
		11000	$7333.34

III. VIRGINIA REMINISCENCES

William Beaumont. The American Philosophical Society. President Jackson. President Madison. President Monroe. Negotiations with University of Maryland. Virginia leave-taking. Virginia reminiscences. Mr. and Mrs. Rives. Sojourn in Washington.

and Physiology in the Écoles de Santé when Dunglison studied in Paris.

[113] Johann Friedrich Blumenbach (1752–1840), of Göttingen, famous for his physiology theory of *Bildungstrieb* (growth impulse).

[114] Karl Asmund Rudolphi (1771–1832), *Grundriss der Physiologie*, Berlin, 1821–1828.

[115] François Joseph Victor Broussais (1772–1832), probably the most impressive medical lecturer in Paris when Dunglison studied there, published *Annales de la Médecine Physiologique*, 1817–1834, and *Traité de Physiologie appliqué à la Pathologie*, 1834.

[116] Sir Charles Bell (1774–1842), first Professor of Physiology at the University of London, published *Animal Mechanics*, 1824, and did basic research in the physiology of the nervous system.

[109] Osler said that his *Dictionary* "was one of my standbys. . . . But the book of Dunglison full of real joy to the student was the 'Physiology,' not so much [for its] knowledge: . . . but there were so many nice trimmings in the shape of good stories" (Harvey Cushing, *Life of William Osler* 1: 238, 1925). It was at once widely accepted as a class book for students (Samuel D. Gross, *Autobiography*, 2: 329–335, 1887) and together with his early efforts in teaching the subject, it earned for Dunglison the title of "Father of American Physiology."

[110] Paul Fitzimmons Eve (1806–1877), Professor of Surgery, Medical College of Georgia, 1832–1850.

[111] 4 vols. 1823. N. P. Adelon (1780–1852) taught physiology privately in Paris about the time Dunglison studied there.

[112] François Chaussier (1746–1828), Professor of Anatomy

WILLIAM BEAUMONT

Early in the year 1832,[1] I received through my friend, Mr. Nicholas P. Trist, who was, at the time, private secretary to Genl. Jackson,[2] an invitation from Dr. Lovell, surgeon general,[3] to visit Washington with the view of instituting certain experiments on Alexis San Martin, a Canadian under the charge of Dr. Beaumont[4] of the United States Army, who had a fistulous opening into the stomach, and whose case is now well known to every physiologist. Circumstances, however, occurred, which rendered it problematical whether I should be able to visit Washington, and, in consequence, I wrote a letter to Dr. Lovell, of which the following is a copy.

University of Virginia
Jan. 11th 1832 [5]

Dear Sir,

My friend Mr. Trist will explain to you the causes that have prevented me from witnessing and assisting at the experiments instituted by Dr. Beaumont and yourself in the interesting and rare case, which is now, and has been, engaging your attention; as well as from expressing in person my high sense of the honour you do me by desiring, that I should in any manner be associated with you.[6] I had seen a detail of certain experiments by yourself and Dr. Beaumont on the subject of this case several years ago; and had fully intended to notice them in the work on *Human Physiology* recently published by me, but the journal was mislaid and I could not refer to it in time. One of these experiments, if I recollect rightly, was on *artificial digestion*, and the result showed, that a piece of corned beef experienced the same changes as when it was inserted through the wound in the interior of the stomach; and another—that the vegetable substances underwent chymification in the stomach more speedily and thoroughly than different kinds of meat, which were passed in at the same time. Both experiments were interesting as confirming the views of

---

[1] This must have been 1833. Beaumont was prevented from coming to Washington early in 1832 by the outbreak of the Black Hawk Indian War. He did not reach Washington until about November of 1832.

[2] Andrew Jackson, President of the United States, 1828–1836.

[3] Joseph Lovell (1788–1836), M.D. 1811, Harvard Medical School, successor to James Tilton, was the first to hold the title of Surgeon General of the Army (1818–1836). The preliminary report of Beaumont's case of a wounded stomach (*Medical Recorder* 8: 14, 1825) was published under Lovell's name but later (*ibid.*, 840) this was corrected and credit given to Beaumont. A further report was published by Beaumont the next year (*ibid.*, 9: 94, 1826).

[4] William Beaumont (1785–1853), M.D. (Honorary) 1833, Columbian University, Washington, D. C., famous for his work on the physiology of gastric digestion, went to Washington at this time especially to consult the literature on the subject and conducted further experiments preparatory to the publication of his book.

[5] Dated incorrectly; it should have been dated 1833; his *Human Physiology* recently published, as noted below in this letter, was published in 1832.

[6] Gout and the weather delayed Dunglison, who wrote to Trist on January 10, 11 and 12, 1833, expressing his interest in the work of Beaumont, saying he felt flattered "to be one of the committee of the 'whole'" and mentioning his offer to Drs. Lovell and Beaumont to report their results for them to the American Philosophical Society.

Spallanzani, and the experiments made at the Hôpital La Charité of Paris, on a female with a fistulous opening in the stomach, and those instituted by Helm, of Vienna, on two similar cases, as regards the effect of the juices contained in the stomach in the solution of alimentary substances. It would have been additionally instructive to have witnessed the effect of saliva on the same aliment out of the body, for the purpose of deducing whether that fluid possesses the sole agency in digestion, as supposed by Montègre, or is merely an adjuvant, as presumed by the best physiologists; and what are the changes effected by it on the aliment compared with those that result from the action of a compound fluid, formed of the various secretions from the supradiaphragmatic portion of the alimentary canal and of the stomach itself—met with in the interior of that organ, and which was the solvent in your experiments.

The results of your second experiment impressed me forcibly, being somewhat at variance with the inferences of Gosse, Montègre, Magendie, and others, from their experiments, to which reference is made in the first volume, p. 455 of my *Human Physiology* (first edition), regarding the comparative digestibility of animal and vegetable substances. Since the period at which your first experiments were made, you have, doubtless, instituted others, which may have led you to confirm or disprove your first obtained results. It would be extremely interesting to me to learn the comparative digestibility, attested by the individual, of the great chemical divisions of aliments—the *amylaceous, mucilaginous, saccharine, acidulous, oily and fatty, caseous, gelatinous, albuminous and fibrinous*, taking as examples, *starch* (arrow root, sago or ordinary wheaten starch) ; *mucilage or gum* (gum Arabic) ; *sugar: acidulous fruits*, with and without the skins ; *butter or suet: cheese*, mild and pungent; *gelatin* (isinglass) ; *albumen* (fluid and concrete as in the raw and boiled white egg) ; and *fibrin*, formed by repeatedly macerating thin slices of muscular flesh in water under 150 degrees of fahrenheit (pure albumen coagulates at 156 degrees; diluted at 212 degrees) as well as the individual articles forming those divisions, when compared with each other. These experiments might be made either in the stomach or artificially; and it might be instructive to adopt both courses.

As regards the nature of the fluid met with in the stomach of fasting animals, and to which, collectively, the term "gastric juice" has been applied, experiments exhibit great discrepancy. It would be gratifying to me to learn the result of your researches. Did litmus paper indicate the presence of any free acid or alkali? If acid, could it be discovered, by burning, whether the muriatic or the Acetic was the one in question or did the fumes of ammonia indicate the existence of either; or a solution of nitrate of silver that of the former? Did the fluid obtained from the stomach when fasting deprive putrid substances of their septic characters? Did it remove the flavour of certain aliments, as of wine? Did you examine it with the microscope—that least satisfactory of all methods for investigating the nature of animal fluids;—and if so, what were the appearances? Did it always coagulate milk?

It would afford me great satisfaction to learn the effects of subjecting pure fibrin, albumen or gelatin to the action of this fluid out of the stomach; and to see how far they experienced mutation in their sensible and chemical properties. The albumen contained in the gastric fluids may be precipitated—when the experiment is made on the fibrin—by a solution of the bichloride of mercury. The precipitate is a compound of the salt and albumen, in the proportion of about one of the former to three or four of the latter; so that, by drying the precipitate, the quantity of albumen in the fluid can be easily determined. If the fibrinous solution be now evaporated, at a moderate heat, until it forms a thick mass, and concentrated Acetic acid be added with the assistance of heat, a tremulous jelly is formed, which is completely dissolved by the addition of warm water, provided the mass be simply fibrin; but if the fibrin has experienced changes during the process of artificial digestion, different results will be obtained.

In like manner, if this artificial digestion has been accomplished on *Albumen*, provided the albumen has experienced no conversion, the solution of the bichloride of mercury will precipitate it, and the quantity of albumen, so precipitated, may be compared with that subjected to the process of digestion;—or the albumen may be coagulated by boiling, and filtered. No other fluid is coagulable.

Lastly: if *gelatin* has been employed, the quantity remaining after digestion may be approximated by precipitating it with tan prepared by infusing an ounce of gall nuts in a pint of water. The quantity of gelatin in the precipitate may be roughly appreciated by considering, that there are somewhat less than two parts of tan to three of gelatin.

In the case of the patient at *La Charité*, the food, during its conversion into chyme, appeared to have acquired an increase of its gelatin, and a substance in appearance fibrinous; but others have asserted, that gelatin has not been met with in the chyme,—which is scarcely comprehensible where gelatin has been the aliment, as the conversion must have been total.

You will pardon me for the length of the preceding detail, every topic of which has probably suggested itself to you already. It will afford me great pleasure to leran any facts, which the case has taught you and Dr. Beaumont, of a physiological character; and I do not abandon the hope of being in Washington in the course of a few days. At present, the roads, from their roughness are almost impassable; but if the frosty weather continues, they will be in order, probably by the commencement of the next week. The whole journey has now to be performed by land and our Virginia roads are proverbial for their badness. Should I be prevented from visiting Washington, I may be perhaps permitted to request an account of your experiments and observations, in order that I may make use of them in a second edtiion of my work on *Physiology*, should one be demanded, or of communcating them, in your name, to the American Philosophical Society, should such be your desire. I am, dear sir,

With great respect,
Obediently yours,

Dr. Lovell [7]   Robley Dunglison.

Dr. Beaumont was neither by nature nor his stars well fitted for the investigation of any profound physiological subject. He was most persevering, however; and his zeal and industry compensated in some degree for his deficiency in ability. It fortunately happened, that I was able to proceed to Washington, and to be associated with him in the prosecution of numerous interesting experiments; it is impossible to imagine a more disagreeable journey from Fredericksburg by the way of Dumfries to Alexandria,—unknown to those of the present day, but lively in the recollection of the traveler of a quarter of a century ago. The results of the experiments were communicated, in a private let-

ter,[8] to Dr. Hays, of Philadelphia, editor of the *American Journal of the Medical Sciences,* dated University of Virginia, Feb. 5, 1833; and by Dr. Beaumont, in his *Experiments and Observations on the Gastric Juice and the Physiology of Digestion* published the same year; and since then they have been incorporated in the different works on physiology and dietetics. At the conclusion of that letter, I remarked,

I cannot help feeling gratified, that the results of this case should harmonize so well with the deductions, drawn from less evidence, in my *Physiology* on many points connected with digestion,—a circumstance which had impressed Dr. Beaumont before I had the pleasure of being introduced to him. It appears, indeed, manifest to me that the great digestive act effected in the stomach is of a physical character, to fit the food for the formation of chyle by the chyliferous vessels, in the small intestine. The results do agree largely with those of Tiedemann [9] and Gmelin—the most accurate writers on this part of physiology. They found, like ourselves, that the secretion of acid begins as soon as the stomach receives the stimulus of a foreign body, and that it consists of the muriatic and Acetic acids. When flints were swallowed, the acid character of the secretion was as distinct as in our case when the elastic gum catheter was introduced; and the result, in both instances, destroys the objection, made by Bostock, regarding the detection of the muriatic acid by Prout,—that as there did not appear to be any evidence of the existence of this acid, before the introduction of food into the stomach, it might rather be inferred that it is, in some way or other, developed during the process of digestion.

The following letter was received by me, about a month after my return to the University of Virginia from Washington.

Washington, Feb. 19, 1833

Dear Sir,

Agreeably to promise I have now the satisfaction of sending you another bottle of *gastric juice* for further investigation. I hope it may be safely received, and will afford you an opportunity further to extend the highly important analysis so happily commenced with the first.

I have also, with peculiar pleasure carefully experimented upon the *masticated meat* and *gelatin* (isinglass), as suggested in your last letter, and herewith send the results of my observations. I hope you will excuse the awkward and unscientific manner in which they were made, and are here attempted to be described. To develop useful facts is my chief aim and ambition; and if I communicate such to you, even in my *own style* and language, I am confident they will be duly appreciated, and *kindly* disposed of to advantage. And I have the *peculiar* satisfaction; and not only this *satisfaction;* but am happy and *proud,* also, to assure you, so far as I have read and understand your very valuable and excellent treatise on *Human Physiology,* that the principles, reasonings and suggestions on the subject of the *gastric functions* and *fluids* are *generally* most indubitably confirmed by my experiment, so far as I have been capable of observing. I am well pleased indeed, to have it in my power to *assist* even in verifying the correctness of *your*

---

[7] This letter was published by Jesse S. Myer, *Life and Letters of Dr. William Beaumont,* pp. 156–158, St. Louis, 1939, in which it is dated 1833.

[8] Published by Dunglison in his *Elements of Hygiene,* pp. 216–222, Phila., 1835; republished by his son in *College and Clinical Record* 2: 241, 1881.

[9] Friederich T. Tiedemann (1781–1861), German physiologist, author of *Physiologie des Menschen,* 1830, also translated into French and English.

views and suggestions, and *irrefragably* establishing physiological facts, which have for ages divided the opinions and distracted the views of physiologists; and, in recent times, become fashionable sources of speculative errors or professional obstinacy.

(Here follows a statement of the experiments referred to, the results of which are given in his book).[10]

I shall be happy to learn the results of your inquiries. I shall continue most cheerfully to receive and faithfully attend to any suggestions you may be pleased to make, while I remain here, and at New York or elsewhere. . . . I am, Sir, with much esteem

Very respectfully yours, etc.
William Beaumont.[11]

My letter to Dr. Hays was strictly private as I did not wish to forestall Dr. Beaumont in his forthcoming publication; on which he evidently felt somewhat nervous, as is sufficiently shown in the following note written to me soon after I left Washington.

Dr. Beaumont takes pleasure in acknowledging his high regard and thankfulness to Professor Dunglison for his kind and valuable assistance and lively interest taken in the prosecution of his gastric experiments, and hopes amply to be able to remunerate him for his generous exertions by affording him a more full and satisfactory view of the result of his observations, when he shall have had time properly to collate and classify them.

The bottle of gastric juice is cheerfully submitted for chemical analysis, with a strong hope and expectation, that Professor D. will succeed in obtaining very important and satisfactory results, and communicate them in detail to Dr. B., so soon as practicably convenient. Dr. B. is much indebted and will be ever grateful for the aid and instruction received from Professor D. and is disposed to reciprocate every generous desire mutually to afford subject for physiological investigations and improvement so far as he is capable. Dr. B. hopes his feelings may be perfectly appreciated. Similar confidence he could repose in but few; but he is well aware, that by *Professor Dunglison* he will never be improperly anticipated in his intention, sooner or later, to publish the experiments collectively, by premature communications to any periodical publication of partial results, whereby impressions of undue transfer of *merit or demerit* from its original or collateral source, might be liable to be made upon the public mind.[12] A simple notice of the case and the intention of the publication, may not be improper, and is not objectionable.

All communications and suggestions from Professor D. on this subject, will be thankfully received, and *gratefully*, if not scientifically, appreciated and attended to.

Washington D.C.
1:00 P. M. Friday
Jan. 25, 1833.[13]

---

[10] P. 73 in the 2nd ed.

[11] Published by Myer, *op. cit.*, pp. 164–165.

[12] In a letter to Trist March 14, 1833 (Library of Congress, Trist Papers), Dunglison again refers to this subject and seemed to be irked by a notice in the *Washington Intelligencer* of March 11, 1833, by S. Beaumont, denying a previous notice that Dunglison and Beaumont were planning to write a history of the case together, which, he said, carried "folly on its front" and expressing disapproval of the announcement that Beaumont planned a volume on the subject.

[13] Published by Meyer, *op. cit.*, p. 159, with others on the subject, pp. 159–168.

The suggestions, contained in this note, although savouring somewhat of a want of delicacy, I need scarcely say were needless. Dr. Beaumont had been long observing and experimenting, although in a desultory manner, on the Canadian, and it was but right, that he should have all the credit to which he was legitimately entitled and the advantages that might ensue from his contemplated publication, uninfluenced by the prior statements of others, and, with this feeling, I determined, that my letter to Dr. Hays should not be made public until after Dr. Beaumont's work had seen the light; and, accordingly, it first appeared in my *Elements of Hygiene* published in 1835, two years after the date of issue of his work.[14] It might have been anticipated, however, from Dr. Beaumont's sensitiveness as to his own rights, that he would have been extremely liberal as regarded those of others, and especially of one, of whose services he speaks so highly in the letters referred to above; and with such anticipations I confess, I was surprised, on the appearance of his *Experiments,* to find no allusion whatever to my having visited Washington, by request, in a Virginia winter from a distance of upwards of 100 miles; or to my having made a single suggestion to him, or been associated, in any manner, with his investigations. The only allusion to me, in the experiments made at Washington, is at Page 233,[15] under the head of "Microscopic Examinations," in this paragraph—"the following microscopic examinations were made with Jones' compound microscope, in presence of Professor Dunglison and of Captn. H. Smith of the Army." It would scarcely be presumed, from this paragraph, what was nevertheless the fact, that the microscope was procured by me, and that the results of observations with it were absolutely dictated by me. My friend Mr. Trist was present; but the fact was overlooked by Dr. Beaumont. Soon after the appearance of his book I saw Dr. Beaumont in Baltimore. As he did not present to me a copy, I offered him payment, which, I think, was *Three Dollars.* He said he thought, under all the circumstances, I *ought* to have a copy gratuitously; but took the money. I did not conceal from him my expectation of seeing in it a more detailed account of our joint labours, and reminded him, that the only allusion to my being concerned in the matter was in the paragraph, which I have cited; when he replied, that he did not think I would care to be associated with so humble an individual as himself!

Some years after the publication of Dr. Beaumont's book (in 1839) I had the pleasure of meeting Mr.

---

[14] See note 8 above.

[15] Beaumont made frequent respectful allusions to Dunglison throughout his book—acknowledging his aid in the preface and referring on page 33 and again on pages 35 and 72 to Dunglison's textbook of physiology, the chemical analysis of specimens on p. 73 and expressing deference to him as a scientist on pages 94 and 113.

George Combe,[16] at the house of my friend Mr. Benjn. W. Richards.[17] At table, a conversation occurred between himself and me on the subject of the experiments of Dr. Beaumont, which were considered by his brother, Dr. Andrew Combe, to be so interes:ing that he made free use of them in his own little work on Digestion and republished the work of Dr. Beaumont. I informed Mr. Combe what part I had taken in them; and was afterwards surprised to find, that he had made the following entry in his journal, and inadvertently, no doubt—had misapprehended me—

Jan. 2. Ther. 32° *Sir Walter Scott and the Ballantynes.*— We met several literary gentlemen today at dinner, and Mr. Lockhart's attack on the Ballantynes, and the reply of James Ballantyne's executors was a topic of interesting conversation. One gentleman remarked, that he had read Sir Walter Scott's life very attentively, and had come to the conclusion that Scott had obtained the full proceeds of the sales of his works from both Constable and Ballantyne; he added, that the facts, stated by Mr. Lockhart himself in his Life, appeared to him to be at variance with the idea, that the Ballantynes were the cause of Scott's ruin. Professor Dunglison, in allusion to Dr. Beaumont's experiment on Alexis St. Martin's digestive powers (see *Experiments and Observations on the Gastric Juice and the Physiology of Digestion,* by William Beaumont M.D., surgeon in the U. S. Army, reprinted with notes by Andrew Combe, M.D. Edinburgh, 1838), mentioned, that he had suggested, and also performed the experiments at Washington, which are recorded in Dr. Beaumont's work. (*Notes on the United States of North America during a Phrenological Visit* in 1838–9–40, by Geo. Combe, Vol. 1. p. 303, Edinburgh, 1841).

I could not allow so erroneous a statement as this to pass unnoticed; and, therefore, inserted the following correction in the number for March 1, 1841 of the *American Medical Intelligencer,* of which I was editor.

Mr. Combe. It was with no little regret, that we observed in the volumes of Mr. Combe on the United States, which have just been published, an allusion evidently to a private conversation, which took place at a dinner table, in which Mr. Combe makes the editor of this journal do Dr. Beaumont the signal injustice of stating, that the suggestions and experiments, made at Washington, and detailed in Dr. Beaumont's book, were by the editor. The error is another instance of the difficulty and impropriety of travelers attempting to detail private conversations. Mr. Combe was informed, that certain suggestions were made to Dr. Beaumont, and certain experiments performed by the editor in Washington along with him, whereas the observation of Mr. Combe would lead to the inference, that Dr. Beaumont himself suggested and performed none of them. The editor of this journal has been extremely careful not to detract from the results of the meritorious and persevering investigations of Dr. Beaumont, and it is pain-

ful to him to have the subject brought forward in this manner. It is strange, indeed, that Mr. Combe should not have seen the injurious effect of such a statement to one party, even if the impropriety of attempting to detail private conversations had not impressed him.

The remarks of Mr. Combe gave occasion to no *reclamation* on the part of Dr. Beaumont. The last letter I received from him was the following, asking me to render him a still farther service.

Washington
22 Jan. 1834

Prof. Dunglison
Dear Sir,

Being about to present a memorial to Congress for an appropriation to indemnify me for the extraordinary expenses I have been at for the last ten or twelve years in preserving, sustaining and supporting *Alexis St. Martin,* and in prosecuting and publishing my *Experiments and Observation on the Gastric Juice etc.,* I have taken the liberty of asking your kind services in my favor, if consistent with your just and generous disposition. It is believed, that the expression of your sentiments on the subject will have a very favorable influence, and greatly conduce to the fair attainment of this object. I therefore beg leave very respectfully to solicit a statement of your views and opinions on the merits, general usefulness, and importance of the work, and also of the expediency and utility of a farther prosecution and investigation of the subject, forwarded to me at this place, soon as practicably convenient. I would, also, respectfully suggest the propriety of adding to your own the signatures of such other professional men of your medical institution as you may deem advisable to obtain.

I am, sir,
Very respectfully,
Yrs. most obediently,
William Beaumont.

Such an application appeared to me of very questionable propriety, and not at all likely to be successful; nor do I know, that it accrued, in any way, to Dr. Beaumont's advantage.[18]

THE AMERICAN PHILOSOPHICAL SOCIETY

In the year 1832 I received notice of my having been elected a member of the American Philosophical Society, which was the more gratifying to me as proceeding from the most celebrated society of the country, and one holding its meetings in the city in which seven years before, it had been suggested (see page 25) that the societies of the country would probably visit upon me the offense of occupying the chair of medicine in the University of Virginia. (See supplementary "Ana.")[19]

---

[16] George Combe (1788–1858), follower of Spurzheim, author of a *System of Phrenology,* lectured in the United States from 1838 to 1840.

[17] Benjamin W. Richards, at one time Mayor of Philadelphia, organizer of the Laurel Hill Cemetery (see p. 159), was associated with Dunglison as a member of the American Philosophical Society as well as such institutions as the Deaf and Dumb Asylum and the Blind Asylum (Henry Simpson, *Lives of Eminent Philadelphians,* p. 840, Phila., 1859).

[18] In spite of the unanimous support of the leading scientists of the day, including that of Dunglison, a bill to grant Beaumont $10,000 was defeated. Congress objected to appropriation of money for any scientific or philanthropic purposes whatever (Meyer, *op. cit.,* pp. 216, 219).

[19] Pp. 173–174. See also pp. 130–131.

PRESIDENT JACKSON

It was during a visit to Washington, that I was requested to visit General Jackson—then President—in consultation with his attending physician—Dr. Thomas. He was suffering greatly from pleurodyne [20] or pain in the side, to which he had been subject, and which was probably owing to wounds he had received in different encounters. His usual remedy—for he was *heroic* in the treatment of his own case—was to lose a quart of blood. Bloodletting had been practiced before I saw him on this occasion, but he had not obtained his accustomed relief, owing, he said, to the bleeder not having taken a sufficient amount from him. It appeared to me injudicious at his time of life:—he was then nearly seventy years old—to reduce himself so frequently as he did; and I so suggested to him,—his physician according entirely with me. We agreed to recommend strong counter irritation; and as he was about to proceed on a progress to the eastern states,[21] to direct a warm plaster, *animated* by cantharides, to be worn. The first plaster, applied before he set out, afforded him decided benefits; and immediately before leaving Washington, he sent for a fresh one, which was accidentally made somewhat stronger, so that by the time he reached Baltimore, he felt inconvenience from it. His mind, however, was so much occupied, and his physical endurance so great, that he neglected it; and it was not until he arrived in Boston, that he had it attended to by Dr. Warren.[22] It was then found to have acted as a blister and, of consequence, to have produced intense irritation. The temporary evil proved, however, of permanent benefit: for the prolonged counterirritation ultimately relieved him entirely; and when he subsequently awarded the credit for this, I frankly told him that much had depended upon himself in keeping the plaster applied for a longer period than was ever contemplated, under the irritation it was producing.

When investigating his case, I told him it would be well to examine minutely his side. "Yes, Doctor"—he replied—"that's the way I like to do things." And immediately proceeded to expose it, pointing out to me the direction in which wounds had been received by him in the vicinity of the suffering part. I had different opportunities of witnessing his well known energy of character. He was, then, deeply interested in the conduct of South Carolina; and took no pains to conceal

his regret and indignation at the course that state was pursuing;—remarking to me on one occasion, that such a union as she contended for, was "a rope of sand," and not the one he had fought for.[23] He, at all times, exhibited great solicitude for the health of Mr. Madison, whom he knew I was attending, and who was, at the time he consulted me, much affected with rheumatism.

To show to what an extent constitutional scruples were indulged in it may be mentioned, that on the occasion of an eclipse, whilst General Jackson was President, my friend and colleague—Dr. Robert M. Patterson, Professor of Natural Philosophy there—was desirous of borrowing a telescope belonging to the general government with which to observe it. With this view he applied to his brother-in-law, Mr. George Leiper, who was then in Congress, to get General Jackson's consent to the temporary loan of it. Mr. Leiper suggested, that as General Gordon was the representative in Congress from the district in which the university was situated, the application had better be made through him. General Gordon was accordingly approached on the subject; who expressed his great desire that the university should have the telescope but doubted whether it was constitutional. The application was not made to the President, but when afterwards asked, as I was told by Dr. P., what he would have done had the case come before him, he replied "The university should have had the telescope."

PRESIDENT MADISON

Of the intellectual and moral character of Mr. Madison it is difficult to speak in terms of adequate commendation. His beautiful seat of Montpellier was a few miles from Orange Courthouse, and, as I have before remarked, was much visited on account of the public and private reputation of its distinguished proprietor. Of Mrs. Madison it is needless to say more than that she was eminently worthy of the character she had everywhere acquired, for kindness and elegant hospitality.[24] To my own family, she was always most attentive, and repeatedly desirous of our visiting them, when the vacations at the university admitted of our so doing; and I have now before me a letter from Mr. Madison, in answer to the one I had written expressive of my fears of introducing, with my family, hooping cough into his plantation.

---

[20] Pleurodyne might be equated with present-day intercostal neuralgia.

[21] This must have been about June 1, 1833, just before Jackson set out on his triumphal tour of the Middle Atlantic and New England states. He suffered from 1806 until his death in 1845 from the effects of his chest wounds. Soon after Dunglison saw him, Jackson consulted Physick in Philadelphia; and when he reached Boston on this tour, he had to be confined to bed for several days by severe hemoptysis.

[22] John Collins Warren (1778–1856), political supporter of Jackson, more famous later as the surgeon who introduced ether anesthesia.

---

[23] South Carolina in November 1832 declared the National tariff act null and void and threatened to secede, but Jackson issued a proclamation warning that laws of the United States would be enforced with military power, if necessary.

[24] Dorothy ("Dolly") Payne Todd Madison, popular First Lady, moved with her husband to Montpelier after his term of office at the White House.

Montpellier, June 27, 1826.

Dear Sir,

Your favor of the 22d.[25] did not come to hand till yesterday afternoon. We are so fortunate with respect to the hooping cough, that it may be safely brought in the case of your little patient who may herself be benefited by an excursion. Mrs. Madison and myself will, therefore, with much pleasure, receive the fulfillment of the promised visit from yourself and Mrs. Dunglison, as soon as your conveniency will permit. May we look out for you at the close of the present or commencement of the next week? Meantime, accept, both of you our joint salutation and good wishes.

James Madison.

This letter was written on the day after I was called to Mr. Jefferson in his last illness. The visit, therefore, had to be postponed until after his decease, and I well recollect the melancholy satisfaction we both had in recalling many of the actions and sayings of Mr. Jefferson.[26] I have already stated the great veneration which Mr. Madison entertained for him, and their mutual esteem and regard. This was signally exemplified in the interest he felt in learning all the particulars of the last days of his illustrious friend. Between those two eminent persons there was much that was alike, and much that was dissimilar. Both were highly intellectual; but, if I were to venture upon a distinction, I would say, that whilst Mr. Jefferson had more imagination, Mr. Madison excelled perhaps in judgment. In social intercourse, no one could be more delightful than Mr. Madison. As a host, he was convivial and entertaining,—his table bounteously supplied from his well-stocked farm: the *cuisine* reflecting credit on the excellent hostess: and his wines of the best—madeira being his preference. In my latest visits to him, when confined to the bed or sofa in the next room, he would invite me to take his place at table; and call out, that if I did not pass the wine more freely, he would "cashier" me! Whilst he was President, he had imported a large stock of madeira wine, for which he had paid *five dollars* a gallon, and one of the last five gallon demijohns of which he sent to me at the University of Virginia. During the last years of his life, he was much affected by rheumatic pains of the joints, which interfered greatly with his movements, and ultimately confined him to the setee or bed.[27] Before I left the University of Virginia, in the autumn of 1833, I had been requested, more than once, to visit him for

this;[28] and, although he experienced relief, he was never afterwards entirely free from it. The fact of his suffering from rheumatism being bruited abroad, specifics for its cure were sent to him, but he was cautious in employing them. Amongst these was an anti-rheumatic liniment, invented by Dr. Carr, of Charlottesville,[29] secretary of the board of visitors of the University of Virginia. On the occasion of one of my visits to Mr. Madison, he informed me that Dr. Carr had sent him the liniment, with directions for its use, under the expectation—I doubt not—that if it proved or appeared to prove of any benefit, he would receive a written testimonial thereof from Mr. Madison, which might be used to his advantage. Knowing Mr. Madison's extreme caution in such matters, I asked him if he had used it? He said "Yes"! Had it proved serviceable? "He could not say." Had he written to Dr. Carr? He said "Yes"! and smiled. I then asked him if he would consider it proper to inform me what reply he had given. He said, "Yes! I wrote to inform him, that I had used his liniment,—that I did not know that it had done me any good, but that I thought it not improbable that it might have prevented me from becoming worse." He smiled with me at his reply, which I said the Doctor could certainly not use as a testimonial, and it never was so used.

In May 1834, I was requested to visit Mr. Madison from Baltimore. I found him, at that time, suffering still more from his rheumatic affection, which was accompanied by febrile movements, and more or less dyspnoea, from which I augured unfavorably, but did not consider, that there was any immediate danger.[30] Exactly two years after this, I was again summoned to him,—the letter to me stating that he would not consent to take any remedies, unless they were pre-

---

[25] This letter by Dunglison, June 22, 1826, is now in the Madison Papers, Library of Congress.

[26] July 1, 1826, Dunglison explained his delaying the visit because of Jefferson's diarrhea and poor prognosis (Madison Papers, Library of Congress). When Dunglison arrived at Madison's house after Jefferson's death, he brought Jefferson's gold-mounted cane, bequeathed as a token of friendship to Madison (Irving Brant, *James Madison, Commander in Chief 1812–1836*, p. 456, 1961).

[27] Arthritis crippled Madison, interfering not only with his walking but also with his writing (Stanislaus Murray Hamilton, *Writings of James Monroe* 8: p. 233, fn., 1903).

[28] The Madisons made use of Dunglison's talents almost from his arrival. June 30, 1825, he sent medicine to them for a servant; August 27, 1827, vaccine matter (Madison Papers, Library of Congress); and in 1828, his reputation was greatly enhanced when a child was unsuccessfully operated upon against his advice (Irving Brant, *op. cit.*, p. 459). Confined to his house during the winter of 1833–1834, Madison used a rocking chair for exercise; a violent itching eruption at that time was relieved by a lotion recommended by Dunglison (*ibid.*, p. 512).

[29] Frank Carr, physician, teacher, farmer and for many years a magistrate, editor (about 1829) of the *Virginia Advocate*, the second newspaper of Charlottesville, and active in the Albermarle Agricultural Society, died 1854 (Vera U. Via, Looking Back, *The Daily Progress*, p. 42, Charlottesville, Va., April 28, 1950).

[30] This visit was reported in the newspapers. The bedridden patient greeted Dunglison with a request to pass judgment on a case of sherry wine just received, and reported that the doctor, "a better judge than I could be were my palate in better health, pronounces the wine to be of the first chop" (Irving Brant, *op. cit.*, p. 519). June 8, 1834, Dunglison wrote to Mrs. Madison inquiring after Mr. Madison's health and referring to misrepresentations of his opinions with regard to Mr. Madison's health (MS. 567, New York Academy of Medicine).

scribed or sanctioned by me. He was, then, again labouring under heat of skin, with increased quickness of pulse, and I may remark that for years—from the commencement, indeed, of his rheumatic ailment—his pulse had become greatly accelerated; and, from being somewhat more frequent, as is the case with the aged, and intermittent in its character, had risen to ninety and upwards, and was perfectly regular. His dyspnoea was now great; and although he was still cheerful, he felt, that his end was approaching; and I left him, I believe on the 23d of May, with no expectation of his existence being protracted for any length of time. He died on the 28th of June 1836, at the age of eighty-five.[31]

[N. B. The following letter was pasted to this page of Dr. Dunglison's diary.]

Orange C.H. Friday night May 16th—34.
My dear sir:

You were so kind as to say, that I could write for you when the occasion might require it for Mr. Madison's health. He is now in a more precarious state, according to my judgment than I have ever witnessed. It commenced with a breaking out on the skin and has extended all over the body and limbs causing great suffering especially at night, during which he is for the most part deprived of sleep, very feverish and talkative and nervous. Soon after rising he is more composed but falls into sleepiness during the whole day with very limited intervals of conversation. He is fast reducing, within some days, in strength and I am apprehensive may not be able for long to have the proper remedies administered with effect. Independent of your qualities as a physician you know he has always entertained such a confidence in you individually that probably no one else could induce him to resort to the proper relief. I have therefore to ask of you professionally for a visit. I think the stage leaves Baltimore early in the morning so as to take the steam boat for Fredericksburg which arrives there about three o'clock and from thence after nine at Orange C. H. the same day. Should your avocations permit you to come on, or should they not, please inform me to have a vehicle ready in the former case and suggest someone else to be employed, in the latter, of eminence—I remain my dear sir with great esteem and consideration,

Yours truly,
J. P. Todd[32]
Dr. Dunglison

PRESIDENT MONROE

With Mr. Monroe I had fewer opportunities of personal intercourse. He was not in good health whilst at the university, and consulted me thereon.[33] He was apparently exceedingly amiable, but diffident and reserved; and made a much less favorable impression in regard to his intellectual powers than Mr. Jefferson or Mr. Madison. Yet these gentlemen had a great regard for him and the expressed opinion of the former was, that although Mr. Monroe might not always be able to explain the process by which he arrived at a conclusion, the conclusion was, notwithstanding, generally right. His advent at Montpellier was looked forward to with pleasure—as I had an opportunity of observing; and a small chamber, in immediate communication with the dining room, was called "Mr. Monroe's room."

When I became acquainted with him, his circumstances had become embarrassed; and he was apparently *distrait,* and awkward; although, at all times, full of politeness. An example of his abstraction occurred at one of the examinations at the university, at which he was present. Seated between Mr. Long and another of my brother professors he appeared to be absorbed in the exercises, when suddenly turning round to Mr. Long, and, doubtless, desirous of exhibiting his interest in that gentleman, he said,—"Pray, sir, how is your father?" Mr. Long replied, "My father has been dead, sir, for the last twenty years." Mr. Monroe was, of course, silenced.

NEGOTIATIONS WITH UNIVERSITY OF MARYLAND

In the spring of 1833, I received a communication from Professor N. R. Smith, of the University of Maryland, of which the following is an extract.—

There is some movement of the waters here and changes will probably take place in our school. Should they occur, they will have some relation to you, and, in confidence, I therefore apprize you, without the privity of anyone. I do it because it might be of importance to you, and to us, that you should have an early intimation. (letter dated April 4, 1833.)

This letter detailed the contemplated action of the faculty. In another letter, dated April 25th, 1833, Dr. Smith informed me, that the faculty had resolved unanimously to recommend to the board of trustees, that I should be appointed to the chair of Materia Medica, my fee to be $20.00 from each student, and all candidates required to have taken my ticket twice. "With you in our chair of Materia Medica," he adds,

I am confident, that, with a chair of surgery" (his own chair) "better filled, we should constitute the strongest faculty in America. This fact, I trust, the good people of our community will not be slow to discover. I am confident, that your removal to this city would not mar your pecuniary interests, whilst certainly it would extend your

---

[31] Death apparently resulted from a failing heart due to arteriosclerosis. While the arthritis may have been rheumatoid in type, considering his penchant for good wine, it is also possible the president suffered from gout, which could well account for all the symptoms enumerated.

[32] John Payne Todd (1792–1852), Mrs. Madison's son by her first marriage.

[33] Nearly all of Mr. Monroe's letters in later life allude to his health. A letter he wrote to Madison, February 8, 1830 (in the editor's possession), gives details of his health, saying he was improved but still confined to the house. Soon after, he moved to New York to live with his daughter, and in a

letter of March 28, 1831, wrote to his friend, Dr. Charles Everett, that his cough was constant, tells of his doctor's (Dr. Bibby) treatment and asks for Dr. Everett's and Dr. Carter's (of Richmond) advice (S. M. Hamilton, *op. cit.* 7: p. 231, 1903).

reputation. . . . We do not intend to ask any pledges of you, but, having presented you the chair, to leave you to act according to your interests and feelings. I fear, that we should suffer much, however, if we do not succeed in obtaining you. But I am sanguine, that you *will* join us, because I am sanguine, that it is for your interest, reputation, and, I trust, happiness. (not to disparage your friends in Virginia). . . . You will hear from our Dean officially at the same time that you receive this.

The following was the official letter of the Dean.

Baltimore April 26, 1833.

Professor Dunglison,
Dear Sir,

I have been instructed by the faculty of Physic of the University of Maryland to apprize you of the progress of a negotiation now going on between the faculty and Professor McDowell, relative to his withdrawal from the university. Our object in this arrangement is to effect a new, and more advantageous, organization of our school, under which we are particularly anxious to secure your valuable cooperation, provided we can offer you sufficient inducements to join us.

The following is a part of our plan. Dr. McDowell has consented to retire, provided we will pay him $1,000.00 annually during his life, for which the faculty are to become jointly and severally bound, *as a Faculty,* and are to pledge the graduating fees as collateral security, which, by act of the Legislature, are to be raised to $20.00. Our number of graduates was 54 the present year, last year 62; so that the fees derived from that quarter will more than meet the annuity. Professor Baker has also proposed to resign, on condition, that you shall be appointed as his successor, and will agree to accept the following organization.—The Chair of Institutes to be abolished; special Physiology to be united to the Anatomical Chair;—Pathology to practice; and Medical Jurisprudence and Therapeutics to Materia Medica. Each student will be required to attend two full courses of each of the six professors, and to pay a fee of $20.00 for each ticket;—such students only as have already been members being allowed the privilege of graduation under the old arrangement. My class, for the last winter, was 120, notwithstanding we were much affected by the prevalence of cholera in our city. Should we accomplish our proposed arrangement, and secure your cooperation, it is the opinion of the faculty,—an opinion, which, I think, is well founded, that our class would, in a short time, amount to 180 or 200.

After detailing other particulars in regard to the school, Dr. Geddings concludes:—

The faculty direct me to beg, that you will consider all the bearings of the case, and that you will inform me, whether you would be willing to join us under our present conditions. You will have your chances of practice amongst the rest of us. The field is ample, and your talents and reputation would no doubt secure to you a due proportion. Should you be disposed to unite with us, I can only assure you of the pleasure it will afford myself and colleagues to see you *one* of *us.* It will not be necessary for you to become a candidate, but merely that you should send me a letter expressive of your willingness to accept, if elected. Please let me know your decision as early as possible, as time is precious; and believe me,

With highest consideration, yours,
E. Geddings—Dean.

From Dr. Potter, whose equivocal course on a former occasion has been mentioned, I received the following *mixed* epistle.

Baltimore April 29 1833

Dear Sir,

I have this day finished the perusal of your work on human physiology, and take the earliest occasion to acknowledge my obligations to the author of a system, which I consider the most clear, intelligible, and, at the same time, the most comprehensive I have ever read. While the physiologist will find himself pleased and instructed, the pupil will perceive, that he has, at last, found a safe and easy guide to the fundamental principles of his profession. As a *textbook,* it will take precedence of all other systems.

You have or will receive an official communication from the Dean of the medical faculty of the University of Maryland, in which you will find an unanimous concurrence in a wish to make you our colleague; and should this event (so much to be desired) be the result, we will, with one voice, and one heart, use every exertion to make your time and place as agreeable as possible. No one is more zealous than, dear sir, yours very truly,

Nath. Potter.

To the letter of Professor Smith, I returned the following answer.

University of Virginia.
Apl. 9th. 1833

Dear Sir,

I return you many acknowledgements for your kind communication of the 4th inst. which reached me only this morning. It is pleasing to me to hear again from you on any subject, and not less so on one, which regards the progress of your school. I wrote you upwards of twelve months ago, by my friend Mr. Saml. Leiper, of Philadelphia, but as I did not hear again from you, I conceived, that, as frequently happens in the case of private conveyances, the letter had not reached its destination. I need hardly say that I feel gratified at your attention having been again directed towards me to supply any vacancy, that may occur in your university. I have not heard of your numbers this year; but I feel satisfied, from the specimens I had of the capabilities of the teachers whom I heard when I was in Baltimore, that the number of students ought annually to increase.

I scarcely know, at present, what reply to make to you, connected with the immediate subject of your letter. At one time, I had determined never to connect myself with a school, where nativity was regarded in the selection of teachers. This objection was however removed by the communication I subsequently had with one of your board of trustees; and still more by the appointment of your excellent colleague, Dr. Geddings, whose translation to Baltimore cannot but have added to the reputation of your school. That objection, at one time seemingly invincible, has, then, been removed, and my difficulties on this score are not insuperable. I am not situated, however, in the same respect as when the faculty kindly interested themselves in my behalf, on a former occasion. My school is rising; and my emoluments are somewhat ample. I will, in confidence, state to you exactly the proceeds of my chair simply; and you will then see, that I must necessarily have difficulties in relinquishing a situation attended with numerous advantages, but devoid of several recommendations to an individual fond of science and literature, and who has been accustomed to civic life. The number of students, who have paid their fees into the proctor's hands is 40;

and 10 have joined my class of medical jurisprudence, for which I receive a separate fee. I have, therefore, received this year from

40 full students at $25.00 each	$1,000.00
10 medical jurisprudence at $15.00 each	150.00
Salary as professor	1,500.00

making $2650.00; and if to this I add $350.00 for house and grounds, my emolument for duties of professor alone is $3,000.00; and this without any drawback whatever, unless I reckon as such an addition to my house, for which I pay $32.00 rent.

My professorial avocation occupies me two hours on alternate days. All the rest of the time is my own, and can be devoted to literary, professional, or other purposes. All this constitutes a *sheet-anchor*, which cannot be abandoned without due consideration.

In the plan that you proposed, the chairs of Institutes and Materia Medica are, I understand from you, to be incorporated; but it is proposed, I presume, that both subjects shall be taught in the ordinary period of a session: one of the objects being, I suppose, to diminish the number of professorships. I am doubtful, however, whether justice can be done to either, under such circumstances. I need hardly say that physiology, which would fall under the department of the Institutes, is a pleasing subject to me; but I think, that no one can feel particularly partial to the other (Materia Medica), important—*essential*—as it is. I shall, however, hear from you on all these points; but I should wish it to be distinctly understood, and shall make it so, if any proposition comes officially to me,—That I will in no way go before your Board of Trustees in the light of a candidate. I will not be a competitor for any chair. If a situation is tendered to me, I will give it respectful consideration, and decide according to the best judgment I may form of my interests. I need hardly say, that the kind attention I have received from you, and the fitness, which impressed me, when I was in Baltimore, of the professor of the chair of surgery, for the performance of his difficult duties, would render it pleasing to me to be associated with you; but, whether I can do so *prudentially*, will depend upon the shape and nature of the proposition made to me. In the meantime, be good enough to answer the following queries. What is the amount of the ordinary professional remuneration of the physician in Baltimore? The fees for each visit, and for consultation? The nature of the aid afforded you by the Legislature? The amount of drawbacks on the fees at this time? The amount of fees paid last session to the schools in question? I had almost omitted to remind you, that we have a law of our university requiring, that any professor, who may be desirous of withdrawing at the end of a session, shall send in his *positive* intention three months beforehand, or, in other words, on or before the 18th of April. The nearer this law can be fulfilled the better, in order, that the Visitors may have full time to look out for a successor. Their kindness to me would probably not interfere, in any respect, with my prospects; but I feel equally anxious not to injure those of the university over which they preside. It would, consequently, be agreeable to me to have the matter assume a tangible shape at as early a period as possible, in order that I may come to a decision. . . . In conclusion, all I can say at present is to repeat, that the objections I formerly had to attaching myself to your institution do not now exist, and that if the circumstances can be made such as to induce me to leave the advantageous situation I now hold, it will give me great pleasure to join you. Believe me, dear sir,

Very respectfully and truly, yours
Robley Dunglison.

Prof. N. R. Smith.

The following reply was given by me to the official communication of Prof. Geddings.

Univ. of Virginia, Apl. 30, 1833

Dear Sir,

I have the honor to acknowledge receipt of your letter, of the 26th inst. proposing to me, in the name of the *Faculty of Physic of the University of Maryland*, to permit myself to be nominated to the Board of Trustees of the university, as *Professor of Materia Medica, Therapeutics and Medical Jurisprudence* in that institution, under the conditions mentioned in your letter. Although these conditions are not in every respect, such as I could desire, and the arrangement of subjects not strictly in accordance with my wishes, I do not think it probable, that any difficulty would arise in my acceptance of the situation, should the Professorship be conferred upon me by the Board of Trustees. At present, at least, I do not anticipate any. Farther than this I cannot say. On a former occasion, when I was honoured by my name being presented to the notice of the Trustees, the report of my being a candidate, and competitive for the office—was, I fancied, injurious to the interests of the University to which I now belong. The attachment I feel, and must ever feel, to it as the place in which several happy years have passed away, would prevent me from doing anything which could unnecessarily interfere with its prosperity, which the report of my being a candidate for a chair in any other institution, if not followed by my appointment to it, would certainly do. I must therefore request the faculty not to propose me as a candidate or competitor on this occasion, and, if competition should arise, to drop me altogether. The same feeling towards the University of Virginia would induce me to beg, that what is done may be done quickly, in order that no delay may arise in the appointment of a successor, should I be called upon to leave my present professorship. A law of the university requires, that three months notice should be given prior to the end of the session. This is now impracticable; but the nearer we can approach it the better.

Be pleased to present me, in the kindest and most respectful terms, to your colleagues, and to thank them for this additional mark of their confidence. I need hardly say how much I regret, that the proposed plan will remove from their body any of those that have been so long associated with them. To Professor Baker my acknowledgements are especially due, for the too partial consideration which he has been so good as to extend towards me.

Accept, dear sir, my thanks for the kind manner in which you have conveyed to me your own wishes and those of the faculty, and believe me, with the highest respect,

Obediently yours,
Robley Dunglison.

Prof. Geddings,
Dean of Med. Fac. of Univ.
of Md. etc. etc.

A few days after this, I received the following statement of my election, from the Dean of the Faculty. I may remark, that I had suggested, that the subject of Hygiene should be added to the duties of the professorship.

Baltimore 10th May 1833

Professor Dunglison,
Dear Sir,

I have the gratification to inform you that, at a meeting of the Board of Trustees of the University of Maryland, held yesterday, you were unanimously elected Professor of *Materia Medica, Therapeutics, Hygiene and Medical Jurisprudence,* Professors McDowell and Baker having withdrawn from the school. I need not assure you of the high gratification it will afford my colleagues and myself to have you associated with us, and to be aided by your able cooperation in the important cause in which we are engaged. Our school will now commence under an entire new organization, and under auspices far more favorable than heretofore. We have six chairs; fee for each $20.00; and require two full courses of lectures. The graduation fee $20.00; matriculation fee abolished; and infirmary fee $5.00 to be paid one year only. Dissection is optional, but, when prosecuted will be $10.00. These are the whole expenses of our course. . . . The organization of your chair is such as to allow you ample scope, and, should you desire it, you can lecture on the vital properties (which, I think, would be proper) preparatory to your course on Materia Medica etc. The special physiology will be as much as I shall be able to introduce into my course. I hope, therefore, that the arrangement will meet your approbation, and that you will agree to accept. Please let me know your decision as early as possible, as time is important.

With highest respects, yours,
E. Geddings
Dean Med. Faculty.

The faculty had determined to pay to Dr. McDowell $1,000 per annum on his retirement, during his life; and they had also agreed to establish, at their joint expense, a medical journal, of which Dr. Geddings should be editor. I declined taking part in these matters, to which no objection was made; and accepted the chair, tendered me by the Board of Trustees.

From different members of the faculty I received the kindest letters on my acceptance. In one dated June 2, 1833, my friend Professor Smith thus expressed himself.

We are most sanguine in regard to a class for next session. I have heard favorable news from the summer class in Philadelphia. Our present organization is popular everywhere, and we must succeed. Walsh, I see, has behaved very handsomely for a trustee of the University of Pennsylvania (alluding to a kind notice he took in the *National Gazette* of my appointment). You expressed a hope, that we are now on a peace establishment. We are so most strictly. We might quarrel with —— if we chose, but we do not, and need not. Indeed, we are the most harmonious faculty in America. Geddings is everything that we could desire.

But, by far the most amicable expressions were used by the colleague whose name might have filled the blank in Professor Smith's letter. On transmitting me a copy of the advertisement of the reorganization of the school, Professor Potter wrote me the following kindly expressed letter.

Baltimore July 8th 1833
Dear Sir,

I have understood from my colleague, Professor Geddings, that you propose to pay us a visit in all the month, and I pray you to signify in the meantime, the probable time at which we may expect you. My object for this special inquiry is predicated on the anticipation of having you as my guest. In this expectation Mrs. Potter unites with me, and prays, that you will not travel single. Mrs.

Dunglison's company will be necessary to complete the "circle of our felicities." I cannot receive an apology. We have, and enjoy an elevated cool residence, where we live in a plain, and unceremonious style, and should you find "a beggarly account of empty boxes," you will certainly find a sincere reception.

Our prospects for a class are highly flattering, and we only want your cooperation to maintain an honourable and fearless competition with any other, and all other institutions.

Believe me yours very truly,
Nath. Potter.
Prof. Dunglison

I returned Dr. Potter my acknowledgement for his kindness; but declined his proposition. It so happened, indeed, that I was unable to go to Baltimore at the time I had contemplated. I am satisfied he would have been pleased to receive me in his house; but his course had made it difficult to know where I should place him; and, as in all similar cases, I avoided too close a contact. He bore the character of being insincere, and by no means careful of the accuracy of his statements; and it was affirmed, that he occasionally employed the language of hyperbole in his classroom. Long, indeed, before I became personally acquainted with him I had heard of his reputation in this respect. An old colleague of his—Dr. Gibson [84] of Philadelphia—told me that he was the editor of a most appropriately named journal—the *Baltimore Lyceum;* and I heard many anecdotes told of him, which were certainly—to say the least of them—highly imaginative. He gave, in my presence, a lecture on the seventeen year locust, introductory to his course on the Practice of Physic. Apart from the inappropriateness of the subject, it was sufficiently well managed; and I do not recollect, that any of the statements were of a questionable character, but I was told he had stated in private, that when the young being falls from the tree, it makes its way into the ground with such velocity, that you could not overtake it with a case knife.

On another occasion, when a tremendous rain occurred in Baltimore, which was the cause of much destruction of property, he affirmed that it fell with so much force on the roofs as to be projected for a distance from the eaves; and to allow the people in the street to walk under the jet entirely dry. It is proper to state, that in the intercourse, which I had with him in private—not very extensive it is true,—and as a member of the Faculty, I rarely noticed the failing with which he was charged. On one occasion, however, in my class-room, a ludicrous instance occurred in the presence of all the students. It happened, that I was lecturing on the comparative mortality of the cities of this country, when Professor Potter entered for the purpose of hearing me, and, by invitation, took a seat

[84] William Gibson (1788–1868), M.D. 1809, Edinburgh, native of Baltimore, Professor of Surgery at the University of Maryland, 1812–1820, succeeded Philip Syng Physick at the University of Pennsylvania.

near me. In the progress of my discourse, I gave the mortality of the different cities from a recent number of the *Journal of Health,* published in Philadelphia, and of which Dr. Bell was the editor. This statement made the mortality of Baltimore greater than that of Philadelphia. I saw, that it excited his feeling as a Baltimorean, and was apprehensive, that he might interrupt me. He properly, however, waited until I had concluded, when he came up to me, and I immediately put up my hand to detain the class. I listened to what he had to say, and remarked to them—"Gentlemen! my friend and colleague, Professor Potter, informs me, that the reason why the mortality of Baltimore seems to be greater than that of Philadelphia, is that in Baltimore they include the stillborn, whilst in Philadelphia they reject them." "Yes, Gentleman"! said he "and in Philadelphia, they reckon all those stillborn who die under five years of age"!! It is impossible to imagine the merriment that this exaggerated—totally unfounded —statement produced, not only in the students but in myself. Yet, he did not appear to be aware of his indulgence in hyperbole; and he undoubtedly often expressed his abhorrence of it in others. In a conversation with my friend Dr. Bernard M. Byrne, of the United States Army,[35] he spoke of the Washington Medical College with great contempt, and especially of the mendacity of one of its professors,—adding, that it was utterly impossible for any faculty to prosper when one of its professors was—as he said—for he was not very chaste in his language—"a damned liar."[36]

VIRGINIA LEAVETAKING

I immediately communicated to the venerable Rector of the University—Mr. Madison—and through him to the Board of Visitors, my intention, with their permission, to withdraw from the University at the termination of the Session. From him, and from the Board, I received the kindest and most complimentary expressions on the occasion. When Dr. Patterson and myself had thought it probable, that we should leave the University in 1831, I communicated with Mr. Joseph C. Cabell—at an after period Rector of the University —from whom I received the following kind reply.

Williamsburg, 12 May, 1831

Dear Sir,

Your favour of the 24th April has come to hand and brought me the very unpleasant intelligence of the approaching termination of your official connection with our University, as well as of that of your valuable colleague, Doctor Patterson. I regret exceedingly, that no advantages, in the power of the board of visitors to bestow, should

---

[35] Died at Fort Moultrie, 1860.
[36] Horatio Gates Jameson (1778–1855), with James Henry Miller (1788–1853), founded the Washington Medical College of Baltimore, in 1827, when he was thwarted in his desire for appointment to the medical faculty of the University of Maryland, incurring the wrath of the latter, which included Potter (William Frederick Norwood, *Medical Education in the United States Before the Civil War,* p. 242, 1944).

be sufficient to retain you as a member of the institution, and that we shall be compelled to surrender to other institutions advantages so enviable, and which they have the discernment justly to appreciate. My request to receive an early communication of any movements, calculated to remove you from the University, was prompted mainly by a desire, on my part, to consult again with the other Visitors in regard to any practicable changes that might tend to lessen objections and improve the advantages of your situation. This object is defeated by your commitment at Baltimore being placed upon a condition, dependent solely on the volition of others, and beyond the power of our board to modify or control, for I understand, from the terms of your letter, that, if elected, you are under a promise to accept. In stating this object I am influenced solely by a desire to manifest the willingness on my part to do whatever I might find practicable to reconcile your personal advantage with the other interests of the Institution. I presume I did not express myself with sufficient clearness in announcing my wish to you. But I presume that such efforts, on my part, would have been ineffectual, for I suppose you have, in weighing the matter, taken the good dispositions of all the Visitors towards you as an element in all your calculations. You acted with frankness and liberality in announcing your views to me last autumn; and I believe you will carry with you the high approbation and esteem of all the Visitors. I cordially hope, that you may prosper to your utmost wishes in your new situation, and that whatever portion of the country may have the good fortune to possess yourself and Mrs. Dunglison, you will find no cause to regret your arrival on our shores. Be pleased to present me, in the most respectful terms, to her, with every good wish, that a friendly spirit could suggest. I am perfectly persuaded of the sincerity of those professions of goodwill to the University, which I receive from both Dr. Patterson and yourself, and it will give me pleasure to enter into farther conferences with both of you, at our next interview, on the general interests of the institution, and on the best mode of supplying the vacancies created by your departure. In the interim, I pray you to accept the assurance of my continued respect and esteem.

Joseph C. Cabell.

The members of my class kindly invited me to a dinner, given on the occasion of my leaving, which I declined,—the expressed intention being equally complimentary to me, whilst the carrying it into effect would necessarily subject them to pecuniary and other inconvenience. At the termination of the Session, I took my leave of them in *A Lecture delivered to the Medical Class of the University of Virginia, at the Conclusion of the Course, and on the Occasion of His* (my) *Leaving the Institution,* which was published by them. The exordium to this indicated the wide field of duties that had to be cultivated by me.[37]

We have now, my young friends, arrived at the termination of the duties assigned to this chair,—duties more multifarious than are attached, so far as I know, to any professorship of medicine in this, or in any other, country. We first inquired into the various functions that are executed by the living body in health, and then applied this knowledge to the elucidation of those morbid manifestations that constitute the pathological or diseased condition. The mode of managing this diseased condition, so

---

[37] He was listed as Professor of Physiology, Pathology, Obstetrics and Medical Jurisprudence.

as to terminate in health, next engaged us, and as collateral subjects of inquiry, the application of medical knowledge to the clearing up of cases that might occur in courts of justice, and, lastly, the art of facilitating the delivery of the foetus, and the attentions to be paid to mother and infant during the childbed or puerperal state were considered. These various subjects, embracing the departments of the healing art, respectively termed *Physiology, Pathology, Therapeutics, Medical Jurisprudence,* and *Tocology* or *Obstetrics* are usually assigned, in medical colleges, to at least four professors, but there is this convenience in having them taught from the same chair, that they are more likely to form one harmonious and consistent whole, and, consequently, to be better understood by the student than when his mind is distracted, as is too often the case in literary institutions, by the discordant sentiments of different teachers. It has happened, too, that where Medical Science is divided into various professorships, the duties of which are expected to occupy the incumbents for a certain and a like period, undue extension has been given to certain branches, and the student has been occasionally doomed to hear the same subject treated from different chairs. This is doubtless one of the causes of the minute attention that is required to many insignificant points of anatomy, the recollection of which may be a good exercise for the memory of the student, but is of no advantage to him in after life. The sphenoid bone, for example, seated at the base of the skull, has various projections from its circumference, to all of which names have been assigned. Seven and twenty of these processes are pointed out by the anatomist, but to the physician or surgeon, in the practice of his profession, the knowledge is useless. No case can arise in which this minute topographical division can assist him. The same minuteness has been extended to other points of osteology, and to anatomy in general. For a like cause, the physiology of obstetrics has been unnecessarily expanded, and many topics, belonging properly to Pathology and Therapeutics, have been incorporated with it. It was with reason, therefore, that the Board of Visitors of this Institution considered, that, during the long session of ten months, three professors might be enabled to discharge competently the onerous duties that are usually divided among six. It has, I know, been objected to the medical school of this place, that the teachers are too few, but, in indulging the objection, the fact of the unusual length of the session has probably been overlooked, whilst it has not perhaps been known, that the medical schools of New Hampshire and Vermont, the former of which has 98 students, with shorter sessions, have the same number of professors as the Medical School of this Institution.

These remarks were written when I was about to associate myself with a medical school, in which there were six professors; and all my subsequent experience has but confirmed the views I then expressed; and it will, I think, be admitted that the students of the University of Virginia, who have attended there one session, are uncommonly well prepared, when they enter a medical college elsewhere, and exhibit their qualifications in a most creditable manner, at their examinations for graduation. The late action of the medical association held at Richmond, excepting the University of Virginia, only, from a proposed regulation in regard to sending delegates to the Association "in consequence of its peculiar mode of instruction and ex-

amination," [38] and this in consequence of the high encomiums passed on its pupils by Professors Huston [39] and Wood [40] more especially, is sufficient evidence of the estimation in which it is held by those who are best acquainted with it. (*The Medical News* for June 1852, p. 51 & p. 56)

My peroration was couched in the following language.—

The most painful part of my duty is yet unaccomplished;—to speak for a moment,—for such topics do not admit of detail—of matters that chiefly concern myself, and to refer to the severance of those ties, that have bound me to you, and, for a much longer period, to this noble institution, in whose continued prosperity I have always felt, and *must* feel the most signal solicitude. Eight summers have elapsed since I first set my foot upon these shores;—gratified at having been selected to fill the chair of Medicine in this University, and elated with the prospect of being enabled, in such a field, to employ my humble exertions for the advancement of my profession; but doubtful,—as everyone must be, who is about to take up his residence amongst strangers,—as to the reception I should meet with, and as to the favour with which my efforts would be received. A brief experience sufficed to dispel these doubts, and almost from the moment of my arrival amongst you, I felt, that Virginia was my second home. In this home, the happiest years of my life have flitted away, and friendships have been formed which must ever endure. Can it then be supposed, that I avail myself of my appointment to the University of another state—an appointment as unsolicited as it is unmerited—without numerous regrets, and that I can, without a pang, tear myself away from scenes of happiness, and from friends in whose hearts I hope to retain a portion of that space, which they will ever occupy in mine!!

VIRGINIA REMINISCENCES

On the whole, my residence at the University had been exceedingly agreeable. I had experienced the

[38] To avoid distinction between professors and general practitioners, it was proposed at the Richmond Meeting of the American Medical Association that membership be opened to all physicians, but that the business of the association be conducted by a house of delegates chosen by various organizations, including medical colleges whose faculty, among other stipulations, numbered at least six professors and required three years of study, but the University of Virginia was to be represented "notwithstanding that it has not six professors and that it does not require three years of study for its pupils, but only so long as the present peculiar system of instruction and examination practiced by that institution shall continue in force" (N. S. Davis, *History of the American Medical Association,* p. 110, Phila., 1855).

[39] Robert Mendenhall Huston (1794–1864), M.D. 1825, University of Pennsylvania, Professor of Obstetrics and Diseases of Women and Children at Jefferson Medical College (1839), survived the reorganization of the faculty in 1841 to emerge as Dean and Professor of Materia Medica and Therapeutics until retirement in 1854 (Frederick P. Henry, *History of Medicine in Philadelphia,* p. 201, 1897).

[40] George Bacon Wood (1797–1879), Professor of Materia Medica and Pharmacy, succeeded Chapman in the Chair of Practice at the University of Pennsylvania; member of the American Philosophical Society, President of the College of Physicians of Philadelphia and of the American Medical Association; a delight to his intellectual equals, he loved books and science even more than society.

distress of losing my second child. Others had been born to me, so that, when I left the University, I had three children,—a daughter and two sons, the younger of whom was 16 months old. In the year 1829, soon after the death of my child we travelled north, and in company with Dr. Emmet—my colleague—and his wife, spent the greater part of our vacation on the Long Island side of the Narrows, one of the most interesting localities I have ever visited. Professors Key and Long had returned to England. Dr. Patterson, as I have already remarked, had accepted the Chair of Natural Philosophy; and a very competent pupil of my own, Dr. Gessner Harrison, who had graduated in the school of medicine, and was, also, proficient in Modern Languages, had been recommended by Mr. Long to fill his chair. I also had much pleasure, when called upon before the Board of Visitors, in recommending him, and he has since given entire satisfaction. Mr. Madison asked me, in presence of the Board, if he was a man of *genius*. He smiled when I replied, that I scarcely knew what that was, but that he certainly was an excellent scholar, and possessed of those mental qualities which would enable him to be eminently useful as a professor of ancient languages, whilst his moral character was entirely without reproach.[41]

These were the only changes that took place in the Board of Professors at the early period; but, as I have before remarked, the resignation of Judge Lomax from the chair of Law made way for the reception into the Faculty of Professor Davis, and the appointment of Dr. Johnson to the post of Demonstrator of Anatomy and Surgery, and his subsequent elevation to the rank of Professor, modified the corps of Professors somewhat.[42]

The society, thus formed, for all the professors were married, was, in the main, agreeable. Between certain of the families the greatest intimacy existed, but, as in every case in which families are thrown together heterogeneously, entire harmony of sentiment could scarcely be expected, although harmony of action was assuredly practicable, yet this was greatly interfered with by the singular conduct of one of the members of the Faculty—Professor Bonnycastle. Possessed of an exceedingly nervous temperament, and unaccustomed by his previous peculiar position, or rather occupation, from much admixture with the world, he laid down

rules of conduct for himself, and for others, by no means in accordance with those considered by mankind in general to be established. He was a son of the Bonnycastle so well known to every schoolboy, was an excellent mathematician, well read in general literature; but not accurate in his own language, and without—almost without—knowledge of the languages of antiquity. He was, undoubtedly, a man of great ability, and an excellent physicist. He had been employed as foreman,—I believe—of the Dockyard at Woolwich or Chatham and was strongly recommended to Mr. Gilmer by Professor Barlow of Woolwich,[43] and others, to fill the chair of Natural Philosophy in the University of Virginia. His capability was never questioned, and he was highly estimated by his classes of successive years. He married a Virginia lady from Loudon county, and became morbidly sensitive, not only of too great attention being paid to her by those of the opposite sex, but, also, of apparent neglect on the part of any of her own. Hence arose misunderstandings between him, and certain of his colleagues and others. He was likewise himself exceedingly sore, if he presumed, that less attention was paid to him than to others, so that, if, owing to any cause, he was left out from a dinner party, he did not fail retaliating on the offending person, by giving an entertainment and pointedly leaving him out—maintaining it as a principle, that if he had been invited for a dozen times, and was left out on the 13th, the last act cancelled the former, and he was under no obligation to return the dozen civilities that had been rendered him. So well was this feeling on his part known to his colleagues, that if his name was omitted from the list of invitations in any case, they prognosticated a speedy entertainment from him to mark the presumed slight imposed upon him. It was a common remark with him, that there was a signal want of public sentiment in Virginia, and hence no general means of repressing impropriety of conduct, which was readily kept in restraint in England; and yet, when such sentiment was brought to act on his own peculiarities, no one could bear it with less equanimity. It was, indeed, in consequence of the notice I took of his outrageous behaviour to one of my colleagues, that he became exceedingly hostile to me. so that I left the University on by no means amicable terms with him, and never afterwards was thrown into communication with him. The case to which I allude was the following. No intimacy whatever had existed between Mrs. Patterson and Mrs. Bonnycastle. but, on one or two occasions, Mr. Bonnycastle fancied. that Mrs. Patterson studiedly slighted his wife. This notion was sufficient to induce him to openly insult Dr. Patterson, who, he presumed, must be in league with his wife, and, in a conversation with me, he expressed

[41] Gessner Harrison (1809–1862), one of the first students of the University of Virginia, was enticed away from medicine to ancient languages; he taught from 1828 to 1859 (Adams, *op. cit.*, p. 163).

[42] At first Professor Dunglison was expected to teach anatomy and medicine merely as a branch of a liberal education, but in 1827 the school of medicine was enlarged to a department organized as follows: Robley Dunglison, M.D., Professor of Physiology, Theory and Practice of Medicine, Obstetrics and Medical Jurisprudence; John P. Emmett, M.D., Professor of Chemistry and Materia Medica; and Thomas Johnson, M.D., Demonstrator of Anatomy and Surgery (Adams, *op. cit.*, p. 178).

[43] Peter Barlow (1776–1827), Professor at the Royal Academy at Woolwich, mathematician, physicist, and optician (Richard Beale Davis, *op. cit.*, p. 138, 1946).

his intention to do so. I was fortunately passing by Dr. Patterson's door when I saw Mr. Bonnycastle walking up to it with a note in his hand. Knowing his excited feelings, I at once asked him what his object was with Dr. Patterson. He said it was to give him an offensive note, which he had with him. I remarked, "You may meet Mrs. Patterson. Give me the note, and I will hand it to the Doctor." He replied. "If you will promise to do so." I took the note and entered the house of Dr. Patterson, whom I prepared for the reception of an offensive communication, but begged of him to take no notice of it. In this note, Mr. Bonnycastle stated, that he wished to have "no farther communication with him or his despicable family." The insult appeared to me to be so gratuitous and discreditable to him, that I determined, as one of those who constitute the "public," that I would have no farther communication with him, and to this determination I adhered. What was singular, however, Dr. Patterson and himself, some years afterwards, formed amicable relations and a short time before Mr. Bonnycastle died, which he did in the year 1840, he wrote a penitent letter to Dr. P., in which he made apologies for his conduct, and begged of him to say to me, that he hoped to die at peace with me, and at that time recollected only the many happy days we had spent together. I confess, that whilst I deemed it necessary to shew how much I disapproved of his course, I could not but sympathize with his infirmity, for it was doubtless in part a morbid condition. Although highly intellectual, and never in such a state in those freaks as to sanction the idea of mental aberration, he was liable to strange hallucinations. He has often told me, that, when lying in bed wide awake, he has heard all the doors of the house violently slammed, the windows thrown forcibly up and down, the bells all set aringing; by and by, a person had entered his room, got upon him, laid hold of the collar of his shirt, and endeavoured to strangle him, and yet the whole time he has been satisfied there was no real slamming of doors, no throwing up and down of windows, no ringing of bells, and no actual violence done him by anyone. The whole affair was an illusion and his case afforded a striking example of such hallucinations or "waking dreams," occurring in one whose intellect was sound, and yet equally occurring in the insane,—the difference, however, being, that he was satisfied of the delusion, whilst the insane consider it real.

It had always been an anxious desire with Mrs. Patterson, that her husband should be connected with a medical school, and hence she was greatly in favour of his accepting the chair of chemistry in the University of Maryland, if he had been appointed to it in 1831. The same feeling made her desirous, that, on my leaving the University of Virginia, I should recommend him to fill the chair I had occupied. It would obvi-

ously have been improper in me to interfere, in any manner—unless asked by the Board of Visitors—with the appointment of my successor, and in this matter I took strong ground. Moreover, Dr. Patterson had long devoted himself entirely to the teaching of Physics, in which he was highly successful, and although, in time, he might and doubtless would, become capable to teach departments that had been assigned to me, I could not conscientiously recommend him as then fitted for the duty, difficult under any circumstances, but doubly so to one who had not kept pace with the rapidly progressive course of certain of the branches, and, again, he was already occupied with one department of science—physics—and was unable to give time to the acquiring of that knowledge in the others which would have been indispensable. I knew too, that the Visitors would not, and *could* not, feel themselves justified in assigning the duties and emoluments of two chairs to one person, whatever might be his endowments and excellence of character.

These considerations had not much weight with Mrs. Patterson, and she erroneously considered, that I was less the friend of Dr. Patterson than she had thought me; and she so stated. This impression, as I afterwards learned from Dr. Patterson and Judge Kane, she had made on the mind of the latter, and it took some time after I removed to Philadelphia to entirely obliterate it. I doubt not, too, that the same feeling was entertained by others of the family, who, before this occurrence were attentive to me, but when I came to Philadelphia, ceased to be so. It is due, however to Dr. Patterson to say, that he never, so far as I know, participated in the slightest degree in those sentiments. Nor did the circumstance, in any respect, modify my feelings towards him. He has more than once told me, that he greatly regretted the feeling displayed by Mrs. Patterson, and that he approved of my course throughout. A slight circumstance will indicate how far this feeling was carried. In the most undeviating manner I had striven not only to induce Dr. Patterson's acceptance of his chair in the first instance, but both Mrs. Dunglison and myself had left no stone unturned to facilitate the family in establishing themselves, and in rendering their residence agreeable to them. I attended them, also, in sickness, during the whole time I was at the University, without asking for, and certainly without expecting, any remuneration. On my leaving the University, many of our friends presented us with memorials of affection or good feeling, but Mrs. Patterson was an exception; and the reason, given by her, was, that they were unable to afford it. I remarked above, that this was a "slight" affair, but it indicated the feeling I have described, inasmuch as both she and her husband were liberal and generous, and the apology—which was not needed—sufficiently indicated what would have been her course had it not been prevented by other considerations perhaps than the one that was given. She was an exceedingly kind and most valuable friend to Mrs. Dunglison during a severe attack of illness when her life was almost despaired of. For nine nights and days, she watched by her, and, during the whole of that time, never had her clothes off, so that I felt, and still feel, most grateful to her, and deeply regretted the view she took of my conduct on that occasion, of the propriety of which I never had any doubt. A few months ago (1851), I had the satisfaction of prefixing the following dedication to the eighth edition of my *Medical Dictionary*.

To
Robert M. Patterson, MD
PRESIDENT OF THE AMERICAN PHILOSOPHICAL SOCIETY,
ETC. ETC.
ONCE HIS COLLEAGUE IN THE UNIVERSITY OF VIRGINIA,
ALWAYS HIS FRIEND,
THIS WORK IS DEDICATED,
WITH UNCHANGED AND UNCHANGEABLE SENTIMENTS, BY
THE AUTHOR.
Philadelphia, October, 1851.

Of the Professors who were at the University when I left it, but one remains,—the Professor of Ancient Languages, Dr. Gessner Harrison. Professors Long and Key—as I have remarked—returned to London. Professor Bonnycastle died at the University. Professor Emmet in New York; Professor Blaetterman at his farm, after having been removed from his chair of modern languages. Professor Johnson, after the loss of his position at the University, removed to Richmond, where he still, I believe, resides. Judge Lomax still occupies the honorable post of Judge of the General Court of Virginia having been recently reelected (June 1852); Professor Davis, as I remarked elsewhere, lost his valuable life in the discharge of his duties; Dr. Patterson left the University in the year 1835, and, within the last few years, retired from the post of Director of the Mint, which he had filled since the time he quitted the Institution, and Professor Tucker—the oldest of any—is still, at an advanced age, possessing the same mental vigour he exhibited when I first knew him. Time has certainly dealt lightly with him. He has resided for some years in Philadelphia, and is now engaged in the preparation of a History of the United States.[44] Whilst at the University, he wrote an interesting life of Thomas Jefferson,[45] which, on account of its large size—two volumes octavo—has not been as extensively diffused as its merits entitle it to be.

On my retirement from my chair, every endeavour was made to obtain a suitable successor, but not many prominent candidates offered. The one appointed was Dr. A. T. Magill, of Winchester, Virginia, a son-in-law of Judge Henry St. George Tucker, with whom I had no previous acquaintance. He had written an Essay on Typhus Fever, which had obtained a prize from the

Medical Society of the State of New York, but was not sufficient for the foundation of an exalted reputation. He did not, however, retain his post long, having died of pulmonary consumption, the seeds of which were sown before he commenced his duties at the University.[46] As he was unable to be at his post at the commencement of the Session, and my presence in Baltimore was not absolutely needed before the beginning of November, I consented to lecture for one month and finally left the Institution before he arrived there, and never, consequently, made his personal acquaintance. The following extracts from letters of Dr. Patterson to me exhibit the impression made by Dr. Magill and, at the same time, the feeling entertained towards us by Doctor and Mrs. Patterson.

Oct. 22, 1833

... You have been, now, some days in Baltimore, and we are fairly separated. We feel it most sensibly—more so than we can tell, or than we ourselves anticipated. Mrs. Patterson and myself are both left without an *intimate* at least, if not without a friend at the University. As to Helen, she has not paid a visit since you left here, and seems too much disposed to shut herself up with her own family alone. Dr. Magill made his appearance among us as soon as you had gone. He is exceedingly like his brother Augustine in face and figure, and I suspect, in caliber. I did not hear his introductory lecture, but I am told, that it was a kind of sermon, in which the bible was recommended as the best *text* book in physiology, etc. It had, according to Davis, a right smart sprinkling of commonplace, secondhand quotations, about "looking through nature up to nature's God," —and "the tomb of all the Capulets." At last, he said, that he did not know that he could terminate his lecture better than by a certain remark, which he made accordingly,—and the students thought as he did, and, naturally enough, began their plaudits, and drove him off the ground. But it was afterwards discovered, that he was not yet done, and he was accordingly sent for, read the *balance* or *shank* of the lecture, and received another dose of claps and poundings.— But the lecture is to be published and you can judge of it for yourself.

His next lecture, I am told, was a sort of supplement to the first, and also very *pious*. He has, they say, some turn for originality, which he displays in his mode of pronouncing the King's English. Thus, he says gelatine and gelatinous, with the g hard, and, to keep up consistency and have the *semper durum,* he even spoke of gelly, which, however, he had the grace to say, that *some people call jelly;* this was in speaking of what he calls eyenert[47] substances. You see that he is a *throughgoing* (another of his words) orthoepist.[48] You will think me not very good natured, and, to tell the truth, I am vexed at the fellow, and my reason is, that he has the indelicacy (to say the least of it) to speak of your work on physiology, *to his class, slightingly* and disrespectfully. He says it is a mere compilation, sometimes pretty well put together, but loaded with a great deal of useless and extraneous matter. He greatly prefers Magendie, and regrets, that there are not copies enough at

[44] *History of the United States to 1841,* 4 vols., 1856–1858. For his other writings, see McLean, *op. cit.*
[45] Published in 1837.
[46] Alfred Thurston Magill (1804–1837), M.D. 1824, University of Pennsylvania, with John Esten Cooke and Hugh Holmes McGuire, in 1825 started the short-lived College of Physicians of the Valley of Virginia, was awarded the prize mentioned in 1828 (Blanton, *op. cit.,* p. 251).
[47] Inert.
[48] One skilled in correct pronunciation.

the bookstores here to supply the class, but, as there are not, he will condescend to allow the use of your book for the present session.—Now, all this looks to me like the envy and jealousy of a little mind.[49] Johnson has met with his match" (He, Johnson—had acted in a manner which was regarded by the Faculty as very unworthy,[50] and was undoubtedly an inferior man, both morally and intellectually) "They are bosom friends. . . . I believe I should have been somewhat coldly disposed towards *any* successor of *yours.*

Nov. 6, 1833.

I heard Dr. Magill lecture last Saturday. His subject was Hearing. He said the description of the organ did not belong to his department, and acoustics certainly did not belong to his department,—so that he had very little left. He gave us, however, a touch of acoustics and of anatomy too,—but no figure, no diagram,—nothing but a few written notes,—a brief *compilation.* The sentences were all grammatical enough, and he reads not very impressively, yet not badly. How far he succeeded in communicating any definite notions of the subject to the student, I can only *guess* and so may you. His lecture was between 15 and 20 minutes long. So much for audition. He had accomplished the whole of *vision* in the preceding lecture, and he promised to treat, in the next, of *touch, smell* and *taste.* I told him I thought he was getting out of his senses very rapidly. You see, dear Doctor, that he can beat you at lecturing all hollow. It would take you five times as long to get over the same ground. So, Paddy's little watch from Tipperary went twice as fast as the big Gog and Magog clock at Temple Bar.

Dec. 4, 1833.

I have heard nothing lately about Dr. Magill's lectures. As a man and a physician, I like him very much. We have been attending together a terrible case of inflammation of the eye, and this has made us better acquainted. He is not of the same class as one of his colleagues.

Jan. 29, 1834.

. . . Dr. Magill seems to be getting on very smoothly with his class. He makes no high pretensions. I heard him a few days ago (for the second time only) on fever. It was a much better lecture than that on physiology.

July 4, 1834.

"When in the course of human events" etc.—that is to say, this is independence day, and never was the zeal for liberty to be exhibited by such repletion as is to be witnessed in this neighbourhood today. We are to have a whig festival at the Midway, and a people's dinner at *Horsedevil* or *Hors-de-ville,* as some Frenchified folks call it. Then we are to have speeches *galore.* First comes the student's orator—a *second* Cicero, as some *thousands* have been before him. Then Col. Preston, who loves his country so dearly, that he did his best to break it in pieces, and who is now to smash the Jacksonites. Then up rises the Duke of Empty Barrels, who will, of course, beat Preston *all hollow.* What a pity, that you are not here,—that you and I might take our bottles of Alexandrine and Claret in

peace and quietness together. . . . They have raised a story here, that Dr. Magill has not given satisfaction, and, that he will not have his appointment confirmed. The fact is, that Dr. Magill has two cordial enemies here, a rival, and a lickspittle. In physiology, a manifest falling off was perceived, and would be freely acknowledged by the professor; but in *practice* I am told, that he gives great satisfaction. It is his *forte,* his favourite study. Besides, he is a very estimable man, and has a very estimable family, and I hope and believe, that he will be confirmed in his station.

July 21, 1834

After a thorough investigation, by the Board of Visitors, of the circumstances connected with certain sayings and doings of one of your late colleagues, Prof. Johnson saw fit to give in a voluntary resignation of his chair on the ground of the inadequacy of the income. It is true (I still consider you one of *us,* and therefore tell you secrets which you must keep snug) that the Rector sent him word, that if he did not *volunteer* a resignation four of the five members of the Board were prepared to vote for his removal, and the fifth did not see how he could justify himself if he did not join with them. It is not, however, to be inferred, that this had any influence in producing the note of resignation which immediately followed, since the ground taken for the measure was an entirely different one.—Dr. Magill's appointment has been confirmed, in accordance with the wishes of all the *present* members of the Faculty, and of the great mass of his class. You may rely upon this, whatever reports you may have heard to the contrary. His weakest point was on physiology; and that subject has been transferred to the chair of Anatomy. He has now, therefore, only practical subjects left, and in these he is *generally* considered as excelling. He is, moreover, industrious and zealous, and he will improve. Nor can we, on such an occasion, keep out of view his excellent personal character, or that of his amiable wife. I suppose, now, you think, that our deep regrets on your removal were all because we were losing a distinguished *professor.* Don't flatter yourself, my dear doctor!

At the termination of his letter, after speaking of the resignation of Dr. Johnson, and his having been before the Board of Visitors to give his opinion in regard to Dr. Warner,[51] who was appointed on that occasion to the vacant chair, Dr. Patterson remarks;—

I told the Board, when I was before them, that if the event, which had just occurred, had taken place, as it was like to do, last year, I had reason to believe, that they might have had you, and your Dr. Smith in the two medical chairs. They looked wild with astonishment, nor was their air unmixed with incredulity,—but Mr. Randolph confirmed my assertion.

I have no recollection of having given any ground for this opinion of my excellent friends, but, most assuredly, I never regarded Dr. Johnson as an efficient or worthy colleague.

The mention of the *"Duke"*—or as he was commonly called the *"Earl,* of Empty Barrels" recalls to my mind the pleasant intercourse I occasionally had with Gover-

---

[49] October 29, 1833, Dunglison wrote from Baltimore to his friend Trist (Trist Papers, Library of Congress), saying about Magill " I am afraid he is not the man that ought to be placed in your and my favorite University."

[50] When a request by the students that Johnson's salary be raised was denied, he resigned, and joined the faculty of the Richmond Medical School. He resigned here also, after which little is heard of him as a medical teacher (Blanton, *op. cit.,* p. 46).

[51] Augustus Lockman Warner (died 1847), M.D. 1829, University of Maryland, conducted a private dissecting room in Baltimore, just to the rear of the University, which he left when called to Virginia.

nor Barbour,[52] who was a well informed and agreeable person, but greatly underrated in consequence of the grandiloquent language, which he too often employed, and for which he had always been celebrated. Mrs. Randolph informed me, that one of his brothers spoke of him to her in language scarcely less inflated than that on which he was animadverting. "My brother James," he remarked—"clothes a beggarly idea in robes of royalty, and calls down the lightning of heaven to kill a gnat."

I called in Washington upon Governor Barbour, when Secretary of State, soon after the laying of the foundation of Bunker Hill monument, at which ceremonial he had been present, and, as he stated, had to "make has way through one consolidated mass of human nature."

Mr. Walsh informed me, that in describing his own land, he pointed out "a rivulet meandering down the vale and ultimately disemboguing itself into a ditch." "Peter's mountain," the loftiest of the ridge called the "Southwestern range" of mountains, which may be 600 feet high, he designated as the "Chimborazo of those Lilliputian Andes." I confess I listened to his conversation with great pleasure. He had seen much of men and things, and had seen accurately; and was far more agreeable and entertaining than his brother Philip,— the Judge of the Supreme Court[53]—who, I doubt not, was an able lawyer, and good legal speaker, but, in private conversation, was anything but interesting,— generally indulging in long disquisitions on matters that were personal to himself, as on the extent of his farm, the cost of his buildings etc., etc. The characteristics of the two brothers were well hit off whilst they were both in Congress.

To shave the state two Barbours try
One shaves with froth, the other dry.

MR. AND MRS. RIVES

Before leaving Virginia, my wife and I paid a last visit to Mr. and Mrs. Rives[54] who lived in the exercise of a delightful hospitality at Castle Hill, about 16 miles from the University of Virginia. They had recently returned from France, where Mr. Rives had served as Minister for four years. He was a Visitor of the University, and took great interest in the Institution, and, on one occasion whilst I was chairman of the Faculty, and a rebellious spirit on the part of the Students had dictated a highly objectionable paper reflecting on one of the professors, signed by sixty-five individuals, he

rendered essential service, by addressing them, and thus aiding in bringing them back to the rules of law and order. Mrs. Rives was an accomplished and agreeable lady, eminently pious and zealous in the faith in which she had been brought up. Her collection of autographs was, at that time, good, owing to the numerous opportunities she had had of communication and correspondence with distinguished individuals at home and abroad. In the year 1842, she published reminiscences of her residence abroad, under the title *Tales and Souvenirs of a Residence in Europe by a Lady of Virginia* which I had great pleasure in seeing through the press for her, as it was by no means convenient to have the proofs corrected at such a distance from the place of publication. The work was one of interest, but did not have that extensive sale, which inferior works of fiction frequently obtain. At a subsequent period, she wrote a small volume entitled *An Epitome of the Holy Bible, for the use of Children by a Mother* which was undertaken with the view of obtaining funds for the rebuilding of an Episcopal church in the vicinity of Chapel Hill.[55] This was printed in Charlottesville, but the selection of an appropriate binding, and the superintendence of its execution were confided to me. The following letter from her will explain the elevated object she had in view in this undertaking.

Castle Hill 18 Nov. 1846.
Dear Dr. Dunglison,

. . . I must put in requisition for a few moments, your patience, to offer again my best thanks for the kind and polite manner in which you have met my wishes, and to make a brief explanation, by way of apology for the liberty I have taken. For the three years past, I have had a project in contemplation, of building a pretty little English church on a spot hallowed by many sacred associations. My own plan is to have it Gothic, built of solid stone, and finished with fine oak. This, as may well be supposed, will cost about four times as much as the paltry edifices that go by the name of churches in the country generally, and I have, of course, in so novel an undertaking, had *cold water thrown upon it* in large quantities. Still, we determined to endeavour, with the blessing of heaven, to plant the banner of the cross *firmly* in the wilderness. To do this, without giving offense, has required no little ingenuity. Had the requisite sum arisen chiefly from a donation offered by Mr. Rives, we should have incurred the reproach of vanity;—those who professed to be very willing to lend their aid to a church "quite good enough" would have been jealous, and we might thus have hazarded an interest even higher than the *visible* church of Christ. I have, therefore, gone quietly on, engaging the younger and more enthusiastic portion of the congregation in my plan, and setting them an example of creating a fund by the exercise of their ingenuity in various ways. This has not only added materially to the requisite sum but has had the far more important effect of engaging their best feelings in a holy cause. Some of these young people, however, require the stimulant of an energetic example, and I have undertaken to set one for them. As a matter of curiosity, I have preserved a record of the various works of all sorts (and the variety is very droll) in which I have engaged, and find, that the little book just sent to your

[52] James Barbour (1775–1841), Governor of Virginia, United States Senator, and Secretary of War during Adams's administration.

[53] Philip Pendleton Barbour (1783–1841), Speaker of the House in 1821, was appointed to the Supreme Court in 1836.

[54] William Cabell Rives (1793–1868), schoolmate of Francis Walker Gilmer and Thomas Jefferson Randolph, married Judith Page Walker, edited the *Madison Papers*, and wrote an excellent biography of President James Madison in 1859.

[55] Castle Hill.

kind care, if it has the success I hope for, will make my contribution to the fund, so far, about a thousand dollars. This is a separate affair from Mr. R's subscription, which is highest on the list. You will not think, I am sure, that I have made this long explanation to boast of my humble agency in a good work. I have ventured to make it only because I felt, that if the commission with which I have ventured to trouble you had been only a matter of personal interest, I should have been making an unwarrantable demand upon your valuable time, and only wished to show that I did not "bind burthens heavy to be borne" without taking a due share of them.

With regard to the binding of the little book I have no special fancy for the one indicated. Any improvement, that your good taste may suggest will be gratefully adopted. I should like it to be as pretty, and attractive to the young people for whom it is designed, as the case may admit. May I beg, that Mrs. Dunglison will accept copies for each of her dear little ones? I think her maternal heart will respond to some parts of it:—it was written for my own, some years ago,[56] and has often since been revised by several intelligent and pious clergymen, as well as my husband and myself. With our united and kindest regards to her and yourself, as well as your family, I remain, Dear Doctor Dunglison, most truly yours,

J. P. Rives

In a subsequent letter, dated 11 Jan. 1847, she acknowledges the receipt of some of the books, and remarks :—

they do great credit to the taste and execution, being really beautiful. The execution is their's (Messrs Lindsay and Blakiston's), the taste I attribute to you. They are greatly prettier than I ever thought they could be made, especially for so reasonable a price. In short, we are all delighted with them, and beg you to accept our united and best thanks for your kindness in the matter.

In the last letter I received from her on that subject, dated August 6, 1847, she remarks,—

I am sure you will congratulate me when I say, that our church is fairly underway, and daily growing in beauty and in stature. The work is beautifully executed by the best artisans to be found; but it is, as may be supposed, a costly affair. We are collecting all the means within our reach, in the hope of completing the stone work before the cold weather, as we should hardly be able to secure the services of our accomplished masterbuilder and his men for only a portion of another year. The small sum in the hands of Messrs. Lindsay and Blakiston will be something, and if it is perfectly convenient to you to ask it, and reasonable to them to require it, we may add it to the rest. I would not trouble you with this little matter, but for the kindness with which you offered to arrange it with them for me.

I found that the proceeds only amounted to about *ten dollars.* This I immediately forwarded to Mrs. Rives, which reached her, I hope in safety, for I received no acknowledgment of it. I did not write again, however, in consequence of her announcing to me in the same letter that Mr. Rives had been appointed a delegate to the Episcopal convention to meet in the city of New York in October of that year. "He has accepted"— she says—"the appointment, and I think I shall be

tempted to accompany him in his visit to that city. We shall hope to have the pleasure of calling on yourself and Mrs. Dunglison *en passant."* They did not call, however, and I have never since had the pleasure to meet them. For the last three years indeed, Mr. Rives has been again serving his country as Minister to France.

In the year 1859 Mr. Rives published the first volume of his *History of the Life and Times of James Madison,* of which he sent me a copy to which I have prefixed a letter or two from him. The condition of the country has interfered with the farther progress of the work. To Mr. Allibone, the able author of the excellent Dictionary of Authors, I furnished some materials for a notice of the literary labors of both Mr. and Mrs. Rives, and the accompanying letters have relation to that subject.

Newport, August 27, 1859.

Dear Dr. Dunglison,

I feel very much touched and gratified at the kindly interest you continue to manifest in my humble literary attempts—an interest which contributed so much to their success, and for which I assure you I have always been sincerely grateful.

Mr. Rives and myself were both amused at the idea of such amateur works being classed among the literature of the age but the thought is too flattering not to be accepted with pleasure; and especially to receive commendation from such high sources and in a work of most distinguished merit, cannot but be particularly pleasing.

I have at home any quantity of flattering notices of my literary attempts both in the journals of the day and in letters of congratulations from friends. I could make a satisfactory collection I think if I were there. Will you be so very kind as to drop me a line and inform me if it will be too late on my return home which will be in a few weeks—about the fifteenth of September. If not, I can contrive to procure them through my sister at home,—though the uncertainties of the mails and other possible contretemps would make it more desirable for me to attend to the matter myself, if the time permits.

We are here for the benefit of the sea air and bathing, which does me a world of good, and Mr. Rives no harm. Indeed the agreements of this lovely spot which we find more and more embellished in every visit, contribute much to renew his health and strength. People say he looks as fresh as he did twenty years ago, and I attribute some of this stationary youth to our annual visit here.

He joins me in kindest regards, and I remain, dear Dr. Dunglison,

Most truly yours,
J. P. Rives.

Castle Hill, Feb. 2, 1860.

My dear Dr. Dunglison,

I return the letter of Mr. Allibone with many thanks for the very kind interest you manifest for me and mine.

I am sure I cannot be wrong in attributing whatever kind things were said about me to the editors of the journals in which they were noticed, as from each of them, either Mr. Rives or myself received letters still more flattering. I can certainly be security for the *Home Journal,* the *Nat. Intelligencer,* and the *Lit. Messenger,* I suppose also the others, as no effort was ever made to get up a puff.

I dare say such innocent compliments would be safe from any challenge. Do you remember once telling me your

---

[56] 1840.

opinion of some mysterious globules that had been recommended to me? That I might take the whole set in my cup of tea without the slightest danger. Without venturing to propose a dose to you, I may yet be permitted to thank you for the great kindness you shew me in endorsing what has been said in my favor.

Mr. Rives is sincerely gratified at the kind interest you manifest in his *Life & Times of Madison*. He is getting on with the second volume, though interrupted by the preparation of the Madison papers committed to his charge by Congress. Within a few days past he has been still farther interrupted by requests from many quarters for an expression of his views on public affairs at present, which he has answered in a letter you will find in the Richmond papers next week. As the production of a friend, I think you will find interest in it. He is not at home today, or he would join me in the expression of cordial kindness with which I am, dear Dr. Dunglison,

Most truly yours,
J. P. Rives.

914 Arch Street
Feb. 6, 1860

My dear Sir:

I am this moment in receipt of your note of 4th inst. and beg to thank you for the trouble you have taken in the matter.

I was a little afraid that you might think me a little too particular in the premises: but you know how little value is to be attached to commendations in American Journals as a general thing.

How much more honourable to Mrs. Rives to have the expressed good opinion of her work of Robley Dunglison, N. P. Willis, and Washington Irving than to anonymous notices, however flattering! You doubtless mourn with me the loss of our friend Mr. Gilpin.[57] We expect to hear a review of his career from Mr. J. R. Ingersoll on Thur. next a week (13th inst.) in the hall of the Historical Society, and I have received an eloquent tribute to his character from Mr. Everett which I expect to present on the same occasion. I hope that you will be with us and that Mr. Tucker will accompany you. Excuse this writing, it is my seventh letter to-day; having written six between four and six o'clock this A. M. This, however, is not a common thing with me, I generally retire to bed late. Doubtless you are pleased with the new Worcester Dictionary. He gives you as a *Magister;* citing your Dictionary as one of his text books. One of the most valuable of late publications is the Dict. des Contemporains of Vapereau, Paris, 1858, 7021 sketches of Persons, pp. 1802, and supplement to be published.

Believe me dear sir,
faithfully yours,
S. Austin Allibone.

Robley Dunglison, M.D., LL.D.

In philology (one of your specialties) a remarkable scholar is Charles Short (a graduate of Harvard) teacher of a classical school at 12th and Chestnut Sts. He was highly recommended to Mr. Binney, Senior, by Mr. Everett. How much Irving compresses in his few lines on "Souvenirs"!

As I know not where to direct to Mr. Rives at present, I venture to trouble you to address enclosed letter, (referring to our friend Gilpin's death; he valued Rives!) and to mail it with yours.

---

[57] Henry Dilworth Gilpin (1801–1860), Philadelphia lawyer, author, editor, and Attorney General of the United States.

When I left Castle Hill, on my way to Baltimore, in 1836, Mr. Rives kindly gave me letters of introduction to many gentlemen in that city, who showed me great attention.

SOJOURN IN WASHINGTON

On our route through Washington, we spent some time with our excellent friends—the Trists—who with Mrs. Randolph, were living in Washington. In this way, I had once more, opportunities for visiting General Jackson to whom Mr. Trist was still private Secretary, and a warm personal friend. At the table of the President I had the pleasure to meet Judge Story,[58] next to whom I was placed at table. When I was first introduced to him, he could not exactly catch my name and after we had been conversing for some time, he suddenly asked;—"Is it Professor Dunglison whom I am addressing?" and when I replied affirmatively, he remarked, I thought I was introduced to Dr. Douglas of Virginia. At this entertainment, I had an opportunity of witnessing a little of the *etiquette* observed on such occasions. Major Donnelson, who was the gentleman-usher and prolegomenon, told me, that Judge Powhatan Ellis would hand in to the dinner table Mrs. Dunglison, and asked of me to do the same with Mrs. Ellis, as soon as dinner should be announced. Soon afterwards, he came up, and apologized to me, saying, that he did not know that Major Barry, the Postmaster general, was in the room; and that *he* would take charge of Mrs. Ellis, from which I inferred, that these posts of honor, as regarded the ladyguests, were given, by preference, to the Members of the Cabinet, an arrangement which is, of course, perfectly unobjectionable. It ought, however, to be a great object with the master of the ceremonies to take care, that no *contretemps* like the one that happened to me should occur. I had not the slightest feeling on the occasion, but I can readily see, that a touchy individual might have been annoyed.

During the dinner, Judge Story and myself had a most agreeable conversation in regard to men and things, and particularly to the distinguished British statesmen and jurists with whom he had corresponded. On separating, he told me he would present to me some of the interesting discourses he had delivered to public bodies, and I had the satisfaction, sometime afterwards, of receiving the following kind and complimentary letter from him.

Cambridge, near Boston.

Dear Sir,

I owe you an apology for not having, before this time, sent you the copies of some discourses of mine, which I promised when I had the gratification of meeting you at Washington. The truth is, that, on my return home, I could not find them and it was not until late in the last autumn, that I was able to obtain them, no copies being on sale, and my own copies having (I hardly know how)

---

[58] Joseph D. Story, John Marshall's close friend and disciple on the Supreme Court.

disappeared. I laid aside the copies of such as I could find, that I might take them with me on my visit to Washington last winter. But I was so unlucky as to forget them on the eve of my departure. This is the long story of my apparent neglect, and I hope you will deem me less unpardonable than I might otherwise seem to be. Pray do me the favour to receive them, even at this late day, as a tribute of my unfeigned respect for your talents and character, and also, as a reminiscence of our meeting at the President's Table.

(For an account of our meeting by Judge Story in a letter to his wife, see Vol. VII, p. 831 of his *Letters* etc.)

You are now removed to a different sphere of professional labours; and I hope, that you may, in your new position, reap all the honours and rewards to which you are so justly entitled. Your contributions to medical literature (I rejoice to find) are already stamped with the warm approbation of the public. I trust, that America will long continue to cherish one whose residence among us has been so useful and so gratifying.

Believe me, very truly yours,
Joseph D. Story.

Dr. Robley Dunglison,
etc. etc.
Maryland.

IV. THE UNIVERSITY OF MARYLAND

University of Maryland. Monday Evening Club. Don Ramon de la Sagra. *Elements of Hygiene. General Therapeutics.* Medical journalism. Sir John Forbes and the *British and Foreign Medical Review.* Negotiations with Jefferson Medical College. Appointed Professor in Jefferson Medical College. Changes in the University of Maryland.

UNIVERSITY OF MARYLAND

. In October 1833 I took up my residence in Baltimore, boarding with a respectable person—Mrs. Sutherland—until I was able to obtain a residence which I did in the following spring in Hanover Street, not very far from the Medical College, and on the 31st of the same month I delivered my first Introductory Lecture, which was published by the Medical class, and, as is the custom there, to a large assemblage of ladies and gentlemen, so that the professor has to frame his lecture by a knowledge of that fact. The subject of my discourse was a brief retrospect of the condition of the departments assigned to me, in former times, marked, as they were, by ignorance, credulity and superstition, and of the immense improvement that had taken place in comparatively modern periods, which

encouraged a hope, that as the physical and moral sciences pursue their onward progress, and as the means of observation and experiment are augmented and facilitated, the science may attain a pitch of perfection, of which at the present time we can form no adequate conception, shedding light where all is now obscurity, and tending to dispel doubt and difficulty wherever existent.

In the peroration I thus spoke of matters more immediately personal to myself.

On making my appearance for the first time in this hall, and on presenting myself to a new and a numerous audi-

FIG. 11. The University of Maryland School of Medicine. Reproduced from *Celebration of the Sesquicentennial of the Medical and Chirurgical Faculty of the State of Maryland*, p. 10, 1949.

ence, I need hardly, perhaps, affirm that I labour under emotions which give occasion to an embarrassment, that might, at first sight, appear affected in one, whose duty it has been to deliver public lectures for so long a period. A short allusion to the circumstances under which I have recently been, and am about to be, placed will show, that there is as much foundation for such emotions as there is sincerity in the assertion; and the charge of affectation would rather apply, were I,—from an extension of those feelings of delicacy, that ought to induce us to make all allusions of a personal character necessarily brief—to avoid referring to them altogether. It is probably known to many of my hearers, that, nine years ago, an agent—now no more (Mr. Francis W. Gilmer) to whose high mental endowments I might add my testimony, were it necessary, after the eloquent sketch of his character from the pen of a distinguished inhabitant of this city (Mr. Wirt),[1] and especially after the important trust reposed in him by the illustrious founder of the University of Virginia, and his coadjutors in the government of that institution—was dispatched to Europe, with full powers to select professors for several of the chairs.[2] His choice fell upon me for the department of medicine, and it was under feelings of gratification at selection, and under hopes of future usefulness, that I first took up my residence amongst Virginians; but under doubts, as I had occasion to observe elsewhere on a recent occasion (*Valedictory lecture delivered at the University of Virginia*, and already noticed on pp. 60–61), as to the reception, which, as a stranger, I might meet with, and as to the favour that would be extended to my humble efforts towards the advancement of my Profession. That

[1] William Wirt (1772–1834), who married a sister of Francis W. Gilmer, wrote the preface to a new edition of Gilmer's *Sketches and Essays* published in January 1828, to which Dunglison may be alluding here (John Pendleton Kennedy, *Biography of William Wirt*, rev. Ed., 2: 209, 1851).

[2] See: William P. Trent, English Culture in Virginia, *Johns Hopkins University Studies*, 7th ser., Parts V–VI, pp. 1–141, Baltimore 1889; also Richard Beale Davis, *op. cit.*

generous people extended towards me, from the moment of my arrival amongst them, the hand of good fellowship and indulgence, and made me feel as I have since felt, that the Old Dominion was my second home. To the enlightened Board, to whom the highest offices of the University of Virginia are assigned I owe especial gratitude for their unvarying kindness, and for the manner in which they have been pleased to appreciate my feeble but zealous exertions for the establishment and prosperity of that noble Institution. Their goodness, as well as that of many of my late colleagues, and of the people of my second home in general will ever be regarded as among the elements of the most pleasing portion of my existence; for

> "I cannot but remember such things were,
> That were most precious to me."

Circumstances equally gratifying to me, and somewhat analogous to those that occasioned my residence in Virginia, have induced my removal to this Commonwealth; and proudly and amply satisfied shall I be, if I experience the same liberal indulgence that was vouchsafed me in the state I have just quitted; and if the united exertions of my able colleagues and myself should have the effect of facilitating the acquisition of knowledge in a science so intrinsically interesting but arduous to the student, and so full of important benefits to humanity.

Although the members of the Faculty of the University of Maryland were by no means equal as teachers, or in their general or professional qualifications, the united body presented a corps of efficient instructors; whilst the Baltimore Infirmary [3] altogether under the control of the Authorities of the University, and of which certain of them, myself included, acted as physicians, and others as surgeons, afforded the student an excellent opportunity for witnessing practical medicine. The college building was likewise well adapted for the objects for which it was erected. Facilities, too, were afforded for the prosecution of practical anatomy greater than in the cities to the north, some of the Institutions of which were not unfrequently supplied by it with the *matériel* for dissection.[4] Still, the tide set so strongly towards Philadelphia,[5] which had long been regarded as the great centre for medical education in the United States, that notwithstanding the energy and

ability of the Faculty it was found impracticable to induce the students to stop in their course, and, consequently, the numbers, who frequented it, did not largely increase. Professors Geddings and Smith were men of unquestioned ability, each eminent in his own department. Professor Potter had seen much, and read no little, but he was indolent, fond of a game of whist—to which he gave much time, and neglected that application, which would have enabled him to keep a *porteè* with his important branch,—the theory and practice of medicine. Professor Hall had less talent and acquirement than his other colleagues, and his lectures did not possess that value, which they otherwise might have done, owing to his introducing into them extraneous matters belonging especially to the domain of natural science. He was anxious, too, at all times, to be regarded as a dexterous surgeon, to which, however, he had no marked claims. He was not highly estimated by the students, although always amiable in his deportment towards them, and, as far as I had an opportunity of judging, towards others.

Professor Ducatel was a good chemist, and an amiable man, but not distinguished as a lecturer. So that, taken altogether, the Faculty was able, and all were desirous of elevating the condition of the Institution as far as their abilities and industry would enable them to do.

Dr. Geddings was fully informed of everything connected with his avocation, and especially with Anatomy, Physiology and Surgical Pathology. He was fond of the literature of the profession, was well acquainted with the French and German languages, and had an excellent library, to which he was making constant accessions. My own collection of books had suffered greatly by my repeated removals; and when I left the University many of them were taken into the Library of the Institution on my proposition, in consequence of the defective condition of the collection on medical subjects.[6] When I had the pleasure of frequent intercourse with Dr. Geddings he was exceedingly ambitious in the good sense of the word, and has often told me how much he was surprised that I could live so entirely content away from London, and that he would give much to possess my satisfied disposition. It certainly has been a great happiness to me, that I have always endeavoured to be contented with my lot, and have, at no time, distressed myself with unfavorable anticipations as to the extent of the class when my colleagues have been gloomy and uncomfortable. Perhaps no one can feel distress more keenly for the time when it actually occurs, but I have endeavoured so to discipline myself as not needlessly to anticipate evils,

---

[3] The Baltimore Infirmary (or "University Hospital" as it is usually called), erected in 1823 to fulfill the need for clinical instruction, was apparently modeled after the Edinburgh Infirmary in name and function. It is one of the earliest medical school hospitals in the U. S. The Broadway ("Maryland") Hospital and the Almshouse with over 2,000 inmates also offered clinical material for the medical students.

[4] Warner's Anatomical Rooms, to the rear of the medical school, when Warner left in 1834, were taken over by William Nelson Baker, who lectured to large classes here while Dunglison was at the University of Maryland.

[5] Thomas Jefferson likewise regarded Philadelphia as the American center for the study of the sciences. When he sent his grandson there, he wrote to Dr. Caspar Wistar on June 21, 1807, his opinion that "there are particular branches of science which are not so advantageously taught anywhere else in the United States. . . . We propose, therefore, to send him to Philadelphia to attend the schools of Botany, Natural History, Anatomy, and perhaps Surgery; but not Medicine" (Charles A. Browne, *Thomas Jefferson and the Scientific Trends of his Time, Chronica Botanica* 8(3) : 413, 1944).

[6] Dunglison gathered a great collection of books, which were sold at auction after his death; it took two days to sell the medical books and two more days to sell the books on miscellaneous literature (Toner Collection, Library of Congress).

and have reaped the good results, not only by keeping my own equanimity but in preserving that of others.

Dr. Smith, without the bibliographical knowledge or facility of Dr. Geddings, had a thorough acquaintance with surgery, and was one of the best operators I have ever known. He had been unfortunate in incurring the animosity of certain leading members of the profession in Baltimore, for conduct, which they thought to be unprofessional, or careless of their feelings, but, in all my intercourse with him, I found him disposed to be correct in his deportment towards others, with perhaps, too great a degree of sensitiveness occasionally, which induced him to suspect slights from others that never were intended.

One source of discord, that affected the professor of anatomy more especially, however, was the appointment of an independent demonstrator by the Board of Trustees, who should be, in no wise, under the control of the professor. Dr. Baxley was appointed to this office, and an unpleasant state of feeling existed between him and Dr. Geddings, as long as the latter remained attached to the University. Amongst the Faculty themselves, too, there were elements of discord, which threatened its destruction, and it was no little satisfaction for me to be told, by Drs. Smith and Ducatel more especially, that the conciliating course I always pursued had tended greatly to the peace and harmony of the body, and to learn from others, that I had earned, during my residence in Baltimore, the reputation of being "a peacemaker."[7] The first not very polished introduction, indeed, which I had to Dr. Oliver Wendell Holmes, of Boston, was at the house of Dr. Mütter[8] of Philadelphia during the session of the American Medical Association, in the year 1847.[9] He was talking with Dr. Goddard,[10] who called out as I approached them—"Here comes the great peacemaker," and it was highly pleasing to me to hear one of my present able colleagues—Dr. Mitchell[11] in a

Charge to the Graduates of Jefferson Medical College, delivered March 9, 1850, and published by the graduating class, in referring to the history of the school, speak of my "great address as a pacificator."

My residence in Baltimore, so far as regarded social intercourse, was very pleasant to me. I found, however, that few of my medical brethren went much into society, and that the medical profession did not take the same elevated position as in Philadelphia. Few of them were men of letters, and almost all were occupied with their dignified calling more as a means of sustenance than as a noble science. They preferred, in other words, the trade to the science, on which Sir Benjamin Brodie has made the pointed remark, that "Medicine is a noble science, but a low trade"[12]—a remark that coincides closely with the views I have always taken of the subject.

MONDAY EVENING CLUB

Sometime after my removal to Baltimore Messrs. Robt. Gilmor, Jonathan Meredith, John P. Kennedy[12a] and myself, with others, instituted an entertainment to be held at each others houses, on the plan of the "Wistar Association" in Philadelphia which had been, for so many years, most successful in bringing hebdomadally together those who, without such an opportunity for a reunion, might rarely be thrown in contact with each other. I confess, too, that the very fact I have mentioned—of my professional brethren being but little seen in company—had its effect in making me more zealous for its establishment. It was so established, under the name of the "Monday Evening Club," of which Mr. Kennedy served as the accomplished Dean or Secretary, so long as I was connected with it. It was entirely successful and was continued as long as I staid in Baltimore, and for sometime afterwards, but was subsequently abandoned. On my leaving Baltimore, to reside in Philadelphia, my friend Mr. Kennedy informed me—an "obituary notice" was inserted on the "minutes" of the club, which I have never seen, but I doubt not—from the kindness always exhibited to me by the members and from the excellent feeling and taste of the recorder it was more flattering to me perhaps than was strictly merited.

[7] Henry Willis Baxley (1803–1876), M.D. 1824, University of Maryland, taught at his alma mater 1834–1839. He resented the popularity of the dissection classes of Dr. Baker, and to Dunglison, who was Dean at this time, fell the responsibility of accommodating matters (Eugene F. Cordell, *Historical Sketch of the University of Maryland School of Medicine*, pp. 86–88, 1891).

[8] Thomas Dent Mütter (1811–1859), M.D. 1831, University of Pennsylvania, Professor of Surgery at Jefferson Medical College (1841–1857), established the Mütter Museum, in 1858, at the College of Physicians of Philadelphia (Simpson, *op. cit.*, p. 730).

[9] The first regular meeting, at which Nathaniel Chapman was elected President.

[10] Paul Beck Goddard (1810–1866), M.D. 1832, University of Pennsylvania, Professor of Anatomy, Franklin Medical College of Philadelphia, member, American Philosophical Society, and fellow of the College of Physicians of Philadelphia.

[11] John Kearsley Mitchell (1793–1858), M.D. 1819, University of Pennsylvania, member of the American Philosophical Society, fellow of the College of Physicians of Philadelphia, Professor of Medicine, Jefferson Medical College (1841–1858), father of Silas Weir Mitchell, and friend of many literary

notables, including Edgar Allen Poe, Bayard Taylor, and Oliver Wendell Holmes (Ernest Earnest, *S. Weir Mitchell*, p. 2, Phila., 1950).

[12] Sir Benjamin Brodie (1783–1862), pre-Listerian London surgeon who had an enormous practice that for some years yielded him as much as $50,000 in five dollar fees (Douglas Guthrie, *History of Medicine*, p. 312, 1946).

[12a] John Pendleton Kennedy (1795–1870), lawyer, politician, statesman and man of letters (Charles H. Bohner, *John Pendleton Kennedy, Gentleman from Baltimore*, Baltimore, Maryland, 1961), supported Perry's expedition to Japan and Kane's search for the lost arctic explorer Sir John Franklin. He wrote four novels that are still readable: *Swallow Barn* (1832), *Horse-Shoe Robinson* (1835), *Rob of the Bowl* (1838), and *Quodlibet* (1840).

In the first formation of such an association it was difficult to avoid giving offence. It was impossible to admit all. Twenty six members would be sufficient for half the year, and as more might desire to be admitted, some had necessarily to be excluded. One gentleman of great worth, in many respects, Dr. Richard Stewart,[18] appeared to Mr. Kennedy and myself a very desirable member. In the first instance, he had shown no wish, and rather a disinclination to be admitted, but on its being proposed to him by me at an after period, when it was believed, that the number of the members would be increased, he consented to belong to it, if appointed. The number was not augmented; and although Mr. Kennedy and myself had been most anxious, that he should belong to the club, and used our exertions for this purpose, we were the persons on whom all his excited feeling burst; and I know not, whether to this day his misapprehensions have ceased.

I have often regarded this as one of the many examples I have been able to adduce, of the fact that where I have been especially anxious to do an act of kindness, I have not only failed; but had the illwill of the party ever afterwards. One of the lady relatives of that gentleman—I was informed, had expressed her indignation that "the Society of Baltimore should be regulated by a Scotch doctor," as she designated me, and I have reason to believe, that the occurrence was, indirectly perhaps, at the foundation of a charge of duplicity whilst I was in Baltimore, made against me by Dr. Chapman, of which I had to take notice, and to which reference will be made hereafter.

DON RAMON DE LA SAGRA

Whilst I was in Baltimore, Don Ramon de la Sagra brought me a letter of introduction from one of my learned friends in Philadelphia. He was the Director of the Botanic Garden in the Havana and was the author of *Historia economico-politica y estadistica de la Isla de Cuba:* 4 to Habana, 1831; and of *Tablas necrologicas del Colera-Morbus en la Ciudad de la Habana y Sus Arrabales*—Habana, 1833. He was an agreeable, sprightly, well informed man, desirous of becoming acquainted with everything of interest, and I had great pleasure, in exhibiting to him what Baltimore offered of a scientific or literary character. He travelled through various portions of the United States, and, whilst in Europe, published an account of his visit, in the Spanish and also in the French language. The Spanish version was entitled *Cinco Meses en los Estados-Unidos de la América del Norte desde el 20 de Abril al 23 de Setiembre de 1835. Diario de Viaje,* Paris, 1836. The French version, of which I have the Brussels edition, is entitled *Cinq Mois aux États-Unis de l'Amérique du Nord, depuis le 29 Avril jusqu'au 23 Septembre 1835:*

Journal de Voyage. Bruxelles, 1837. In these works Don Ramon speaks in the kindest manner of myself and family and of the reception he met with from us:

Absorbido en estas reflecsiones, fui á la Casa del doctor que me habia invitado á tomar el té con su interesante familia, de modo que mi imaginación continuó gozando con otro bello cuadro de felicidad doméstica. El doctor Dunglison es inglés, y vino hace años á este país para regir las clases del departamento médico de la universidad de Virginia. Sus talentos, su conducta y sus bellas cualidades, le han grangeado una justa reputación. Es no solo un sabio por sus costumbres, sino además un hombre amable por su trato y un caballero por sus modales. Hablando del estado de la medicina y farmacia en los Estados-Unidos, tuvo la bondad de acceder á mi súplica de estenderme algunas notas. p. 117.

These "notes" are contained in Don Ramon's work; but many of the names given by me as well as others, throughout the volume, are so *estropiés*[14] as to be scarcely recognizable. My venerable friend Mr. John Vaughan, of Philadelphia, is everywhere called Mr. *Waugham,* Mr. Tyson is Mr. *Tison;* Jefferson, *Jeferson;* Geddings, *Geddins;* J. R. Coxe, J. R. *Coke;* Drs. Wood and Bache, *Doctors Wood and Bake,* Mr. Lea, Mr. *Leea,* Dr. Mease, *Dr. Measse,* etc., etc.

ELEMENTS OF HYGIENE

Before I left Virginia I had contemplated to write a work on hygiene, and when the subject, at my suggestion, was added to the duties of my chair in Baltimore it became still more desirable to have a textbook on the subject, none suitable being to be found in the language. Accordingly, I wrote to Messrs. Carey, Lea & Co. mentioning my intentions and asking them to name a price for it, which, in a letter dated Dec. 6, 1833, they declined, adding, "if agreeable to you to make us a proposition, we will give it that consideration, which anything from you merits, and shall be happy if our views shall accord with yours." The following letter from them indicates the arrangement ultimately entered into.

Dear Sir,

Your letter, dated Jan. 30, 1833 (should be Dec. 30, 1833) is received, in which you accept an offer of *seventy cents* a copy for *fifteen hundred* copies of your new work on Hygiene, and, in addition, you to have *twenty five copies* without charge, which, of course, are not to be sold. The work to consist of about four hundred and fifty pages of pica leaded, extra charges for alterations at press to be at the expense of the Author. . . . We are, very respectfully,

yrs. obed.
Carey, Lea & Co.
Philad. Jan. 1, 1834

Prof. Dunglison
Baltimore.

PS. We, of course, to have the refusal of future editions at the same price.

---

[18] Richard Sprigg Stewart (1797–1876), M.D. 1822, University of Maryland, President of the Medical and Chirurgical Faculty of Maryland (1848–1851) and Vice-president of the American Medical Association (1849).

[14] Mangled.

Six weeks after the date of this letter, I received the following, which indicates the state of feeling amongst men of business in Philadelphia in regard to the course of General Jackson towards the Bank of the United States.

Dear Sir;

The distress, which has pervaded the country, and which is daily increasing through every part of it, has for some weeks past, induced us to doubt of the propriety of going on with your book during such an alarming time. After much reflection and consultation we deem it more to your interest, as well as to ours, to defer, for a time, the publication of it until we see what the result of the present extraordinary situation of affairs will be. Should it be published amidst the alarm, which now exists, the sale would be very limited, and the chance of a second edition almost nothing. Our correspondents are diminishing their orders for new books, and we are now withholding our hands from everything new of any magnitude, waiting the result of the hostility, which our government is now carrying on against the Bank, and the business of the country. We believe a more unpropitious time for booksellers has not existed since we have been in business. We shall be glad to hear from you, and doubt not but that, as our interests are very much the same, you will agree with us.

We are, very respectfully
your obedt.
Carey, Lea & Co.
Philad. Feb. 11, 1834

Prof. Dunglison
Balt.

To this letter I gave the following reply.

Baltimore, Feb. 11, 1834

Gentlemen,

The proper period for the publication of my work on Hygiene,—now that it is yours—I must leave somewhat to your discretion. You will see, however, that the sooner you can determine to carry it on the better, inasmuch as we have no treatise on the subject in the language, and some publication might be made in England, which might interfere with its sale. I shall want it, too, for my class in Autumn, and have calculated upon putting it into the hands of my summer students on hygiene. Let me know, as soon as you deem it advisable to print, and I know you will suffer no more time to elapse than you can help.

Believe me, Gentlemen,
very respectfully yours,
Robley Dunglison.

Messrs. Carey, Lea & Co.

In the autumn of the same year, I received the following letter.

Dear Sir,

It being a matter of some importance, now, to decide, whether the "Hygiene" shall be published this fall or in the approaching spring, we address you, under the impression, that you would prefer the former period, we should, therefore, be glad to hear from you by an early mail and be informed if you can go on at once with it, so that the work may be distributed before our rivers close. Business is mending in spite of the "government," and unless another fit of madness should seize upon our "roaring lion," we may look for a moderately good year in the approaching

one. You will, of course, see Mr. Toy, and ascertain from him how fast he can go on with it.

We are, very respectfully, yours
Carey, Lea & Co.

R. Dunglison M.D.
Philad. Sept. 20, 1834

The work was now proceeded with and it appeared at the commencement of the year 1835. Its title sufficiently indicated the subjects embraced in it.—*On the influence of Atmosphere and Locality, Change of Air and Climate; Seasons; Food; Clothing; Bathing; Exercise; Sleep, Corporeal and Intellectual Pursuits, etc. on Human Health; constituting Elements of Hygiene;* and it consisted of 514 pages 8vo. A supplementary chapter contained a deposition made by me, involving questions regarding the effect of draining a malarious soil; a table of the mean temperature and of the seasons in different places of America, Europe, etc.—Tables of the temperature of St. Augustine, etc. during certain months;—the mean temperature etc. of corresponding months in certain winter retreats; the temperature etc. of Campeche, and a Table of the comparative digestibility of different alimentary substances.

The deposition on the malarious soil is especially interesting as containing the answers to questions put to me by both plaintiff and defendant; the parties having mutually agreed, that my answers should be read as evidence, without being put in a legal form.

The first edition of the work was not exhausted until the year 1844; and the publishers, owing to the slowness of the sale, although it was undoubtedly well received by the reviewers, were not very desirous of printing a second edition, excepting with the view of keeping the work before the public.

At this period, too, the first edition of my *Medical Student* was exhausted; but to publishers, who were accustomed to rapid sales in the case of other works, this also presented no great attractions. Both were on specialties, and both on topics that were by no means indispensable, although very advisable to the medical student. Accordingly, I agreed to the terms presented in the following letter from Messrs. Lea and Blanchard, solely with the object therein expressed, for the copy money could not be regarded in the light of compensation.

Philadelphia April 4th, 1844

Dear Sir,

In relation to your volumes on *Hygiene or Public Health,* and that of the *Medical Student,* we, as well as yourself, are desirous of keeping them before the public. It is, therefore, understood, that we are to be at liberty to print an edition of the volume of Hygiene in such octavo form as we may elect, the edition not to exceed fifteen hundred copies,—you to bring it up to the day by a revision and such additional matter as is necessary and, when we wish, carry it through the press.

Should we elect to print an edition of the Medical Student, or such part of it as may be determined on, in a duodecimo volume of not over one thousand copies, you are to furnish the copy and carry it through the press. For the rights thus stated we are to place to your credit *five hundred dollars* payable six months after the publication of the volume on Hygiene. No other editions of either of these works are to be printed or published during the time we shall have copies on hand.

Yours very respectfully
Lea & Blanchard

Prof. Dunglison

The second edition, accordingly, appeared in the autumn of 1844, under the modified title—*Human Health; or the Influence of Atmosphere and Locality; Change of Air and Climate; Seasons; Food; Clothing; Bathing and Mineral Springs; Exercise; Sleep; Corporeal and Intellectual Pursuits, etc., etc., on Healthy Man; constituting Elements of Hygiene;* in 464 pages octavo.

The following is the number of copies and the amount of copy money paid me, reckoning the sum paid for the second edition, to be in the ratio of that paid for the first editions of the *Elements of Hygiene,* and of the *Medical Student* up to June 1852.

A.D.	Edition	Number of Copies	Copy Money
1835	First	1500	$1050
1844	Second	1500	322
		3000	$1372

GENERAL THERAPEUTICS

Whilst the *Elements of Hygiene* were passing through the press I was occupied in preparing a work on General Therapeutics, the only one of the kind in the English language. It always appeared to me, that whilst the properties and mode of administration of the various articles of the materia medica received ample attention, but little was paid to the great principles, that should guide the practitioner in prescribing them; and that the most crying defect, in practitioners in general, is their want of knowledge of general therapeutics. It is easy enough to prescribe, but not so easy to know when and what we ought to prescribe. It was well said by Primerose, in his work *De vulgi erroribus in Medicina,*[15] "propterea sola experientia absque doctrina et ratione incerta est et conjecturalis. Qui enim novit rhabarbarum purgare bilem nescit tamen quando, quibus et cui morbo prosit, nisi sit medicus doctus et peritus."[16] I may be mistaken in forming any value of my own works, but it has always seemed to me, that my *General Therapeutics* was calculated to be of some service by inducing an inquiry into the *modus operandi* of remedial

agents, even if such inquiry were not to lead to the same results as my own. And this idea has been confirmed by the kind expressions of some who had examined the work. The principles, which I taught from my chair, and are contained in this book, I was pleased to learn from my friend Dr. Fonerden,[17] of Baltimore, had exerted a decided influence in modifying the heroic practice, which generally prevailed at the time there, and, very recently, Dr. Bullitt,[18] of Louisville, gave me credit for having had great agency, by the work in question, in diminishing the destructive *calomelization* so universal in the west at the time it was written. Be this as it may, the doctrines contained in it have been confirmed by all my after observation, and I have, consequently, had but little alteration to make in them in subsequent editions. Yet, certain of the doctrines, which nevertheless maintain their ground as firmly as they ever did, were designated by a reviewer—Dr. Elisha Bartlett—in the *American Journal of the Medical Sciences* [19] for August, 1837, as "dangerous and erroneous" and the book was declared to be overburdened with attempts to explain the precise and intimate *modus operandi* of medicines.

The action and operation and effect of every article must be accounted for pathologically, physiologically, philosophically, rationally. It is astonishing to witness the pertinacity, activity and ingenuity of this *"detestable mania* for explanation," as the authors of a recent and excellent French work on materia medica and therapeutics call it. It is not enough, that any given medicine or mode of treatment, *cures.* This knowledge would be mere empiricism, unworthy altogether of the scientific physician. We must know *how* it cures, and *why* it cures; and unless these things are made out, we are bound to believe, that it does not cure at all; that we have been mistaken, and that the two circumstances of the use of the remedy, and the cure, which simple observation had taught us, sustained to each other the relation of cause and effect, must have been only accidentally so connected. If any one circumstance exhibits more strikingly than another the folly and absurdity of this passion, it is that of the multifarious and contradictory explanations that are continually and successively invented and maintained. There is hardly a page of Dr. Dunglison's book, which does not contain more or less paper spoiled, and worse than spoiled by magisterial and confident statements of these *hows* and *whys* and *wherefores* of pathology and therapeutics.

And yet these remarks are from the pen of the author of a "Philosophy of Medicine," which philosophy is to consist of simple observation of phenomena, without an attempt on the part of the inquirer to deduce the laws of such phenomena;—a philosophy, which has been the

---

[15] James Primerose (1598–1659), whose book on *Common Errors in Medicine* went through numerous editions.

[16] Because experience alone, without theory and reasoning, is uncertain and conjectural. For who knows that rhubarb purges the bile, nevertheless does not know when, to whom and in what disease it is efficacious, unless he be a learned and skilled physician.

[17] John Fonerden (1804–1869), M.D. University of Maryland, 1823, Professor of Obstetrics and Diseases of Women and Children at Washington University in Baltimore, 1845–1846.

[18] Henry Massie Bullitt (1817–1880), M.D. 1838, University of Pennsylvania, taught at St. Louis Medical College (1846–1848), Transylvania University (1849), Kentucky School of Medicine (1850), University of Louisville (1886), and founded the Louisville Medical College (1868).

[19] Signed only with initials E.B.

cause of every medical novelty and absurdity, and of every foolish practical system;—for homeopathist and empiric equally and confidently appeal to simple experience as the groundwork of their practice! [20] This review appeared nearly one year after I had removed to Philadelphia to join the Jefferson Medical College, towards which, the Editor, Dr. Hays,[21] was known to possess feeling anything but favourable. Connected with this subject, I may mention, that Dr. Pancoast [22] informed me, that he was told by Dr. Stewardson,[23] then of Philadelphia, now of Georgia, that he was asked to furnish to Dr. Hays a review of my Therapeutics, which should animadvert severely upon it, and through it, upon its author. The work was, of course, public property, and subject to any censure it might deserve, or appear to the reviewer to deserve, but it is doubtless objectionable to indulge in personalities, and the only fair ground of objection to the review by Dr. Bartlett was that it brought the author more prominently forward than was necessary.

I hope there was some mistake in the affair as regarded Dr. Stewardson; yet it came to me in a very direct manner. Dr. Hays himself informed me, that he had received a much more severe notice of the Therapeutics than that by Dr. Bartlett, which had been withdrawn,—I forget whether he said at his suggestion, —and Dr. Stewardson, many years afterwards, informed me, that he, Dr. S., had written a notice of it, which he was glad he had not published, and one reason he candidly gave was, that he did not find that my views were so objectionable as he then thought them. Drs. Bartlett and Stewardson were both pupils of the School of Observation of Louis, and, therefore, likely, in many respects, to accord in their views.

In the number of the *"American Medical Intelligencer"* for August 15, 1837, I inserted the following article:

Review of Dunglison's Therapeutics. In the number of the *American Journal of the Medical Sciences* for August— just issued—there is a review of the *Therapeutics* of Dr. Dunglison, in which the reviewer very properly examines freely into the different views of the Author, adding,—"We have spoken of Dr. Dunglison's treatise with freedom; it is very probable, that the Author and his friends may think, that we have spoken, in some instances, with undue license and unnecessary severity." We assure the reviewer, in all sincerity, that no such feeling has been excited in us by the perusal of his strictures, and we think we can answer, that the impressions of our friends will be the same as our own. It will be obvious to everyone, who will take the trouble of reading the work, and the review in the *American Journal,* that the two authors are in their views, as to therapeutical facts and inquiries, wide as the poles asunder, and there is but little probability that they will ever approach nearer. Conceiving, as we do, that every review should contain the honest, fearless exposition of the opinions of the writer, we should be amongst the last to object to such productions, even when, from inadvertence, or other causes misstatements may have been made as to the sentiments inculcated in the work reviewed. The opinions of the reviewer are those of one individual, the opinions in the work reviewed are clearly those of the author. When, therefore, they are widely apart, the difference must be regarded merely as difference of sentiment between two individuals. In the very number of the *American Journal,* in which the review appears, this kind of discrepancy is perceptible between the reviewer of the *Therapeutics* and the author of a bibliographical notice of *Colles on the Venereal,* written, we believe, from the initials, by an estimable physician of this city (Philadelphia)—the former retaining the ancient belief,—as he seems to do on most questions—that mercury is an "antidote" to syphilis, the latter being inclined to ascribe the removal of syphilitic symptoms, under the mercurial plan of Dr. Colles, to the affection getting well "in spite of the action of that agent," and he concludes, that with such views as Dr. Colles possesses upon fundamental points, his book is "calculated to retard rather than promote our knowledge of the pathology and treatment of the venereal disease." Yet, by others, the work has been considered as one of the best practical treatises, not only upon this but upon any subject that has appeared in modern times!

Six years were required to exhaust the edition of fifteen hundred copies of the *General Therapeutics* and when the question occurred as to the propriety of reprinting it, it was suggested by my publishers that, to adapt it for a textbook, to accompany the courses of lectures on Therapeutics and Materia Medica in the Schools, Materia Medica ought to be incorporated with it. Accordingly, the following letter was written to me by them, and agreed to. I may remark that the *General Therapeutics* appeared in September, 1836, and was dedicated as follows:—

To the Gentlemen,
who have honoured the Author
by their attendance on the Lectures
delivered by him in the
Medical department of the Universities of Virginia
and Maryland
this work is respectfully inscribed.

---

[20] Elisha Bartlett (1804–1855), another of the peripatetic medical teachers from New England, wrote *An Essay on the Philosophy of Science* in 1844, which led Osler to dub him the "Rhode Island Philosopher." He condemned all hypotheses, distrusted inductive reasoning, desiring only "facts" and carrying French empiricism to its logical extreme (Richard Harrison Shryock, *Medicine and Society in America,* p. 129, 1960); see also Chap. V fn. 23.

[21] Isaac Hays (1796–1879), M.D. 1820, University of Pennsylvania, protégé of Nathaniel Chapman, whom he succeeded as editor of the *American Journal of the Medical Sciences* (1827–1879), was a benevolent gentleman of the old school with many friends, a frequent guest at Wistar Parties and a Fellow of the College of Physicians of Philadelphia (Howard A. Kelly, *Cyclopedia of American Medical Biography* 1: 393, 1912).

[22] Joseph Pancoast (1805–1882), M.D. 1828, University of Pennsylvania, Professor of Surgery, Jefferson Medical College (1838–1841), and of anatomy (1841–1874), Fellow of the College of Physicians of Philadelphia, and member of the American Philosophical Society.

[23] Thomas Stewardson (1807–1878), M.D. 1830, University of Pennsylvania, on the staffs of Pennsylvania and Philadelphia General Hospitals, lived for a number of years at Savannah, Ga.

Philadelphia, April 27, 1842

Prof. R. Dunglison,

Dear Sir, Our object is to publish a System of Therapeutics and Materia Medica, that would answer for general use, as well as for a Text book for students. This would form two volumes 8vo. of about 550 pages a volume, the page not quite so large as that in your "Practice." In this work, you would mould in your late work on therapeutics. For an edition of *fifteen hundred copies* we to pay *Twelve hundred dollars* at six months from publication. The right to publish future editions to rest with us on the same terms, —you to revise and bring them up to the day. As you have expressed your willingness to undertake such a work on the terms stated, will you acknowledge the receipt of this, assenting to the arrangement, and, at the same time say when it may be expected for the press and for publication.

We are, very respectfully,
Lea & Blanchard.

This edition appeared in December 1842, and consisted of 1500 copies. It was entitled *General Therapeutics and Materia Medica, adapted for a medical text book.* In February 1846, a *third edition* was issued, of *seventeen hundred and fifty copies,* to which were added numerous xylographic illustrations of medicinal plants, etc. This was exhausted in 1850, and, in June of that year, the *fourth edition* of *fifteen hundred copies appeared.*

In the Preface to the second edition, published in 1842, after making some observations on the course I had pursued in the preparation of the part of the work devoted to Materia Medica, I remarked—

The views of *General Therapeutics* are essentially the same as in the first edition. The author has subjected them, however, to a careful revision, and has been pleased to find, that the period, which has elapsed since their first promulgation has but strengthened his belief in their general accuracy, so that he has not deemed it necessary to make many or great modifications.

and these expressions were reiterated in the edition of 1850.

Of the second edition a notice was given in the same Journal, in which the strictures of Dr. Bartlett appeared five years before, of which the following is an extract. It was from the pen of Dr. Joseph Carson,[24] at this time (1852) Professor of Materia Medica in the University of Pennsylvania, and is contained in the number for April 1843.

Among the prominent writers, who have contributed to place this subject (Materia Medica) upon a proper footing —is Professor Dunglison, whose ready pen has, to a considerable extent, been exercised upon it. The present work may be regarded as the third that has been put forth by him, viz. his *General Therapeutics, New Remedies* and the *Treatise* under consideration. The first may be said to be only the prototype of the last, as they can hardly be looked upon as successive editions of the same work, for as indi-

[24]Joseph Carson (1808–1876), Professor of Materia Medica and Therapeutics, Philadelphia College of Pharmacy (1836–1850) and the University of Pennsylvania (1850-1876), wrote a fine history of the first century of the Medical School of the University of Pennsylvania, 1869.

cated by the change of title, the whole production has been so modified and amplified, as to answer more comprehensive and useful purposes than were designed in the first instance, yet, as nothing has been curtailed of the original, but, on the contrary, such improvements made as the author's more mature reflexion and experience suggested, it is presented in the twofold character of a revised edition, and of a new book. It now constitutes two large sized octavo volumes. It is not within our limits to enter upon a critical examination of the contents of these volumes, or to comment upon any peculiarities contained in them. To stand upon their merit or demerit in the estimation of his readers, the author is fully competent, as he has been long accustomed to the praise or censure, which arise from community of views or difference of opinion as regards facts and deductions in medical science. In this age of independent thought and free expression, great latitude of judgment must be allowed, and where numerous topics are treated of, it is unreasonable to require, that all should tally with our preconceived ideas, or be in unison with our prejudices. From his clear style, and long habit Professor Dunglison is an adept in presenting a summary of the precise information possessed upon any subject that he may handle. Of this, the book before us is an example. The chapters upon general principles contain all, that the student should know, without embarrassing his mind and taxing his memory with extended disquisitions, while the account of the individual articles under each class is succint, but sufficiently extended to convey accurate notions with regard to their origin, source, qualities, uses and modes of application.

The tone of this notice—it must be admitted—is very different from that of the first edition.

The following is the number of copies, and the amount of copy money, paid for the several editions to June 1852.

A.D.	Edition	Number of Copies	Copy Money
1836	First	1500	$ 800
1842	Second	1500	1200
1846	Third	1750	1400
1850	Fourth	1500	1200
		6250	$4600

MEDICAL JOURNALISM

Whilst I was in Baltimore I furnished different articles to the *Baltimore Medical and Surgical Journal* —a quarterly journal, edited by my colleague, Dr. Geddings; and to the *North American Archives of Medical and Surgical Science,* which succeeded it and appeared monthly under the same editor. I also continued to be a collaborator of the *American Journal of the Medical Sciences,* and furnished to it various reviews and bibliographical notices. I was, also, one of the collaborators to the *American Cyclopedia of Medicine and Surgery,* which was edited by Dr. Hays; but lived only until the letter A was finished. There were many excellent articles in it but it did not answer the purpose of the publishers to continue it. Two only were furnished by me—an elaborate one on "Asphyxia," and one on the "Atmosphere," therapeutically and hygienically regarded.

SIR JOHN FORBES AND THE BRITISH AND FOREIGN MEDICAL REVIEW

In the summer of 1835, I received the following letter from Dr., afterwards Sir John Forbes [25] in regard to the establishment of the excellent Review, which he published for many years, with so much credit to himself and advantage to the profession. It was the commencement of a correspondence, which was highly agreeable to me, and gave me an opportunity of not only serving him disinterestedly, but the cause of medical science, of which he is admitted by all, to have been an energetic and able supporter.

Chichester July 15, 1835.

My dear Sir,

Although without any personal acquaintance with you, I am sure you will take in good part the present communication, for the sake of our common profession, and which is, moreover, only one of the ordinary debts, which you have incurred by your literary celebrity. The inclosed prospectus will explain the nature of the work in which I am about to be engaged, and the object of the present communication is to ascertain, whether you are in anyway disposed to cooperate with the Editors in its execution, or, if not, whether you will, at least, favour us with your best advice and opinion as to that portion of it, which will have reference to America as a *foreign* country. Besides giving Reviews of all the *good books* published in *every* country, where there is a Medical press, we wish to give *Annual Reports* of the state of medicine in them during the preceding year, noticing the discoveries and improvements, the new theories and practices, the books published, the obituaries of eminent men etc., etc. As *preliminary* to each series of Reports, we should, moreover, like to have one good *Statistical Report*, giving an account of the Medical Schools, Hospitals, Medical Societies, State of the Medical Press, Lists of Journals, Transactions of Societies, Etc., general character of the profession, mode of remuneration, grades etc. Now, my dear Sir, will *you* undertake to supply either or both these Reports (i.e. one Statistical Report and a literary Report *annually*) or, failing this, will you indicate some gentlemen in your new country, able and willing to do so?

We should, also, desire one or more gentlemen as *Correspondents*, and they might be either the *Reporters*, or not, as might happen,—although, of course it would be preferable to have all the work executed by the same parties. The duty of the Correspondent would be to transmit, from time to time, lists of works published in different parts of the States,—to indicate those worthy of immediate notice,— to endeavour to organize a plan for the transmission of books, and the *exchange* of Journals; and, in short, to communicate anything that might be interesting to the English medical reader. I feel much difficulty in establishing anything like an interchange of Journals through the American booksellers in London.—Another point on which I wish to consult you is the *republication* of the *Review* in America. Of course, as proprietor, I would prefer its being *not* republished, but as I can have no doubt, from the nature of our plan and the character of the writers engaged, that the *British and Foreign Medical Review* will be the most interesting Medical Journal *for America* ever published in England (whatever may be its success at home) its republication may, I presume, in the amazing fertility of your press, be pretty safely calculated on. Now,

would it interest you to be, in anyway, connected with this republication? Could we make any arrangement that might be mutually beneficial to us? If not,—would you have the kindness to name the thing to any influential publisher in one of your great cities. Of course in the event of a republication, *beneficial* to the proprietor, the *proofsheets* would be transmitted to America by post. In explanation of a preceding statement as to the interesting character of our Review, in regard to America, I will merely state, that being of the largest size of Quarterly Reviews (18 sheets) and containing nothing but *Reviews* and *Foreign Intelligence*, it will give a much fuller account of books than any other publication and *especially of the medical literature of the Continent of Europe*. Our corps of Reviewers for *German, French, Italian*, etc. is most numerous and respectable. Another ground of anticipated superiority is the system on which the Review is founded, *every page of writing will be paid for* and therefore the Editors can ensure, at the same time, the very best writers, and their own independence. This is a plan, which has never been attempted in England before, and it is one, which can only be effected by the outlay of a large capital in the first instance, and supported by a *very large circulation* in the end. I do not conceal, that upwards of 1000 copies must be sold, before the expense of print, paper and contributions is covered, but I confidently anticipate, after a few years, that the sale will be such as to make the work more profitable than any hitherto published in this country. Of course, any communication, that may be made by you or any other friend in America will be paid for in the usual manner.

Believe me, my dear Sir,
Yours very truly
John Forbes

Original Reviews of works published in America by yourself or men of first rate talent will be acceptable,— they will be paid for at the rate of from 4 to 6 guineas per sheet, but such reviews must be sufficiently *early* to anticipate other Reviews here:

J.F.

I don't see, why the B. & F. M. R. might not be made a sort of American Journal, by having an *American* Editor, and containing reviews of all the best publications of America. By speedy transmission of proof sheets piracy could be effectually prevented. Think of this.

I embraced the cause of the Journal very warmly, and, certainly, took great trouble to fall in with the wishes of the Editors in regard to it. In the first instance, it was proposed, that my name should be associated with that of the Editor on the title page. The proposition appeared to me to be, in many respects, objectionable, and a very important obstacle was, that it was impracticable for me to exert any editorial control over the work, whilst I should, of course, be jointly responsible for everything that appeared in it. The plan was, indeed, abandoned before my letter to the Editors reached them, as appears from the following communication, in which other forcible objections are contained.

Chichester Dec. 15, 1835

My dear Sir, —I cannot say how much I am obliged by your very kind letter, so liberally offering yourself as our ready benefactor in any way. I can only thank you sincerely. Although considerations, that could not be withstood have forced us to forego the gratification of seeing your name actually associated with ours on the title page,

---

[25] See chap. I, footnote 65.

on further consideration and consultation with many friends we had come to the conclusion before your letter arrived, solely on the score, that the mere existence of your name as Co-Editor would infallibly destroy the character of Independence and impartiality which we are most anxious to maintain. The English would see in every favourable notice of American Works the partial judgment of the American Editor, and the Americans themselves would lose the *comfort*, such as it might be, of thinking that the opinions of their writings were at least the opinions of men who had no connexion with them either national or political. You yourself, doubtless, will see the force of these considerations, and to be ready to wonder with me, how I could have ever overlooked them in the first broaching the subject to you. I trust, however, that this arrangement will not, in any way, deprive us of all the services we have hoped for from you, and which we have taken care to notice in Advertisement affixed to our first No. The chief of these services will be, 1. The sending a Review now and then (not many yet as we are really overstocked—but the one you name on Smith's book) 2. Supplying us with a report on Medical America (See Dr. Hecker's first Report in No. 1) 3. Supplying us or rather instructing our Agents to supply us with books and Journals; and last but not least, kindly assisting our American publishers in their efforts to extend the circulation of our book. . . . I really do hope this said B. & F. Rev. may assist in bringing all the honourable cultivators of our science, in all countries, more together. I trust your new countrymen will not be displeased with what we have said of them in more than one place of our first number. It is said sincerely, I assure you. Both Conolly and myself are, I hope, *Liberals* in every sense of the word, and for the people of America I have from my cradle been taught to have a warm admiration. Believe me, my dear Sir, ever most truly and sincerely, yours.

John Forbes

The allusion to their having noticed the services they hoped for from me, in the advertisement to the first number, induces me to give the following extract from it.—

One of the most gratifying circumstances connected with the preparation of the present work (the British & Foreign) and one to which they cannot but again refer, has been the prompt and highly valuable assistance afforded to the Editors by their brethren in foreign countries. Amongst those to whom they are under the greatest obligations in this respect, they must be permitted here to name DR. HECKER, of Berlin; DR. GERSON and DR. OPPENHEIM, of Hamburg; DR. CHELIUS, of Heidelberg; DR. PONFICK, of Frankfort; DR. LEFEVRE, of Saint Petersburg; DR. LOMBARD, of Geneva; DR. SEOANE of Madrid; and DR. DUNGLISON, of Baltimore. To the first and last named of these gentlemen, the Editors are under especial obligations. Some of the fruits of Dr. Hecker's kindness will be found in the present number, but the Editors are indebted to him for numerous favours equally important, though of a more personal kind. To *Dr. Dunglison*, the distinguished professor in the University of Maryland, the obligations of the Editors are, if possible, even greater. Animated by the liberal desire of promoting the general progress of medical science, and its readier and fuller interchange between two nations in every way allied, he has not only lent his aid to secure the circulation of this Journal in America, but has enabled the Editors, to make such arrangements with authors, editors, and publishers in the United States, as will ensure to them the speedy communication of every work of interest that makes its appearance in that country. Among the contributions which they are led to expect from *Dr. Dunglison*, they hope shortly to be favoured with a report of the present state of Medicine and Medical Institutions in the United States, which cannot fail to be generally interesting to European readers.

This report appeared in the fourth number of the *British & Foreign* for October 1836. It contained an *account of the Medical Colleges*, as regarded their history, organization, professors, etc., opportunities for instruction, fees, etc., and to the 5th number, for Jan. 1837 I communicated an article on the *Medical Associations for the regulation of practice*, and another on the *Remuneration of the Medical Profession* with a comparative table of the then *medical, surgical and other fees* of New York, Baltimore, and Charleston, and another of the *Charges for Medicine supplied by the Druggists* of New York and Charleston; For the fourth number, I believe, I wrote the review of Dr. Harlan's *Medical and Physical Researches*, and these, I think, were the only communications I sent for insertion in the *British & Foreign*.

The course of the "American edition," as it was called, which consisted of a certain number of copies printed in London, but marked on the cover as destined for American use, did not however "run smooth," owing to insufficient agencies; and all my exertions failed in making the American portion of the undertaking profitable to the Editors. In the year 1840, it was deemed by Dr. Forbes, desirable, that an advertisement should be inserted in different Journals, which should contain, amongst others, the opinion of the *Review*, entertained by distinguished members of the profession on this side of the Atlantic, and I had no hesitation in asking some of my professional brethren to write with me in the expression of sentiments as to its merits. I stated unequivocally, that I regarded it as the best *Medical Review* in existence, and recommended it with all my heart,—and it was in acknowledgment of those services that I received the following letter from the senior Editor:—

12 Old Burlington St.
London, Nov. 18, 1840

My dear Sir,

I write purposely to thank you for the new and splendid proofs of your kindness exhibited in the "Testimonials" from Philadelphia, which I know are derived from your kind exertion. I have written to thank Dr. Lee for his great kindness in this matter, and I have taken the most effectual step to show my appreciation of the value of the whole Testimonials, by reprinting them, and circulating them in every Medical Journal published in this country. I expect substantial benefit from this step,—high honour I am sure of, as I think such a compliment was never paid any work before.

In the year 1844, being desirous of obtaining a good English microscope, I wrote to Dr. Forbes, who entrusted the superintendence of its formation by Messr.

Smith & Beck to Dr. Carpenter,[26]—himself an excellent microscopist. I received it safely, and it gave me great satisfaction. In acknowledging the receipt of the money expended by him, Dr. Forbes wrote me the following letter.

Old Burlington St.
Dec. 28, 1845

Dear Dr. Dunglison,

I duly received your letter of Nov. 27 with its enclosure of £18 in repayment of my outlay for your microscope. Whenever I learn from you the outlay for Advertisements (made at Dr. Forbe's request, at my discretion, in different medical Journals) I shall gladly pay the same to Wiley and Putnam here, on your agents' account. I am extremely obliged to you for this troublesome work you have so kindly undertaken. I hope we shall see the advantage of it in increased sales bye and bye. As yet I have not heard of this being the case. Chapman's book has been lying by me—or rather an Article on it—for some time. It will appear in my April number. There is one Article in my present (January) number, which I am anxious all my *experienced* friends should read, and favour me with their comments thereon. It is on "Homeopathy and Allopathy," and what I have called "Young Physic." The article is written by myself. If my conclusions are just, it is high time we were thinking of doing something in the way of *reform*, quite different from what has, of late years, excited the cry among us doctors. I should not be surprised if any of your publishers might find himself repaid by printing this article as a pamphlet. *Its* name and *my* name together might probably sell it—and I cannot help thinking, that it might be of some use to your heroic druggers as well as ours. At any rate, I shall be much obliged if, when you have read my paper, you will send me your opinion thereof. . . . Believe me my dear Sir,

Yours faithfully,
John Forbes

PS. As I am having some separate copies of my article thrown off, I will send you, and a few of my other friends in America, some. These copies will have a *title page and mottoes*, which should be given if it were thought worth while republishing.

In regard to the allusion to "Chapman's book," I may remark, that this was a review of the *Lectures on the Diseases of the Thoracic and Abdominal Viscera* of Dr. Chapman, of Philadelphia. With that review—as the letter of Dr. Forbes sufficiently shows—I had nothing whatever to do, nor do I know who was its author. It was evidently, however, intended to be kind, and was certainly as complimentary as the work merited, perhaps more so, yet, in the absence of all evidence and of fact, Dr. Chapman believed, that I wrote the article, and so informed Dr. Meigs, who mentioned the circumstance to me, on returning from the funeral of Dr. Randolph[27] in his company.

[26] William Benjamin Carpenter (1813–1885), M.D. 1839, Edinburgh, Licentiate of the Apothecaries Company 1835, wrote many popular works on physiology (George Johnson, Address, *Medico-Chirurgical Transactions*, 2nd ser. 51: 27–30, 1886).

[27] Jacob Randolph (1796–1848), M.D. 1817, University of Pennsylvania, son-in-law of Philip Syng Physick, surgeon on staffs of Pennsylvania and Philadelphia General Hospitals, Fellow of the College of Physicians of Philadelphia, and member of the American Philosophical Society.

The Article of Dr. Forbes on "Homeopathy, Allopathy and Young Physic" I was most happy to have reprinted in Philadelphia, and applied, accordingly, to Messrs Lindsay & Blakiston, who undertook it. It was greatly objected to, especially by the "druggers," and small fry of the profession, who were afraid that the promulgation of the truths uttered in it might interfere with the mysteries of the craft, and thereby injure the *trade*; but the philosophical searcher after truth, who properly considered, that all mystery should be discarded in the pursuit of a science so noble, hailed its appearance with satisfaction. For myself, I had no hesitation whatever in replying, most favorably, to its distinguished author; who published my letter, along with others from eminent physicians, not only in this country but in every part of the world, without, however, the names of the particular writers being appended to their respective compositions. In the number for July 1846 of the *British and Foreign Medical Review* they appear under the head of "Extracts from Correspondence." "They are made public"—says Dr. Forbes—

in the hope that they may tend to promote the object with which the article was written—viz. the improvement of practical medicine. Emanating, as they almost all do, from men not only of reputation, but of long experience—from men in every rank of the profession, and living in different countries, they will shew, that the opinions promulgated by the editor are neither novel nor singular, but are in strict accordance with those entertained by the most eminent members of our body. All the merit the writer of the article in question has ever sought to claim for it is, that it openly avowed what the writer knew to be the sentiments of the wisest and best among his brethren. The correspondence and the other documents now published will, it is hoped, spread more widely the knowledge of the fact—that such are, in reality the sentiments of a large portion of the physicians and surgeons of the present day. Such a knowledge cannot fail to give greater confidence to the younger members of the profession not merely to declare their opinions, but to modify their practice according to the views they may conscientiously entertain. It is mainly to the younger members of the profession, that the writer of the article in question looks for the consummation of the reformation in therapeutics, which he is desirous of promoting, and which he believes to be absolutely necessary and inevitable.

It certainly required no little moral courage in Dr. Forbes to wage battle against such a host of routinists as occupy the ranks of the profession. According to his desire, I gave him my opinion of the article in a letter, dated, Philadelphia Feb. 28, 1846, from which he published the following extract, in the number of the *British & Foreign* above referred to:

The article has created quite a sensation here, and knowing well, that it could be laid hold of by the homeopathists, and, garbled as it has been, I was myself anxious, that it should be reprinted in full, so that no permanent misrepresentation might exist. The favourable portion of your remarks has already been extracted by them (the homeopathists), yet they have not concealed, that you are no homeopathist, and have endeavoured to show, that you are not consistent, by contrasting your admissions in regard to the reform produced by the practice of Hahnemann with your exposition of its absurdities. The whole article accords signally with my own views. In regard to the "agenda, cogitanda etc." I have scarcely an objection to make. Whilst I lived in—(Virginia), I was generally regarded as an "inert" practitioner, because I did not practise the energetic and heroic treatment universal there; and, since then, my remedial agencies have been considered to belong to a "masterly inactivity." I apprehend, that in the progress of life, every one becomes less and less active, is more and more disposed to attend to "the divinity that stirs within us," is less and less disposed to believe in the special adaptation of drugs to special morbid conditions; and more and more in the great principles of hygiene and therapeutics. With one single admission only would I hesitate to accord. You ascribe immense influence to Hahnemann as a reformer of regular practice. In this country his doctrine, and course of treatment, have had but little effect on the "regulars." In the cities they have long become less active, but if anyone is entitled to the credit more than another here, it is Broussais.[28] Nowhere, not even in France, were his views so extensively embraced; and under their adoption the excessive bleedings of the Rush school, and the hypercatharsis in use every where, were abandoned, and a more rational and milder system introduced. The good sense of observers of the day has, also, I apprehend, had much to do in bringing about this salutary reformation p. 246.

P.S. August 1865. Early in February 1857 Sir John Forbes published a small volume in which he embodied his views on the treatment of disease. It was entitled *Of Nature and Art in the Care of Disease* & in the Preface he stated his "intention on some future occasion, to publish another volume consisting of the original article "On Homeopathy, Allopathy and Young Physic," together with a selection from the correspondence elicited by it at the time from his medical friends. Sir John died, however, in 1861 and the work has not appeared. His "Nature & Art" was reprinted in New York in 1858 from the second London edition.

Dr. Jacob Bigelow,[29] of Boston, with views similar to those of Sir John Forbes wrote a few years ago on *Rational Medicine* and in 1859 on *Nature in Disease.* This work he did me the honor to dedicate to me. (See the volume, with letters to me prefixed, in my library. Dr. Bigelow is a learned and accomplished physician, a clear writer, and accurate thinker. Would that there were more of them in the profession than there are).[30]

Dr. Forbes continued to be the Editor of the *Review*

---

[28] François Joseph Victor Broussais (1772–1838), who was at the height of his teaching fame in Paris when Dunglison studied there, pushed bleeding, purging and puking to such extremes that a revulsion set in, led by Chomel, Louis, and Laennec, which led to general abandonment of reckless bloodletting and purging.

[29] Jacob Bigelow (1787–1879), M.D. 1810, University of Pennsylvania, Professor of Materia Medica, Harvard Medical School, eminent botanist, read a discourse on *Self-Limited Diseases,* in 1835, which had an important influence on moderating medical treatment in this country and was published in various collected editions of his lectures, including *Nature in Disease,* mentioned here by Dunglison.

[30] This "P.S." was inserted thirteen years after the diary was originally penned in 1852.

to the 48th number inclusive, which was published in October 1847. In a letter, dated June 22, 1847, he asks me—

Has the fame of my abdication of the throne critical yet reached Philadelphia? With the October number my labours cease, and my Review dies! It will, I believe, be resuscitated in a somewhat different form and under a somewhat different name by the amalgamation of it and Johnson's. Both may pay as one, while both failed to pay as two. I believe the Editor's name will not appear, but I shall have no connexion with it in any way. It will be the joint property of the joint publishers. Knowing as you do the labour of such a task as mine has been during the past 12 years, you will not be surprised, that I desire to relinquish it on reaching the grand epoch of *Threescore.* Had considerations of a pecuniary kind influenced me, I should have taken this step long since as I have been for many years convinced, that the book will never *pay.* In making out my account (and I shall treat my readers to this in my *finale*) I expect to find, that my *outs* and *ins* are nearly balanced, leaving *entirely unpaid all my* own labour both as a contributor and editor. I shall complete the book by a large volume of *Index,* which is now in preparation.

In a subsequent letter, dated August 16, 1847, Dr. Forbes says:—

Many thanks for your very kind note of the 29th of July. I am very sensible of your uniform good will towards the Review, and cherish your high appreciation of its merits as it deserves. I have just written a *postscript* for my last number, in which I give a simple history of the whole affair. In this, I assure you, I assume the *superbiam quaesitam mentis,* authorized by such testimony as yours and many other most estimable friends. I expect that you will have a really good Journal in the future *British and Foreign Medico-Chirurgical Review,* the first number of which representing the *two* old Quarterlies, will appear on the first of January.

In that "postscript," Dr. Forbes gives a most interesting account of the *Review,* and the reasons, which interfered with its not having succeeded better as a commercial speculation, and they whose opinions are of any value, will fully accord with the sentiments contained in the following passage:

It succeeded in establishing for the first time, in medical literature, a high critical tribunal analogous to those which had previously conferred so great benefits on general literature. It investigated the literature as well as the scientific merits of books, and boldly pronounced its opinions of them, whether favorable or unfavorable, without regard to the name or station of the authors. It made the medical profession of this country (England) as familiar with the production of foreign countries as with their own. It allowed no improvement, or discovery, or new fact of interest or curiosity to be promulgated in any part of the world, without conveying it speedily to its readers. In giving its judgments, it was altogether beyond the influence of authors or publishers, or of any individuals, whatever, and if its judgments were not always right, they were never prompted by unworthy motives. A few respectable individuals intolerant of fair criticism from overweening vanity or self conceit, several weak men, whom nature never intended to meddle with the pen, and sundry intra-professional quacks of various rank and complexion have occasionally attempted to raise the cry of partiality and injustice against its decisions;

but I here boldly appeal to all disinterested men and good judges, whether, taken as a whole, the opinions promulgated, and the critical awards made by the writers in the *Review*, from its beginning to its termination, have not been remarkable for their honesty, accuracy and justice. That they were *always right* no man of common sense could for a moment, pretend, but that they were as nearly right as the means at the Editor's command could make them, is confidently asserted. As consequences of this character, it is still farther maintained—*first* (in reference to the past), that the *Review* has, of late years, been regarded by the best informed members of the profession, in this and other countries, as the highest authority on medical subjects; and *secondly*, (in reference to the future) that, by showing what a learned, independent and liberal review is, and can effect, it has not only prevented, in time to come, works of inferior character and merit from finding acceptance with the profession in this country, but has assured the perpetuity of a work or works of the same high stamp as itself. *Brit. & For. Med. Rev. Oct. 1847*, p. 571.

On this side of the Atlantic, there were examples of the dissatisfaction referred to by Dr. Forbes, in the above extract. I have already alluded to the case of Dr. Chapman. Dr. Horner's [31] book on *Anatomy and Histology* had been critically noticed, and had been stated to be written "in the American variety of the English language." He was therefore, dissatisfied, and unwilling—I was told—to aid in the testimonial subsequently given to Dr. Forbes, after his retirement from the *Review*. A severe criticism had also been passed on the *Operative Surgery* of my colleague Dr. Pancoast; and the wounded vanity of Dr. Martyn Paine, of New York, led him to reply to a critique on his *Medical and Physiological Commentaries*,[32] which by no means added to his reputation. This critique he ascribed to Dr. Carpenter, who had nothing to do with it. It proved to have been written by a Surgeon of the name of Morgan, who wrote, I believe, the first part of a work, entitled *First Principles of Surgery*, and afterwards went to Australia.[33] He imposed upon Dr.

Forbes an outrageous plagiarism from Channing's Essay on Milton. Dr. Carpenter was led to address me a printed letter, disavowing altogether the authorship of the article, which disavowal was endorsed by Dr. Forbes, and I hope was ultimately as satisfactory to Dr. Paine, as it was at once, to every unprejudiced person. The letters are contained in the *American Medical Intelligencer* for December 1841, of which I was Editor, and in justice to Dr. Carpenter, I took pains to have them disseminated as widely as I could but I took no part in the controversy.

The "large volume of Index," of which Dr. Forbes speaks in one of his letters, was published. It is a valuable termination to a most valuable series; and one which was of inestimable advantage to me as a teacher, as it must have been to every one similarly circumstanced; yet, as Dr. Forbes remarked in his Postscript —its very excellences were impediment to its extensive reception.

Those high qualities, which obtained for it the suffrages of the best judges, were direct obstacles to its success with a large portion of the medical profession. It was pitched on too high a strain for the less educated practitioners. Its physiological and philosophical discussions, its investigation of principles, its generalizations, its correct but abstract and philosophical style, were, in a great measure, thrown away on this class of readers; they desiderated something lower, humbler, simpler, something they could at once grasp and understand. The mere "practical men," also,—men, who look upon medical literature as a sort of manufactory, storehouse, shop, or waggon, for investigating, detecting, promulgating, and conveying "improvements in practice" by which they mean new or improved application in some unusual case etc etc.; in a word some fresh addition to the already boundless stock of empirical conventionalities—were sadly disappointed at finding in its pages general principles taking the place of individual cases, and therapeutical doctrines giving the go-by to bran-new specifics and charming formulae without a chemical or pharmaceutical defect.

On the retirement of Dr. Forbes from the editorship of the *British and Foreign*, the high compliment was paid him of a very large number of the most eminent members of the profession—*two hundred and sixty five* —subscribing one guinea each "to testify their respect for his literary and scientific character, and their sense of the value of his disinterested services, in establishing and carrying on for a period of twelve years, the *British and Foreign Medical Review*," and to present him with "a lasting Memorial of their Approval and Esteem." In the list are the names of several distinguished American physicians, but far fewer than there ought to have been,—Drs. John Rhea Barton,[34]

[31] William Edmonds Horner (1793–1853), Professor of Anatomy and Dean of the University of Pennsylvania, whose work in the untrodden field of microscopical anatomy has original observations but earned him little fame because his literary style was said to be stilted (William Shainline Middleton, William Edmonds Horner, *Annals of Medical History* 5: 40, 1923).

[32] Martyn Paine (1794–1877), M.D. 1816, Harvard, Professor of the Institutes of Medicine and Materia Medica, University of the City of New York, instigator of the New York Anatomical law, published *Medical and Physiological Commentaries*, 3 vols., 1840–1844, and *An Examination of a Review, Contained in the British and Foreign Medical Review of the Medical and Physiological Commentaries*, 1841, in which he said "it is extremely difficult to separate Dr. Carpenter from the reviewer," pointing out the plagiarism in the *British and Foreign Review* and wisely urged that all reviews should be signed or initialed unless written by the editor himself.

[33] George T. Morgan, a great resurrectionist, lecturer at Marischal College, Aberdeen, Scotland, author of *An Outline of Inflammation; being First Principles of Surgery*, London and Edinburgh, 1837, republished in Dunglison's *American Medical Library*, Phila., 1838, left Aberdeen, when Marischal College was absorbed by Aberdeen University in 1854, settled in

London, and a few years later, in Mauritius, a British colony east of Madagascar in the Indian Ocean.
[34] John Rhea Barton (1796–1871), M.D. 1818, University of Pennsylvania, famous Philadelphia surgeon and originator of the Barton bandage, has been memorialized by the Barton Professorship of Surgery at the University of Pennsylvania (*University of Pennsylvania*, p. 353, Boston, 1901).

82   RADBILL: ROBLEY DUNGLISON, M.D.   [TRANS. AMER. PHIL. SOC.

W. Gibson, C. D. Meigs,[35] Thomas D. Mütter and myself being the only Philadelphia contributors. To all
these gentlemen, Dr. Forbes, in the volume of "Index"
dedicated the whole work, "with a deep sense of the
honour conferred upon him."

NEGOTIATIONS WITH JEFFERSON
MEDICAL COLLEGE

It was during my residence in Baltimore, that I first
became personally acquainted with Dr. Granville Sharp
Pattison,[36] who was, at the time, Professor of Anatomy
in Jefferson Medical College. In travelling to the
South, he heard me lecture in the University of Maryland, and was kind enough, at that time, to express a
hope, that we might become associated with each other,
in medical instruction. I also became personally known,
whilst there, to Dr. Revere,[37] the Professor of Practice of Medicine in the same Institution, who imbibed,
I believe, the same desire. In the year 1836, specific
and definite propositions were made to me by them, and
I, at length, consented to accept a Chair of Institutes of
Medicine and Medical Jurisprudence in the College if
it were tendered me by the Board of Trustees. At
that time, there were but six Professorships in the
school, and to procure my services, it was proposed to
add a seventh. Under the then existing Faculty,[38] and
especially under the energetic action of Dr. Pattison,
the school had increased its numbers, so that the class
of 1835–36 amounted to 364 students. A portion of
that "energetic action" consisted, however, in a most
objectionable system of announcing the assumed advantages of the Jefferson Medical College to the dis-
paragement of all other Institutions, and especially of
the University of Pennsylvania. The "Announce-

[35] Charles Delucena Meigs (1792–1869), M.D. 1817, University of Pennsylvania, Honorary M.D. 1818, Princeton University, Professor of Obstetrics and Diseases of Children,
Jefferson Medical College (1841–1861), fellow of the College
of Physicians of Philadelphia, member of the American
Philosophical Society and the Society of Swedish Physicians
(Thomas G. Morton and Frank Woodbury, History of the
Pennsylvania Hospital, p. 508, 1895).
[36] Granville Sharp Pattison (ca. 1791 or 1792–1851), controversial anatomist, Professor of Anatomy, Physiology and
Surgery, University of Maryland (1820–1826), Professor of
Anatomy, Jefferson Medical College (1831–1841), Medical
College of University of City of New York (1841–1851),
notorious for his feud with Chapman (William Snow Miller,
Granville Sharp Pattison, Johns Hopkins Hospital Bulletin
30: 98, 1919).
[37] John Revere (1787–1847), son of Paul Revere (of Midnight Ride fame), M.D. 1811, Edinburgh, Professor of Medicine, Jefferson Medical College (1831–1841) and Medical
College of the University of the City of New York (1841–1847)
(Gayley, op. cit., p. 46).
[38] The faculty of Jefferson Medical College in 1836 consisted of: G. Sharp Pattison, Anatomy; John Revere, Theory
and Practice; George McClellan, Surgery; Samuel Colhoun,
Materia Medica; Jacob Green, Chemistry; and Samuel Mc-
Clellan, Medical Jurisprudence, Midwifery and Diseases of
Women and Children. It remained intact from 1833 until 1839.

ment" of the College for the year I have instanced, was,
indeed, so improper and unprofessional that it gave occasion to severe animadversions by Dr. Caldwell of
Lexington, as well as by myself, in the American newspaper of Baltimore, and in the North American Archives of Medicine etc. edited by my friend Dr. Geddings; and I frankly mentioned the circumstance not
only to Dr. Pattison, but to Drs. Revere and Green.[39]
The vain boast, that the Jefferson Medical College was
not only "not surpassed, but not to be surpassed" was
certainly unworthy of any respectable school. I never
doubted, however, that, should I be connected with the
school, the offence would not be repeated with my
protest.

APPOINTED PROFESSOR IN JEFFERSON
MEDICAL COLLEGE

After different communications had passed between
us,[40] I received the following letter from Dr. Pattison.

Philadelphia, June 24, 1836
My dear friend,
It is with sincere gratification I enclose you the Commission of your appointment as Professor of the Institutes of
Medicine, and Medical Jurisprudence in Jefferson Medical
College!! Associated, as I trust we shall be, in the same
Institution, for many years, I feel assured, that the sentiments of regard we have, since our acquaintance, entertained for each other, will continue to increase, and that
we shall zealously be fellow-labourers in the great cause of
medical improvement in the United States. Your last letter
in which you consented, in the event of your being elected
to accept of either professorship, of Materia Medica or of
Physiology etc. (on consideration, I thought the title of
"Institutes of Medicine & Medical Jurisprudence" better
than that of Physiology, and it allows of more lattitude)
I kept sacred. I did not even mention the fact to my wife,
or my intimate friend, Dr. Revere. But, in proposing the
creation of a seventh chair to my particular friend, Judge
King,[41] who signs the inclosed letter, I found it necessary,
in the strictest confidence, to pledge to him my word, that
I would ensure your acceptance in the event of your election. This pledge he required before he would consent to
move in the matter, justly observing, that, to create a
seventh chair, unless under the certainty of a gentleman of
your distinguished talents and reputation filling it would
most seriously injure the Institution. You will oblige me
if you will send your official acceptance of the office, enclosed to me, by return of post, addressed to the Hon. Judge
King etc. etc. Dr. Revere and myself would have, in person, presented to you your appointment but as I have been,
long after the time I intended, detained in the city on this
business, and as I propose to shake hands with you in
Baltimore, on Tuesday next, on my way to Virginia, I could
not possibly leave the city. Moreover, the Faculty meet

[39] Jacob Green (1790–1841), M.D. (hon.) 1827, Yale Medical
School, Professor of Chemistry, Experimental Philosophy and
Natural History, Princeton University (1818–1822) and of
Chemistry, Jefferson Medical College (1825–1841), where his
father was a member of the Board of Trustees.
[40] An invitation from Pattison to Dunglison, December 24,
1835 (ms: 506, New York Academy of Medicine), indicates
that Dunglison may have visited Philadelphia about this time.
[41] Secretary to the Board of Trustees of Jefferson Medical
College.

tomorrow to prepare the annual announcement and to inform the Profession of your being now one of *us*. . . .

Yours most faithfully,
Granville S. Pattison

The letter of appointment from Judge King, as Secretary of the Board of Trustees, was answered by me immediately, and, after a full understanding of the position I should occupy, as the incumbent of a new chair, I accepted the appointment. To this I was led by various considerations. There did not appear to me to be the deep interest in the affairs of the school in Baltimore that ought to have impressed the Board of Trustees; and, hence, they not unfrequently acted on important matters without consulting the Faculty, and at times, in opposition to the wishes of the latter; and, moreover, as I have already remarked, and experience proved its truth, the tide set so strongly towards Philadelphia, which had been so long esteemed the great centre of medical education, that it appeared impracticable to stem it, so that I despaired of seeing a permanently large school established in Baltimore. I decided on leaving it, however, with regret, for all my associations—professional and others—had been exceedingly agreeable; and, I believe, sincere sorrow was experienced by my colleagues at my withdrawal. In evidence of this, I may adduce the following official letter from Dr. Geddings, who succeeded me as Dean of the Faculty, as he preceded me in that office:

My dear Sir,

Your letter of yesterday, informing the Faculty of your determination to resign the Professorship of Materia Medica etc. in the University of Maryland, was laid before that body last evening, whereupon I was instructed to communicate to you the regret felt by the Faculty, that circumstances should have rendered it expedient for you to withdraw from the Institution, and deprive it of the able and zealous support, which it has derived from your exertions during the last three years. While they look back with pleasure upon the harmonious cooperation which has characterized their actions as a body since your connection with the school, and your steady exertions to maintain a feeling of mutual regard between those who were engaged in the same cause with yourself, they feel sensibly the loss they are about to sustain by your resignation. Although the Faculty are aware of the magnitude of this loss, they are too sensible of the justness and force of your motives to suffer their private interests to weigh against your future prosperity, in which they all feel interested. Permit me, on my own behalf, to indulge the hope, that although, in future, our field of action may be different, the same private feelings and relations may exist between us as heretofore, and be pleased to accept every assurance of my highest esteem and regard.

Prof. Robley Dunglison   E. Geddings
Balt.   Dean, Med. Faculty.
July 1, 1836

CHANGES IN THE UNIVERSITY OF MARYLAND

My withdrawal from the school was soon followed by that of Dr. Geddings, who, in the succeeding spring accepted an appointment to a new chair of "Pathological Anatomy and Medical Jurisprudence," which had been added to the Medical College of the State of S. Carolina. It had always been maintained by some of my colleagues, that the Board of Trustees of the University of Maryland were not constitutionally entitled to hold their places; open hostility now broke out between them and the faculty, and, ultimately, the latter succeeded in ousting the former; and in restoring a government, of which the Faculty constituted an important part, and of which they had been deprived by an illegal act of the Legislature. Hence it became necessary to legalize all the acts that had been executed during the existence of the Board of Trustees from whom I received my appointment; and to many young gentlemen the honours of the Institution.

Whilst I was connected with the University of Maryland, I officiated as one of the Physicians to the *Baltimore Infirmary*, became a Licentiate of the *Medical and Chirurgical Faculty of Maryland*,[42] as required by law; a member of the *Medico-chirurgical Society of Baltimore*[43] in 1834; of the *Philocretan Society of Baltimore*,[44] in 1835, of the *St. George's or St. Andrew's Society of Maryland*[45] about the same time; of the *Massachusetts Medical Society*[46] in 1836; and of the *Medical Society of the State of New York*,[47] in the same year.

The only change that occurred in my family, during my residence in Baltimore was in the birth of my son Richard James, which took place on the 13th day of November, 1834.

---

[42] The Medical and Chirurgical Faculty of the State of Maryland was chartered in 1799 "to prevent citizens of Maryland from risking their lives in the hands of ignorant practitioners or pretenders to the healing art" and could license by examination or presentation of a satisfactory diploma from a medical college (Eugene F. Cordell, *Medical Annals of Maryland*, pp. 20–22, 1903).

[43] The Medical-Chirurgical Society of Baltimore, founded by Samuel Baker about 1830, in 1832 drew up a *Code of Medical Ethics* and in 1848 a medical fee schedule (Cordell, *ibid.* p. 94).

[44] One of the early cultural organizations of Baltimore devoted to discussions and lectures on various academic subjects (*American and Commercial Advertiser*, September 28, 1835, quoted in a personal communication from Miss Elizabeth C. Litsinger, Enoch Pratt Free Library, Baltimore, Md.).

[45] Organizations of descendants of English and Scottish ancestry, respectively, for the relief of poor and distressed countrymen.

[46] Founded 1781, an outgrowth of a Medical Society in Boston in 1736, to establish a library and publish *Medical Communications* (Henry R. Viets, *Brief History of Medicine in Massachusetts*, p. 103, 1930).

[47] Organized in 1807, two honorary members could be elected from outside the state each year.

V. JEFFERSON MEDICAL COLLEGE

Philadelphia and Nathaniel Chapman. Isaac Hays. Jefferson Medical College. George McClellan. Reorganization of the faculty. Pattison and Revere. Reorganization of 1841. Progress of the School. Pattison and Revere. New York University seeks a medical professor. Death of Dr. Pattison. Transylvania University. Charles Caldwell. *The American Medical Library and Intelligencer.* Dr. Samuel Forrey. Athenian Institute. Insane poor of Pennsylvania. Vaccine virus.

FIG. 12. Jefferson Medical College when Dunglison was there.

PHILADELPHIA AND NATHANIEL CHAPMAN

In September 1836, I left Baltimore with my family for Philadelphia. I was entirely aware of the odium in which the Jefferson Medical College had been held, by those especially who were connected with, or allied by interest or affection to, the University of Pennsylvania. The medical department of that Institution had been so long in existence and had so long been regarded as the only avenue to distinction in medical teaching, that it held supreme sway over the minds of most perhaps, of the practitioners of Philadelphia; and every attempt to interfere with it was looked upon with jealousy, if not contempt; so that, when Dr. Drake joined the Jefferson School in 1830, some of the professors of the University of Pennsylvania treated him with marked coldness. He was well known to Dr. Chapman; yet, as he himself told me, Dr. Chapman did not even speak to him whilst he occupied the chair; and others equally manifested coldness towards him, and when, after that, Dr. Pattison was appointed to the Chair of Anatomy, the hostility which had been previously felt towards him, especially on the part of Dr. Chapman, and which had led to a duel between Dr. Pattison and General Cadwallader, the elder,[1] was certainly not diminished. Had the fight taken place between Dr. Chapman and Dr. Pattison a possibility might have existed of a subsequent reconciliation, but when the former permitted Gen. Cadwallader to suffer in his cause, all approximation of the original contending parties became impracticable. Dr. Geo. McClellan

---

[1] Brother-in-law of Dr. Chapman.

also, had rendered himself extremely disagreeable to many of the profession by conduct that certainly was not always defensible. So that neither the past history of the College, nor its present condition was in good odour, when I joined it, and it was prognosticated in Baltimore, and affirmed by those whose wishes were fathers to the thought, that my position in Philadelphia would be made exceedingly disagreeable to me. I had no fears of this, however. I determined so to conduct myself to every one, that there should be no just cause of offence; and if, notwithstanding this, my course should not be approved of by others, I should have the consolation, that it would be satisfactory to myself. I had, moreover sufficient self respect to feel, that if anyone of my medical brethren did not seek my society, or avoided it, the loss, if any, would be mutual. I must candidly say, however, that my feelings were but rarely tried. Before a year had expired, all the Professors of the University of Pennsylvania had called upon me, with the exception of Dr. Chapman. Before I took up my residence in Philadelphia, I had visited him at his own house; and he had been entertained by me in Baltimore. When I removed to Philadelphia, it was for him to decide, whether he would continue to have my society, and it was not for me to repine, should he determine in the negative. I never did repine. We were often thrown into each other's society, and it was not until I accidentally heard, that he assigned a reason for his not having called upon me, which had no foundation in reality, that I took any notice whatever of his conduct. This was first accidentally communicated to me by his colleague, Dr. Gibson, at a party at Mr. Nathan Dunn's,[2] some years after I had removed to Philadelphia. On that occasion Dr. Gibson informed me that a distinguished gentleman of Philadelphia, who was very partial to me, had stated to him, that an eminent member of our profession had not called upon me, in consequence of my having been guilty of duplicity whilst in Baltimore. I told him it was impossible for me to reply to so vague a charge; and begged of him to obtain for me farther particulars, which he promised to do. The next time I saw him, he stated, that Judge Hopkinson[3] was his informant, and that he said I had been guilty—on the authority of the gentleman referred to—of duplicity, by writing against the Jefferson Medical College, whilst Dr. Granville Sharp Pattison was staying with me in Baltimore. I replied, perhaps it would be enough for me to say, that Dr. Pattison had never staid with me whilst I was in Baltimore. This Dr. Gibson thought sufficient, but afterwards he told me that it was perhaps, whilst I was staying with Dr. Pattison in Philadelphia. To this, I replied, that I had never staid with Dr.

---

[2] Listed in the Philadelphia directory as: "gentleman, 3 Portico Square."

[3] Joseph Hopkinson (1770–1842), Judge of the U. S. District Court, author of "Hail Columbia," counsel for Benjamin Rush in the famous libel suit against Cobbett, was Vice-President of the American Philosophical Society.

Pattison more than a single night, I believed; and it was not likely that I should occupy myself in writing against the College in his house; and that there was in fact, no truth in the assertion. I told him, however, what I *had* done, that when the Jefferson Medical College had published, as I have already mentioned (p. 82), a most obnoxious "announcement," I had both spoken and written against it, and that I still maintained the same sentiments in regard to it, but that I had communicated all this to Drs. Revere and Green, so that the charge of duplicity was an absurdity.

Accidentally, some time afterwards, Dr. Gibson let it escape from him that Dr. Chapman was the "eminent professional gentleman" in question. As, however, he had mentioned the affair to me in confidence, I could take no notice of it to Dr. Chapman, but on communicating the whole affair to Dr. Patterson & Judge Kane the former told me, that he had heard Dr. Chapman mention the same thing more than once;—that on the Thursday preceding, he had dined with Dr. Chapman, —Mrs. Chapman, Judge Hopkinson and Mr. Gordon, of Baltimore—Dr. Chapman's son-in-law—being present;—that Dr. Chapman had expressed his most friendly feeling toward me, stating, that he only waited for an opportunity to do me a solid service, but he regretted, that I had been guilty of duplicity in Baltimore,[4] that I was a gentleman, and so forth. I asked Dr. Patterson if he made no reply to this. He said he did not think it required or deserved one. Judge Kane and I both agreed, however, that the matter must be met; and I urged upon Dr. Patterson, as my friend, to see Dr. Chapman, and inform him, that there was not a word of truth in the assertion, and that it must not be repeated. A few days afterward, he called to say, that he had done so, and that Dr. Chapman replied, that he had been so informed by five or six persons, but if I said there was no truth in it, he believed *me*. I have not heard, that he subsequently repeated it.

If Dr. Chapman were so informed, it must have been by malignant and mendacious persons. Duplicity is far from being one of my sins, and the whole story bore too much thè character of having been got up for effect. It was believed, I doubt not, that it was breathed into ears not unwilling to receive it, and it was probably founded on baseness and malevolence. It was idle, however, in Dr. Chapman, to assign this as a reason for not having called upon me. I cannot but ascribe his conduct to my having become associated with the Jefferson Medical College, and I fear the assigned grounds were suggested at a subsequent period, and although they were derogatory to me, they might seem to others more valid than the true reason.

I could not, however, after this, accept of any invitation from him. He had declined calling up me, and had assigned reasons for this, which were as unfounded as they were disparaging, and I consequently, could

hold no social relations with him. Having wronged me, however, he began—as is common—to suspect and dislike me; hinted that I was ambitious of obtaining the highest posts in the American Philosophical Society, and this at one of its meetings—and, as I have before remarked (p. 79), that I had severely criticized his "lectures," published some years ago, in the pages of a foreign Journal—all of which were mere coinages of the brain. I confess, that my knowledge of him neither led me to a high appreciation of him as an intellectual or a moral man. On my earliest acquaintance with him he made an appointment with me to meet him at the wharf on the following morning, to accompany him to visit the Count Survillier (*Joseph Bonaparte*). When I went at the appointed hour, he was not there, but fortunately, Dr. Gibson, who had told me he would be one of the party, as Dr. Chapman was not to be depended upon, was, and we went together. He was a good table companion; spoke to the purpose, but so slowly that, as he himself told me, he could think as he went along; but he was hollow; made protestations, I fear, which he did not feel; and was by no means tender, in his observations on others. In the teeth of the most positive documentary evidence, he pronounced, before me, to several gentlemen, that my venerable friend, Mr. Duponceau, who had died a short time before, and whose executor I was, had stated falsely, that he held a commission in the army of the United States. I, at once, affirmed, that my friend's statement was true.

Nor was I more impressed with his intellectual powers. He had no judgment whatever; and I have been repeatedly ashamed at the figure he cut on the floor of the Philosophical Society. It is true, that I saw more of him in the after part of his career—from 1836 to 1845, inclusive—and, that he might have fallen off; but his published *Lectures*, two volumes of which were published in 1844, ought properly to be esteemed indexes of his intellectual endowments; and they are, most assuredly, very inferior in all respects, both in matter and manner. Strange, that so great a weight should have been hung upon so small a wire, and that such a person should have been regarded by men of intelligence as the "unquestioned head of the profession," or, as one of his colleagues designated him, the "unquestioned head, by position." In no other way, ought he to be so regarded; and whether in this, admits of contest,—that "position" being Professor of the Theory and Practice of Medicine in the University of Pennsylvania! And that colleague the gentleman, who has succeeded him in the office![5] I well recollect, some years ago, walking home from the "Club"[6] at Dr.

---

[4] See Chapter II, fn. 35.

[5] George B. Wood.

[6] Probably the "'Tea and Toast," or Monday Evening Club, founded in 1835 (Walton B. McDaniel, 2d, *Fugitive Leaves from the Library*, n.s., *No. 33:* 93, College of Physicians of Philadelphia, 1959), although it may have been one of the other social groups such as Mütter's medical party or club described by E. R. Squibb (*Journal*, p. 179, 1930), or the

Wood's with Dr. Caspar Morris [7] and his prognosticating the evils, that might result from the death of Dr. Chapman, observing that, in such event, in place of there being one "unquestioned head," there would be several, and that the harmony of the profession might, in this way, be disturbed. I told him I did not participate in those views;—that, granting for the nonce, that Dr. Chapman was such "an unquestioned head," I thought, that the looking up to one man was apt to lead—and I considered there was ample evidence to shew, that it had led—to servility amongst the members of the profession, and that I confessed I preferred a republic of science to anything else; but undoubtedly an oligarchy to a monarchy. Dr. Chapman has now (July 1852) been virtually dead for two or three years and I have yet to hear of any such evils as those that were apprehended by Dr. Morris.

ISAAC HAYS

Dr. Chapman's example in not calling upon me, when I came to Philadelphia, I doubt not influenced some of his subordinates; but as I had no special anxiety to be much acquainted with them, and had no opportunity of testing their social virtues, if they were possessed of any, I never felt the loss of their Society. Dr. Hays and myself had been better acquainted. As I remarked before, I was a collaborator of and was published as such in the *American Journal of the Medical Sciences*, but it was an established rule with the Editor, that no one officially connected with any other Journal could remain a collaborator to it,—not a very liberal rule, it must be admitted, but nevertheless one that was vigorously enforced.

When I became Editor of the *American Medical Library and Intelligencer*, my name was of course, removed from the list of collaborators to the *American Journal*, and I do not believe it was the wish of either party—it certainly was not mine—that the connection should be resumed. Dr. Hays left his card with me when I came to Philadelphia.

I returned his visit, and that is all the communication, of a social cast, which has taken place between us during the last sixteen years. I do not, indeed, think, that there is anything between us in common; certainly not that sympathy of feeling which attracts those of kindred sentiments towards each other. For a quarter of a century, he has conducted the *American Journal of the Medical Sciences*, and it must be admitted with great industry. The "Periscopic" department has usually been well selected. The reader may

however, look in vain for any amount of matter furnished by Dr. Hays's own pen. As a writer, he is by no means happy; and it is to be deplored, that so many inaccuracies are permitted to escape correction in the Journal especially when the Latin language is employed. A warm friend, if not a bitter partisan, of the University of Pennsylvania, the Journal contains numerous eulogies of that school, and of those who are connected with it whilst it has been a complaint, and, I think, one that is well founded, that the same or some measure has not been extended to other institutions and to those connected with them, so that the epithet "American" has been regarded as improper, when applied to the Journal.

JEFFERSON MEDICAL COLLEGE

I had not been long in Philadelphia, before I discovered, that not the best feeling existed between the members of the Faculty in the Jefferson Medical College. They had divided, indeed, into two parties,—the one, consisting of Dr. Geo. McClellan; [8] his brother Doctor Saml. McClellan,[9] and Dr. Calhoun; [10]—the other of Dr. Pattison, Dr. Revere, and Dr. Green; and hence, as the Faculty consisted only of six persons, on important questions they were often tied, and there was real danger of the wheels of the Institution being so clogged, that it might not be able to proceed. My appointment to a seventh chair gave occasion to Dr. Pattison's party counting on my vote to decide in their favour, and this circumstance interfered with the cordiality of my reception by the other party. I had good authority, indeed, for the belief, that Dr. Geo. McClellan was originally much opposed to the institution of a seventh chair, and to me as its incumbent. I determined to pursue an independent course, and when this was seen by Dr. McClellan he overpowered me with his attentions. Both parties then looked towards me; and both were disappointed. On all occasions, I endeavoured to divest myself of undue prepossessions, and to judge of every question on its merits; a course

Saturday Evening Club, which reminded Samuel D. Gross of the Wistar Parties (*Autobiography* 2: 225, 1887).

[7] Casper Morris (1805–1844), M.D. 1826, University of Pennsylvania, Fellow of the College of Physicians of Philadelphia, member of the Historical Society of Pennsylvania and The American Philosophical Society, was a founder of the Philadelphia Medical Institute and of the Episcopal Hospital of Philadelphia.

[8] George McClellan (1796–1847), M.D. 1819, University of Pennsylvania, in 1821 established a private medical school which became (1825) the Jefferson Medical College of Philadelphia, where he was Professor of Surgery until 1838, when he left to form the Medical Department of Pennsylvania College. "'Some of his best friends would say that he was impolitic and unwise, and at times, even inconsiderate and imprudent" (W. Darrach, *Memoir of George McClellan*, p. 39, 1847), but his death was considered a public loss (James F. Gayley, *op. cit.*, p. 30).

[9] Samuel McClellan (1800–1854), M.D. 1823, Yale, studied medicine with his brother, George McClellan, with whom he was associated as pupil (1820–1823), professor at Jefferson Medical College (1828–1839) and Pennsylvania Medical College (1839–1843). (Wm. H. Pancoast, *Introductory*, p. 22, 1874.)

[10] Samuel *Colhoun* (1786 or 1787–1841), M.D. 1808, University of Pennsylvania, associated in teaching with the McClellans at Jefferson Medical College (1831–1839) and Pennsylvania Medical College (1839–1841), Fellow of the College of Physicians of Philadelphia, changed his name from Calhoun to Colhoun in 1832.

which met so completely with the approbation of Dr. Green, that he generally coincided with me; and there was, consequently, not much danger of any injudicious measure being carried in the Faculty. Dr. Pattison had a most earnest desire for the success of the school, and was indefatigable in urging it onwards. His great fault, indeed was, in being too unscrupulous in the means he adopted for this purpose. He was the real author of the offensive announcement of which I have spoken (p. 85), and was firmly of opinion, that the true policy of the Jefferson school was to wage an active warfare with the University of Pennsylvania. Hence he lost no opportunity, particularly in his Introductory Lectures, of disparaging it, and especially its Professor of Surgery, Dr. Gibson, who was too fond of the same, and even of worse, illiberality; and has gained himself an unenvied notoriety for his attacks on many of his estimable professional brethren. Dr. Pattison had the conviction, that mankind are to be forced into beliefs, and that it is idle to be too delicate, if the object is to make merit known. People—he urged—will not appreciate it, unless it is over and over again, and strongly, placed before their attention. I combated his views on these subjects, and ultimately succeeded in inducing him, for a time, to omit all allusions to another Institution in his Introductory discourses. He subsequently, however, indulged in severe animadversions on the University of Pennsylvania, after Drs. Huston and Pancoast had become our Colleagues; and his remarks were so offensive to the former, who participated with me in my sentiments regarding such performances, that I think, had Dr. P. remained in the Institution, he would not have repeated them. I was informed, however, that when he had determined to leave the College, and go to New York, he stated, that he did not believe the "Dunglison Policy"—as he termed it—of peace and goodwill to all, would ever succeed. Yet how triumphant has such a policy exhibited itself since; for, after Drs. Pattison and Revere left the school, not a word of censure or unkindness was ever heard from any chair in it; and the numbers frequenting the school have risen to the unexampled height of over *five hundred students* for a succession of years!

The unpleasant state of feeling that existed more particularly between Drs. Pattison and McClellan tainted the whole Faculty. I was soon brought into unpleasant collision with the latter, owing to what I conceived, at the time, to be—and subsequent reflection has not changed my sentiments on the subject—a most wanton and malicious attack on the private character of both Dr. and Mrs. Pattison.

GEORGE McCLELLAN

On returning from dining with Mr. Nicholas Biddle at Andalusia, my half brother, John Atkinson,[11] now of

Saint-Louis, exhibited great anxiety to shew me a letter, which Dr. McClellan had that day put into his hands; which letter was in answer to one, that Dr. McClellan had written to a Mr. Jones—I think that was his name—who had formerly been a pupil-resident in the house of Dr. Pattison, but had left his preceptor, for some cause or other, with very hostile feelings. The reply was evidently to a letter anxious to obtain testimony unfavorable to Dr. Pattison. Jones affirmed, that Dr. Pattison and his wife had lived together before they were married, and reflected severely on the conduct of Dr. Pattison. I have had no means of judging of the accuracy of this statement; but nothing could be more improper than for such a letter to be put into the hands of my brother, who was, at the time, a student of the Jefferson Medical College, and, therefore, of Dr. Pattison. I immediately took the letter from him, and, at nine o'clock in the evening, called upon Dr. McClellan, and stated how shocked I was at his having shewn a letter, containing such charges against one of his colleagues, to my brother;—adding, that I could not answer for the consequences, if the transaction were to come to the knowledge of Dr. Pattison. He replied, that two parties would be concerned, if Dr. P. were to act offensively in regard to him; meaning, I presumed, that he had no fears of Dr. Pattison. To which I replied, that Dr. P. would probably not challenge him, but would shoot him, and that the world would hold him justified, under the circumstances.[12] After this, I had every reason to believe, his feelings towards me, which had never been highly favorable perhaps, became much colder; but still, at no time, even to the last, was he ever personally offensive to me; nor, indeed, did I even at that very time, notice anything objectionable in his behaviour to Dr. Pattison, when they were together. In the meetings of the Faculty—which he did not always attend—he appeared generally to yield his opinions; and on the very day on which he had designated Dr. Pattison to me as a "swindle"—connected with certain speculations into which he had entered, and which proved disastrous to him—I have seen him put his arms round Dr. Pattison's neck, with expressions of the greatest familiarity and friendship, and call him "Granville"! Dr. Pattison had, however, a great contempt for him, and now that Dr. McC. is in his grave, and that years have passed since his decease, I cannot look back with any sentiment of respect for him. By many, he was regarded as a man of ability or "genius"—as it was generally said, and he had a certain degree of smartness; but I must confess he is the only man, from whom, after an association of three years, I never received—so far as I can recollect—a solitary new idea; and yet, I fear, the qualities of his

---

[11] After the death of Robley Dunglison's father, his mother married James Atkinson, by whom she had one son and two

daughters. John Richard Atkinson, Dunglison's stepbrother, M.D. 1839, Jefferson Medical College, practiced in St. Louis, Mo.

[12] This famous scandal was quickly resurrected in every quarrel with the beleaguered "dueling anatomist."

head were superior to those of his heart. Notwithstanding this, there has never been a medical man, who passed away, who received more, or as many, eulogies. As far as regarded the *United States Gazette* this was owing, in part, to an ardent friend of his, Dr. Bird[18] being one of the Editors; but the warm opinions of him were mainly perhaps owing to an apparent frankness, which led the hearer to infer openness, candour, and perhaps goodness of heart, which were certainly far from discoverable in him at all times, when the veil was withdrawn.

As regarded the College, his dissatisfaction became so great, without any assignable cause, that he began to disparage it on all occasions. He had taken a great dislike to Judge King, or, as he generally designated him, "Ned King," who, beyond all question, has at all times, been a most steadfast friend and promoter of the school; and was in the habit of stating far and wide, that it was ruled by a parcel of politicians, that he wondered how any respectable person would send his son to an Institution, governed by such a "blackguard board of Trustees"; and that the Institution was rotten, "was going to the dogs," and must go to them. Nor were his feeling improved by an addition, which was made to the number of the Board of Trustees, when, in the Session of 1837-8, the Jefferson Medical College was separated, by the Legislature from all connection with the parent institution at Canonsburg, and was erected into an independent establishment. Under the new charter and organization of the Institution, thereby effected, all the officers ceased to exist as such, except the old trustees, who were continued by the new charter. It became, therefore, necessary to appoint professors to the vacant chairs, when all the professors were unanimously re-elected. For a long time, Dr. McClellan declared, that he would not accept a re-election; but, when required to give in his decision by a certain day, he answered in the affirmative. I confess, that after this I had some hopes—but they were not of a sanguine nature—that he would cooperate actively with his colleagues and abandon the random and injurious observations, which he had previously made in regard to the officers of the School; but I was mistaken. Most unaccountably, he appeared to derive pleasure from undervaluing it;—stating, that he had never received *twelve hundred dollars* from it during any Session, when I know, that the first year of my professorship, my income from it was only fifteen dollars short of *four thousand*. Matters at length came to a crisis; and I determined to address my colleagues frankly and fearlessly, and to appeal to them to use all their exertions to modify a course of conduct, which, if persisted in, could not fail to be most injurious to the school, and to all those connected with it. Accordingly, I sent to them the following letter, to be read at a meeting of the Faculty, from which I purposely absented myself.

Philadelphia Mar. 9, 1839.

Gentlemen,

A month or two before the last medical Session of the College, the undersigned was desirous of addressing the Faculty on matters which appeared to him to concern, in the most vital manner, the prospects of the College, but the expectations which he was led to indulge, that these matters might be of less influence than he, at one time, believed, and the impracticability—as it appeared to him—of obviating them before the medical Lectures began—induced him to postpone his intentions.

As the same causes appear, however, to have continued in operation, and to be no less active at this, than at the former period, he feels it a duty to himself, and to the Institution, to state frankly and respectfully, the views which impress him.

It will be recollected by the Dean of the Faculty (Dr. Calhoun), that the undersigned—in the month of August or September last—informed him, that he—the undersigned —had received communications from Richmond, the University of Virginia and Baltimore, stating, that reports were rife, that, owing to dissension amongst the members of the Faculty, and between them and the board of Trustees, and to faulty administration of the affairs of the College on the part of the latter, it was impossible for the College to succeed,—that it was sinking and must sink, and that the authority for this view was reported to be the statements of a member of this Faculty. The undersigned consulted with the Dean as to the best method of counteracting these reports, who agreed, that a carefully drawn up advertisement should be placed upon the cover of the *American Medical Intelligencer* of which the undersigned is known to be Editor; but, subsequently, the Dean wrote to the undersigned, stating his opinion, that the matter had better be left entirely at rest. Feeling, however, that the very fact of no advertisement of the College appearing upon the cover of a Journal edited by a member of the Faculty might encourage the reports everywhere abroad as to dissensions —of which the undersigned is happy to say, from his own personal experience, he knows nothing—he determined, should the advertising on the cover of the Journal not be approved by the Faculty, to copy, at his own expense, the ordinary advertisement as nearly as advisable; which was accordingly done, and this was the only course adopted at that time.

The Session approached; and, during this interval, the undersigned received from every quarter repetitions of the same reports, many of which, doubtless, came to the ears of his colleagues, and all were calculated to have a disastrous influence on the prosperity of the school;—for it is not surprising, that candidates for the honours of their profession should hesitate to connect themselves with an Institution reported to be in a tottering condition. Soon after the commencement of the Session, these reports were again common, not only amongst the students, but amongst the inhabitants of the city, and members of the Faculty were grieved to be told, by some of the latter, that they had authority for stating, that the institution was "going to the dogs," and that, for certain reasons, it must do so.

It is far from the object, as it is from the province, of the undersigned to lay charges against anyone of his colleagues of desiring to injure an institution to which he is

---

[18] Robert Montgomery Bird (1803-1854), M.D. 1827, University of Pennsylvania, professor of materia medica, Jefferson Medical College (1841-1843), playwright and novelist, became part owner in 1847 of the *North American and United States Gazette* and joined its editorial staff (Robert L. Bloom, Robert Montgomery Bird, editor, *Pennsylvania Magazine of History and Biography* 76: 123-141, 1952).

attached. Such a desire would be discountenanced—if on no other ground—by the fact, that the Professors, on the occasion of their recent re-appointment—engaged to exert all their influence to promote the welfare of the Institution as at present constituted. It is the report, that these sinister statements rest on the authority of a member or of members of the Faculty, which is to be deplored, and which is calculated to exert as baneful an influence on the Institution as if it were founded in truth.

Against such reports—as vividly existing at this, as at any former period—it seems to the undersigned of vital moment for the Faculty, individually and collectively, and, in the manner that may appear most fitting to them, to exert all their influence. Of the acts of the Board of Trustees [14] the Undersigned has but one opinion, and this was openly expressed on the back of a communication, which—it was recently proposed—should be laid before the students on the contested subject of the Graduation Fee. The acts of the Board have met with the most unqualified approbation of the undersigned; and it is not less gratifying to himself than it is just to that Body, to attest the devotion and disinterestedness, which they appear to him to have exhibited in the cause of the College. But even were their acts other than they seem to the undersigned to be and to have been, the whole history of similar institutions has shewn that it signifies not how wise even may be the administration of Trustees, hostility on the part of the Professors cannot fail to destroy the institution in which it prevails. The power of managing the affairs of the institution is vested in the hands of the Trustees, and all experience shews, that they will not be driven from its exercise by any hostile movement on the part of the Faculty. It is scarcely to be expected, that, in directing the complicated machinery of an extensive institution, the Board of Trustees can *always* act in such a manner as to give entire satisfaction, but where this is not the result, it is with the Faculty to respectfully represent the matter to the Board, and not to allow their objections to become public;—still less the impression to go abroad, that the acts of the Board are objected to by the Professors, and likely to interfere with the prosperity of the school. In like manner, it is scarcely to be expected, that entire *harmony of sentiment* can exist amongst all the members of a Faculty accidentally brought together, owing to their possessing certain intellectual requisites, yet *harmony of action* is as essential as it is practicable, and this is all that can be meant, when the importance of harmony amongst the members of an institution is spoken of. Nothing is more calculated to produce excitement amongst a body of readily impressible and enthusiastic young gentlemen than an assertion resting —or presumed to rest—on any competent authority, that differences of opinion and action exist amongst the Faculty on any measure which materially affects the interests of the Class.

In the preceding statement, the undersigned has restricted himself to topics, which appear to him after full, and mature, and—he is persuaded—unprejudiced consideration, to be exerting a most malign influence on the onward course of a great Institution; and, feeling as he does, he would regard himself culpable, did he not express, in the most frank and explicit manner, his sentiments to his colleagues. So satisfied, indeed, is he, that no exertion of talent, or of zeal, or of both combined, on the part of his colleagues or himself can counteract the bad effects of the influences to which he has referred, unless they are firmly and energetically repressed by every member of the Faculty,

that, should this course not be adopted, the question may, and he feels will, arise on his part—and he will regret, for many reasons, the occasion—whether his continuance in an Institution, so circumstanced, can be productive of advantage either to it or to himself? The undersigned has always thought, that under an energetic Faculty, activated by ordinary prudence and judgment, there is ample space, in the city of Philadelphia, for two noble Institutions and he sees no reason whatever to modify that opinion.

Robley Dunglison, MD.
Professor of the Institutes of
Medicine etc.

To the Faculty of Jefferson Medical College

It may be well to remark, as exhibiting the singular conduct of Dr. McClellan, in disparaging the Jefferson Medical College, of which he had been one of the originators, that when—as I shall hereafter mention— an offer was made to me of the chair of Theory and Practice of Medicine in the Medical department of the Transylvania University, he called in my absence from home, at my house, and urged to my wife and my brother—who, of course, did not consider his urgency complimentary to me, that I should accept the appointment; adding, that the emoluments there were greater than in the Jefferson Medical College, and that the latter *must* go down;—that it could not prosper with such a rascally Board of Trustees. My brother was so struck with these strange remarks from a colleague, that he made a note of them, which he gave to me. This application to me to join the Transylvania College was on the death of Dr. Eberle,[15]—a similar application—as I shall state hereafter had been made to me before his appointment. Dr. Eberle died on the second of February, 1838; and, being desirous of preparing an obituary notice of him for the *American Medical Intelligencer* for March the first of that year, I was anxious to obtain information in regard to him, and especially as to the part he took in establishing the Jefferson Medical College. The Revd Dr. Ashbel Green, the president of the College, ascribed to him the greatest share; and from Dr. Calhoun, I received the following note in answer to one which I addressed him.

Philadelphia Feb. 14, 1838.

Dear Sir:

Dr. Eberle, Dr. Geo. McClellan and Dr. Jacob Green, assisted by the influence of the Revd. Ashbel Green DD, were the prominent founders of the Jefferson Medical College on Tenth St. Philadelphia, in the year 1825, for which they obtained a charter from the legislature of Pennsylvania . . .

Yours truly
S.C.

---

[14] This was the independent Board of Trustees, created when Jefferson Medical College was completely separated from Jefferson College of Canonsburg, Pa.

[15] John Eberle (1788–1838), M.D. 1809, University of Pennsylvania, co-founder with George McClellan of Jefferson Medical College, author of several medical works, editor of the *American Medical Recorder*, left Philadelphia with Daniel Drake for Ohio in 1830 (Samuel X. Radbill, John Eberle, Pennsylvania Dutch Pioneer in American Medical Education, *Bulletin of the Institute of the History of Medicine* 4: 121–136, 1936).

Dr. G. McC. will send you a more detailed account of his life, in which I have given dates, facts etc.

This detailed account, however, I never received. Since Dr. McClellan's death more especially, and before that, when he was left out on the reorganization of the school, it had often been described as a peculiar hardship that he should have been omitted from a school, which he had founded. He was doubtless, one of those who were actively and efficiently engaged in obtaining a charter for it, and in founding the same; yet the testimony of the President of the College, and of Dr. McClellan's intimate friend and undeviating supporter during his lifetime,—Dr. Calhoun—only ascribed to him a share, and by no means the largest share in the work, and such also was the testimony of Dr. Jacob Green, who was associated with him in the undertaking.

REORGANIZATION OF THE FACULTY

But to return to the letter, which I wrote to my colleagues. It was laid before them at a called meeting, held when I was purposely absent, and at which Drs. Pattison, Revere, Green and Calhoun were, I believe, the members of the Faculty present. It was immediately moved, and carried, that it should be transmitted to the Board of Trustees. This was done, without any consultation whatever with me; so that I was not a little surprised on the Thursday following—not having met with any of my Colleagues in the meanwhile —to be told by Mr. Vogdes—the Secretary of the Board whom I saw accidentally, that it had been handed over to him, and would be presented to the Board of Trustees, at a called meeting. I thus became, as it were, a prosecutor before the Board, although most undesignedly, as my communication was addressed to and only intended for, my own colleagues. I have never, however, had occasion to regret that the affair took this direction. It was impossible for the College to go on without a disruption, under the course that was pursued; and there was every reason to believe, that it would not be modified by the party whose conduct was the most objectionable. The Board of Trustees met; a committee was appointed to investigate the affairs of the College, who summoned the different professors before them; and, after a full investigation, recommended, that all the chairs should be again vacated, and that, in this manner, a new organization should be effected. This was adopted, and, in the reorganization, Dr. Geo. McClellan and Dr. Calhoun were left out. Between the two, however, as I always maintained, there was a marked distinction. The former was actively injurious to the Institution by the course he pursued; the latter executed his task faithfully, and to the best of his abilities. My colleagues, however, were exceedingly dissatisfied with him and the Board of Trustees concurred.

Although such was my course of feeling and of action, he ever after appeared to ascribe the main agency to

me; or if not, he at least believed, that I could have prevented it; and on meeting me in the street was disposed to treat me with disrespect, had I taken the least notice of him, which I did not. He was so stolid, it would have been impossible to convince him on any subject on which he had formed an opinion; and as he had been virtually—by his non-re-election—removed by the Board of Trustees, I was willing to make allowance for feelings of irritation, even if directed against one, who had taken no active part of his removal.

Prior to this decision on the part of the Board of Trustees it had always been the expressed determination of Dr. Geo. McClellan to have nothing to do with medical teaching, which—he affirmed—interfered with time that could be devoted more profitably to the exercise of his profession practically. His views now, however, underwent a change, and he actively set about the formation of the Pennsylvania Medical College in connection with a parent institution at Gettysburg. which went into operation the following autumn. To supply the vacancies in the Faculty of Jefferson Medical College, Dr. Pancoast was appointed to the chair of Surgery; and Dr. Huston to that of Materia Medica; Dr. Saml McClellan retaining his old chair of Obstetrics and the Diseases of Women and Children. For a time, the Faculty of the College were uncertain as to the course the latter intended to pursue; and it was a well founded source of objection to him, that he held his position in the Jefferson School, until within a short period prior to the commencement of lectures. and then gave in his resignation; to avail himself of the same position in the Pennsylvania Medical College. The faculty of the Jefferson College were then compelled to make their arrangements without time being permitted to think of another appointment, and, accordingly, it was determined, that there should be but six chairs; and, at the earnest request of my colleagues, and, I need scarcely say at great inconvenience to myself, I consented to embrace, in my course—of five lectures a week, instead of four—as heretofore—Materia Medica and Therapeutics, in addition to Institutes of Medicine and Medical Jurisprudence; and this without any additional fee.

The amalgamation was not as difficult as I thought it would be; and I managed to go through with all the subjects by the end of the session. The plan I pursued was—first of all, to lecture on the functions of an organ; next on its pathological aberrations; passing to the therapeutical methods for bringing back the pathological to the healthy condition, and, lastly, to the remedial agents, that are capable of fulfilling the therapeutical indications. In this manner, I made a connected and consistent whole, and I do not think I ever gave two courses of lectures which were better received or more satisfactory to myself. So impressed, indeed, were my colleagues, Drs. Pattison and Revere, with this arrangement, that, when they went to the Uni-

versity of New York, the union of the branches was
adopted there. In my own case, the union became
necessary, owing to Dr. Huston succeeding Dr. Saml.
McClellan in the chair of Obstetrics for which he considered himself better adapted than for Materia Medica
and Therapeutics.

In the Annual Announcement of the College for
1840–41 the Faculty express their opinion of the value
of this arrangement; and publish resolutions of the
class in favour of the new professors of Surgery and
Obstetrics; and one to the following effect—

Resolved, that we consider the present arrangement of
teaching the Institutes of Medicine and Materia Medica in
combination, to facilitate greatly the progress of the Student, and to present many advantages over the modes in
which they are usually taught.[16]

Owing to the violent disruption of the Faculty, and
the formation of a fresh medical college into which
nearly one half of them had passed, the class of 1839–40
was reduced to 145 students, as many as under all the
circumstances we had, perhaps, any right to expect.
In the "Announcement" of the following year, the
Board of Trustees adopted a letter written by me, in
which they expressed their conviction that "at no time
had the school been more effective, or more in a condition to fulfil the warmest hopes and expectations of its
friends"; and that it then presented advantages unsurpassed by those of other institutions; concluding
with the remark, that they felt satisfied "The institution was admirably calculated for a career of extensive
and enduring prosperity."

PATTISON AND REVERE

In the year following—1840–41—the number rose to
163, but notwithstanding this increase it was manifest,
that Drs. Pattison & Revere, especially the former,
despaired of eminent success in the College, and it was
suspected, from the absence more than once of Dr. P.
at New York, that he had some intention of moving
thither. Such suspicions were the topic of conversation
between my colleague Dr. Green and myself a day or
two before the lamented death of Dr. Green, which
took place in the month of January 1841; so suddenly,
that he was dead before I could reach his house not
two squares distant, although I obeyed the summons
immediately; and the conversation was more immediately suggested by a visit, which Dr. Pattison had, a
short time before, paid to New York. No communication was, however, made to any of the Faculty until
the day of Dr. Green's funeral, on returning from
which, Dr. Pattison told me that all the arrangements
had been made for commencing the medical department

of the University of New York. I may here remark,
that, some years before, Dr. Matthews, the chancellor
of the University of New York had come to Philadelphia for the purpose of enlisting my services, as a
Professor, in the formation of a medical faculty, but,
before meeting with me, he saw my friend Dr. Robert
M. Patterson, who informed him it was useless to see
me, as he knew I would not then leave Philadelphia,
and hence he abandoned the notion altogether, as he
subsequently told me, and as Dr. Patterson informed
me at the time.

I confess, that the announcement of the intended
resignation of my two colleagues, following so closely
on the death of a third, startled me not a little, but I
had no doubt whatever that an energetic corps of instructors could now be got together, and that the College might be destined to attain a greater degree of
prosperity than ever. On the day following the report
of their approaching secession, I so stated to the class,
and succeeded in restoring confidence in them, in regard
to the permanence of the Institution which could not
fail to have been shaken. In the number of the *American Medical Intelligencer* for 1841, I inserted the following notice.

Medical School of the New York University. The public
prints have announced the reformation of the medical
school of the University of New York; the former professors of which—dissatisfied with the acts of the council—
resigned their situations about two years ago, and before
even they had entered upon their duties as teachers. The
list of the present Faculty includes the names of Drs.
Paine and Bedford, of the former faculty, and has the
additional names of Professors Pattison, Revere, Mott, and
Draper. The institution will go into operation in October
next. This arrangement will deprive the Jefferson Medical
College of two of its able and zealous teachers. By a
singular coincidence, the sudden death of Professor Green,
and the appointment of his two colleagues to the new university, occurred almost simultaneously. We shall part
with our friends and colleagues with regret, and with a
sincere desire, that their exertions may be profitable to
themselves, and to the profession, in the new sphere, which
they have selected. For ourselves, there has ever been but
one rule of thought and of conduct—founded on a desire
to see every honourable association of teachers—no matter
where situated—flourish according to their merits, and no
paltry inducement of self-aggrandisement has ever led, or
could lead, us to interfere, by word or deed, with their
useful labours. A spirit of emulation is of course desirable;
but it should be an emulation as to which institution can
best promote the dignity of the profession, and through it,
of the community. To effect this, it is never necessary to
disparage the efforts of others; whilst it is indispensable
to be honourable and energetic ourselves. With these
views, as honestly entertained as they have ever been
openly expressed, we desire to see the new University of
New York proceed onwards, and at an equal pace with the
different elevated medical institutions of the country. The
loss to the Jefferson Medical College by the death of Professor Green, and by the secession of Professors Pattison
& Revere, is sincerely deplored by us; but we should not
express our honest conviction did we say it was irreparable. We are fully satisfied, indeed, that the board of
trustees of the institution will look with a single eye to

---

[16] According to his son, Richard J. Dunglison (1834–1901),
physiology of the various organs was taught concomitantly with
the therapeutic action of drugs upon these organs (*College
and Clinical Record* 1: 10, 1880); a popular *Lecture on Death*
annually terminated the course.

the supplying of the present vacancies, in a manner, that will add to the harmony, stability, reputation, dignity and honour of the school, and should such determination be rigidly carried into effect, they cannot fail to succeed.

REORGANIZATION OF 1841

Similar confident remarks were made by me in my valedictory at the termination of the course. The unpopularity of Dr. Pattison with the profession was an obstacle to the supplying of any vacancy; but when three professorships had to be filled, of which his was one, and objectionable members of the former faculty had left the school, I felt satisfied it would not be difficult to obtain the accession of the most prominent individuals of the profession. I was exceedingly desirous that my old colleague and friend, Dr. Robert M. Patterson, should be the Professor of Chemistry; but his office as Director of the Mint threw—what were considered, at the time,—insuperable obstacles in the way of his being a Professor likewise. The chair of Surgery was proposed to be divided between two Professors, Drs. Randolph and Mütter; but, fortunately I may say, the former who was, at the time, in Europe, and did not design to return soon enough, declined. His name was, however, inserted in the first announcement of the organization of the Faculty, which was made by me in the *Medical Intelligencer* for March the first, 1841, in the following terms.

Jefferson Medical College—Reorganization. It is with the greatest gratification, that we announce the reorganization of this college, with a corps of professors, whose names and professional acquirements are known over every portion of this country. At a late meeting of the Board of Trustees, the following professors were unanimously appointed to their respective branches.

Dr. Dunglison, Institutes of Medicine and Medical Jurisprudence.
Dr. Huston, Materia Medica and General Therapeutics.
Dr. Pancoast, General, Descriptive and Surgical Anatomy.
Dr. J. K. Mitchell, Practice of Medicine.
Dr. Randolph, Practice of Surgery.
Dr. Mütter, Institutes of Surgery.
Dr. Meigs, Obstetrics and Diseases of Women and Children.
Dr. Franklin Bache,[17] Chemistry.

Of these Gentlemen, Drs. Dunglison, Huston and Pancoast are medical officers of the Philadelphia Hospital; Drs. Randolph and Meigs of the Pennsylvania Hospital; and Dr. Mütter is surgeon to the Philadephia Dispensary.[18] With the college, thus fitly organized, the effect must be to render Philadelphia still more the centre of medical education in the Union. The higher the reputation of the schools, and the more harmonious their cooperation in the

[17] Franklin Bache (1792-1864), M.D. 1814, University of Pennsylvania, grandson of Benjamin Franklin, Professor of Chemistry until 1864, co-author of Bache and Wood's *Dispensatory*, was Vice-President of the College of Physicians of Philadelphia and President of the American Philosophical Society.
[18] The first institution of its kind in this country, established in 1786 to serve the poor; merged with the Pennsylvania Hospital in 1923.

great work of medical instruction, the more certainly must this result be accomplished. Unworthy rivalry should be abolished, but an honourable competition, as to which institution can be most extensively useful to the profession and the public, should endure.

With the session of 1841–2, the new Faculty commenced their eminently successful career. The school now consisted again of seven professorships, and I abandoned the department of Materia Medica and Therapeutics, which I had taught for two years, and which was assigned to Dr. Huston, whilst Dr. Pancoast was transferred to the anatomical chair, which he was well adapted for filling, having been up to the time of his appointment to the Jefferson Medical College, a successful private teacher of that department. It was made my pleasing duty to introduce the new Faculty at the opening of the Medical Session, which I did in the following language, first premising the circumstances, that had given occasion to the reorganization of the school.

The revolution of a single year, gentlemen, has wrought sundry changes in the organization of this school. Whilst its exercises were proceeding energetically and in apparent harmony, death deprived us of a respected colleague, who fell from the tree of existence, not as a sere leaf, but full of the freshness and vigour of maturity. In the last lecture, which I had the honour to deliver from this place, I remarked, that the loss of a colleague so zealous and competent; of a teacher so beloved; of a husband and father so affectionate; of a member of society so correct in all respects, was a sad calamity, and was so regarded by us. Almost simultaneously with this affliction, the plans of two of my former colleagues were matured; and on the very day on which the remains of my lamented colleague were consigned to the tomb, it was announced to us, that they had transferred their services elsewhere. The blow came in the midst of the session, and, for a time, it seemed, I doubt not, to the Class, to threaten the stability of the Institution. But this feeling was momentary, and as it speedily passed away, and reason usurped the place of emotion, the conviction, that the loss was not irreparable, impressed every bosom; and the class of the last session dispersed to their homes in every part of this widespread country, satisfied that the institution was destined for a career of prosperity greater than it had ever enjoyed. As for myself and my remaining colleagues, we had no apprehensions, no misgivings. We felt satisfied, that a corps of instructors could be brought together of congenial sentiments, ample qualifications, untiring industry, and whose honorable and upright deportment and elevated position amongst their medical brethren and the community, would entitle them to the loftiest consideration. Such was the honest opinion I expressed on the occasion of taking leave of my class at the termination of the last session. After alluding to the death of an estimable colleague, and the secession of others, and expressing my wishes, that the new institution in a sister State, might succeed as it should merit,—I used the following language, at parting, to my young friends,—"It has ever been a part I will not say of my policy, but of propriety, to entertain the kindliest sentiments towards every respectable institution, that is engaged in the great object of medical instruction. It becomes us not to harbour an ungenerous thought towards any honourable member or association of members of our profession. It is easy for two or more Institutions to move on harmoniously towards one great goal; and to mutually feel and act—not as bitter rivals endeavouring to injure or destroy each other, but as energetic and devoted members of a profession, which one of the first of the Romans conceived to elevate man nearer to the gods than any other avocation,—competing with each other as to which can render the greatest amount of benefit to mankind. It never can be godlike or proper, or even politic, to disparage the useful efforts of others. All experience indeed, proved the impolicy of such a course; and were it even shewn, that the number of students in any institution, could be augmented by it, success would not be desirable, because it would not be honourable. With these feelings on the part of myself, and, I am sure, I may add, of my colleagues, it will be an object of anxiety with us to have such an organization of the school, as will enable us to persist in our career of usefulness, indulging the best feelings towards our honorable competitors. To the Board of Trustees of this Institution belongs the responsible office of effecting this desirable consummation. My estimable colleagues and myself are desirous of seeing established a strong and efficient corps, composed of individuals whose very names will be a presage of success. To aid in this great object, I may add, that no unworthy consideration of self comfort will be indulged on our parts; and that, so far as I am concerned, I am willing to accept the most humble position—if there be any one in the school which is more humble than another. With this spirit of self sacrifice for the good of the whole, success, it appears to me, is inevitble, under the judicious determination of our Board of Trustees, and I feel justified in assuming, from what I know of the sentiments of several of its members, that their desire, in making the new appointments and arrangements, will be—to ensure, as far as may be in their power, the harmony, stability and reputation of the institution, so that hereafter, and in the lapse of years, her numerous alumni may be enabled to exclaim with pride and exultation—'It was from this flourishing and distinguished school that we received the highest honour of the profession!'" Such was the language I held, and such the sanguine anticipations that were cherished by my colleagues and myself, in regard to the reorganization of the school. And how have these anticipations been fulfilled? In a manner, which has afforded the most unbounded satisfaction to us all. Most nobly have the Board of Trustees discharged their duty to the School, to the Profession, and to the country. They were, to adopt their own language, desirous to obtain the services of gentlemen, known throughout the country as practised teachers, and who possess a widespread reputation as writers on different subjects of their profession, whose very names would be a source of confidence. With this view they banished all personal feelings, and, in the appointment of Professors, endeavoured to keep singly in view, that which appeared to them to be most conducive to the continued prosperity of the School. I may be permitted to congratulate the Board of Trustees, my colleagues, the alumni of the Institution, and the profession on the selection of our new professors. Most cordially do I welcome them to these halls; the theatre, I trust, of a long career of usefulness to others, and of distinction to themselves. Of their eminent qualifications, it becomes me not to speak in their presence; and, fortunately, this would be a work of supererogation. The voice of the profession has pronounced in language not to be misunderstood, their peculiar fitness; and the fact of the increased numbers, thus far, who have joined this institution, sufficiently exhibits, that its voice has been appreciated in the proper quarter. Neither is it necessary that I should expatiate on the history of our school. Its annals are short but eventful. We cannot pride ourselves on a long line

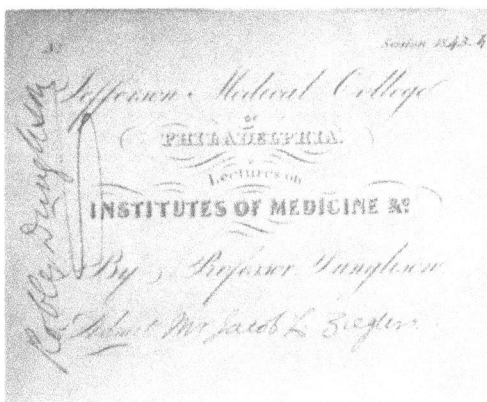

Fig. 13. Student admittance card to Dunglison's lectures. Courtesy of the College of Physicians of Philadelphia.

of distinguished ancestry; for our existence does not date more than seventeen summers, yet, like the country which we inhabit, and which now stands forth amongst the foremost of nations, we appear in the vigour of manhood determined to permit none to excel us in our honourable endeavours to be useful. Possessed of extensive facilities for teaching the science demonstratively, by the numerous anatomical, pathological and obstetrical preparations, and paintings in the museum of the College; of diversified specimens of genuine and spurious articles, plates, and drawings in the Cabinet of Materia Medica, and of a most valuable apparatus in the department of chemistry;—having, withal, through the clinical instruction connected with the Dispensary of the College, the Philadelphia Dispensary, and similar eleemosynary institutions; and still more, by the lectures on clinical medicine and surgery at the Philadelphia Hospital, and the Pennsylvania Hospital, ample opportunities for enabling the student to become familiar with disease, it will be his fault, should his progress be unsatisfactory. Through the liberal arrangement of the Board of Guardians of the Almshouse, the students of this College can be instructed clinically by the Professor of Anatomy, and the Professor of the Institutes, and they enjoy every privilege accorded to those of the sister institution. The Board of Trustees, in the annual announcement of this college on the occasion of its reorganization (written, by the way by myself, but signed and adopted by the President of the Board) have remarked, that with a Faculty thus organized, and bent on harmonious and effective action, they entertain no doubts as to the signal success of the institution; and the Faculty have resolved, that so far as may rest on their exertions, the sanguine anticipation of the Board shall not be disappointed.

PROGRESS OF THE SCHOOL

The success of the reorganization was, indeed, signal; so much so, that in my introductory lecture to the session of 1844–5—or three years afterwards, I was enabled to speak thus: [19]

In concluding these preliminary remarks, it will not, I trust, be considered wanting in good taste, if I say a few words in regard to the present condition of this Institution.

[19] Delivered November 4, 1844.

At the termination of the session of 1840–1, when two of my colleagues and myself were left alone as teachers, and a new organization became necessary—in a valedictory address to my class, I ventured to state, from what I knew of the sentiments of several of the members of the Board of Trustees, that their desire, in making the new appointments and arrangements, would be to ensure, as far as was in their power, the harmony, stability and reputation of the institution; and I foretold, that, in the lapse of years, her numerous alumni would be enabled to exclaim with exultation, "It was from this flourishing and distinguished school, that we received the highest honours of our profession!'" And how has this prediction been fulfilled? In the session of 1840–1, the number of students was 163. In 1841–2, the first session of the new organization, it rose to 209; in 1842–3, to 229; and in 1843–4, to 341,—an unprecedented increase of *one hundred and twelve* in one session. The number of graduates augmented in a still greater ratio; from 47 in the session 1842–3 to 117 in that of 1843–4. (*Published Lecture, p. 12*)

Two years after this, the class had increased so much, that it became necessary to remodel and enlarge the building into the elegant and commodious structure, which now exists; and, since that period, the number of students has not varied essentially, whilst that of the graduates has greatly augmented. In testimony of this I copy the following paragraph from the *Annual Announcement of Jefferson Medical College for 1852–3*, written by myself,—

The increase in the last few years, in the proportion of graduates to the class in attendance, whilst the regulations for graduation have not been in the minutest particular relaxed, is owing to the circumstance, that annually an increasing number of physicians and of students, who have received their education elsewhere, resort to Philadelphia for the purpose of completing their medical education. This is strikingly manifested in the following table of the students and graduates for the last six years:—

Session	Number of the Class	Number of the Graduates
1846–7	493	181
1847–8	480	178
1848–9	477	188
1849–50	516	211
1850–1	504	227
1851–2	506	228
1852–3	556	223
1853–4	627	270
1854–5	565	257
1855–6	510	215
1856–7	488	212.

PATTISON AND REVERE

The resignations of Drs. Pattison and Revere having been made in the middle of the Session, their position was not exactly agreeable to them. The young gentlemen did not sympathize with them in their withdrawal from the school; and although myself and colleagues used every exertion to prevent any annoyance to them, they were occasionally ruffled;—generally, however, more owing to morbid sensibility on their part than to any action of disrespect on the part of the students.

In respect to one, Mr. Neale,[20] the son of a presbyterian minister, their conduct was certainly indefensible. He had spoken disparagingly of Dr. Revere in the dissecting room, and the fact had been communicated to him. He immediately sent for Neale, and as Dean of the Faculty reprimanded him, which ought to have ended the matter. Neale, before lecture one day, came to my private room to inform me of the transaction, and to complain of the harsh conduct of Dr. Revere to him. I told him to take no notice of it; that we were very desirous, that nothing but the greatest respect should be paid to our seceding colleagues, and that he had done wrong in expressing his feelings. The fact of Mr. Neale's being seen entering my room was, however, reported to Dr. Revere, and he at once concluded, that Neale's course was approved of by me. Accidentally, I entered the office of Dr. Huston on one occasion, when I found Dr. Revere and himself engaged in conversion on this very subject. I told Dr. Revere, that I had a better opinion of him than he had of me, and would not have believed him capable of the conduct of which he suspected me. He regretted, I believe, his suspicions, and I took no farther notice of the affair. When Mr. Neale however, presented himself for graduation at the end of the session, Dr. Pattison refused to examine him although he had the usual note from the Dean—Dr. Revere—that the young man was entitled to be examined,—the ground assigned being,—that he had not paid for his private dissecting ticket. —Dr. Revere then declined to ballot for him because he had not been examined by Dr. Pattison,—thus giving to any professor the power, by his own whim and caprice, to refuse the degree to anyone against whom he had any dislike, founded on real or imaginary causes. My colleagues and myself determined, that no such act of injustice should be committed. We postponed the balloting until within an hour of the time of the commencement, and then recommended him—Drs. Huston, Pancoast and myself voting in the affirmative—to the Board of Trustees to receive his degree. On proceeding to the Musical Fund hall, I was surprised to find Dr. Pattison addressing the Board of Trustees on this very case. As soon as he had finished, I explained the matter to them, and our grounds for action, and Dr. Huston followed, after which without any deliberation, they unanimously sanctioned the view we had taken. I confess I have never been able to find the slightest shadow of excuse for this attempted gross act of injustice, which gave me a worse opinion of my friends than I had ever had. A similar case occurred, however, afterwards, when, in the first announcement of the Faculty of the University of New York, to serve their own purpose, they sanctioned the totally unfounded assertion, that the greater part of the Museum of Jefferson Medical College belonged to them,—the matter of fact

---

[20] Benjamin T. Neal, Jr., of Pennsylvania, graduated in the class of 1841.

being, that when their share was allotted them, it was so insignificant, that the greater part—I was told—was sent to the apothecary's establishment of Mr. Geo. W. Carpenter of Market St. for sale!

These annoyances, originating with them more than with others, soured them, however, with their colleagues who remained connected with the Jefferson School; and when they left Philadelphia finally, to proceed to New York, Dr. Pattison—although I was then in affliction—neither called nor left his card at my house. It was gratifying to me however, to find subsequently, that both gentlemen were satisfied, that, as regarded my conduct, they had been in error, if they had suspected anything improper on my part towards them or others. In the year 1842, I had occasion to write to Dr. R. on sending him a copy of the new Pharmacopoeia of the United States to which he was entitled as a member of the Faculty, which, in the year 1840, had sent delegates to Washington with the view of framing a new edition of the same. To this I received the following answer.

New York, Oct. 24, 1842.
Dear Sir,

Your kind letter, with the copy of the Pharmacopoeia, was duly received a few days since, I having but recently returned from the country, where I had been for some time with Mrs. Revere, on account of her health. Please to accept my warm acknowledgments for the book, and the kind spirit of your letter. I am happy to inform you, that Mrs. Revere's health is, I think, now permanently restored, though, during the summer, her situation was most perilous. Please to present our kindest respects to Mrs. Dunglison, who we can never think of, but with respect and affection. Though shadows have passed between us, I perceive now they were but shadows,—and believe me to be

very truly, and respectfully, yours
J. Revere

He had previously, however, written to me the following letter, to which I gave a prompt and favorable answer in the direction of his wishes.

New York June 29, 1842.
My dear Sir,

Perhaps I cannot offer you a higher proof of my estimation of your professional character and my confidence in you as a man, than in the following request. An election is about to take place of Physician to the New York City Hospital,[21] for which I am a candidate; and which I must in all frankness state I am particularly anxious to obtain from my connection with the University of New York. I believe, that a favourable opinion of my qualifications for this situation from you would be very useful to me. If then you can express such an opinion, and will do so, you will confer a very particular favour. Mrs. Revere is at present very ill. She and Helen ask to be most kindly remembered to Mrs. Dunglison. An early answer enclosed to me, will be desirable as the election takes place in a few days. I am respectfully and truly yours,

John Revere.

[21] Probably the New York Hospital, open to all medical students for a fee of six dollars. Bellevue was officially connected with the University of New York later. I find no evidence that Dr. Revere was ever on the staff of either.

Between these dates and his death, which occurred in the year 1847, I received letters breathing the same kind of amicable spirit.

NEW YORK UNIVERSITY SEEKS A
MEDICAL PROFESSOR

The like gratifying return of proper feeling was observed in Dr. Pattison, to whom, as to his colleague, my own conduct had been uniform, and uniformly kind and considerate. Having occasion to forward him his share of monies, which had come into my hands, since Dr. Revere and himself had left the College, I took occasion, at the request of Dr. Clymer,[22] who was a candidate for the chair made vacant by the death of Dr. Revere, to speak of his qualifications, and, in reply, I received the following letter, which, with the others given hereafter, exhibit his feelings towards me to be of the kindest and most complimentary character, and, at the same time, the correctness of his views in regard to the best mode of filling up such vacancies, when they occur in the higher medical schools.

84 University Place,
New York June 2, 1847.
My dear Sir:

I have to acknowledge your favour enclosing a check etc. . . . I have, at the same time, to thank you for your kindness, in favouring me, as a member of the Faculty of the University of New York, with your views, as to the qualifications of Dr. Clymer for filling the vacant chair of the Theory and Practice of Physic in our school. Coming, as they do, from a gentleman of great distinction in the Profession, they cannot fail to have their influence with my colleagues and myself. The duty we have to perform, in suppling the loss we have sustained in the death of our late friend is one of great importance, and as the choice is fortunately left entirely to the Faculty, their sincere desire is, to perform it faithfully, and elect, without reference to personal considerations, the candidate whose talents and reputation as a teacher of the important department of the Theory and Practice will best secure for the service of their Institution the man best qualified to discharge the duties of the professorship. A number of applications have already been received, but as we are bound to advertise for three months before we nominate a Professor, we have determined to keep our minds undecided, until we shall have ascertained who all the candidates may be. So soon as this term shall have expired, we shall then examine most carefully the claims of all, and make a selection.

Believe me, my dear Sir,
Yours most faithfully,
Granville S. Pattison

Robley Dunglison, M.D.

About ten days afterwards, I received the following letter from him, which was marked "Private and Confidential."

[22] Meredith Clymer (1817–1902), M.D. 1837, University of Pennsylvania, pioneer American neurologist, taught at the Philadelphia Medical Institute (1843 and 1849), Franklin Medical College of Philadelphia (1846), and succeeded in obtaining the chair at the University of the City of New York in 1851.

My dear Sir,

Feeling very anxious to make a wise selection to the vacant chair of the Theory and Practice of Physic, I am induced to write you to ascertain whether you have any personal knowledge of Dr. Bartlett,[23] who, at present fills that situation in the University of Transylvania. This gentleman has a good reputation, but having no acquaintance with him myself, and all I know being derived from hearsay, I have thought I might possibly obtain some information from you in reference to him, on which more reliance could be placed. You will, therefore, confer a favour on me if you can supply me with any information on the following points. *First.* What is his character socially, and as a gentleman? *Secondly.* What is your estimate of his talents and professional acquirements? *Thirdly.* What is his reputation as a teacher? In one word, what are your views as to his fitness for filling the vacant chair. There is one question more I would ask you. Were you in the situation of an elector, is Dr. Bartlett the man you would select to the office? If he is not, who, of those who *may be obtained,* would you prefer? You will observe, that the election is in the hands of the Faculty every member of which is most desirous to select the best man that can be obtained, without reference to personal or local considerations. I am aware, that there is some delicacy in your answering the inquiries I have made, but you may depend on whatever you write, in reference to this matter, being strictly confidential. Your letter in reference to Dr. Clymer will have its weight with the Faculty, but I must tell you, in confidence, that I individually, think we ought to have a higher man to fill the Chair. In a pecuniary point of view it is probably as valuable as any professorship in the country, and as an introduction to by far the most valuable field for the Practice of Medicine, it is decidedly the best in the United States. *There is one man in the Profession who, could he be obtained, would, at once, place our Institution far above all others; but unfortunately that man is not to be obtained. I mean Dr. Robley Dunglison. Most sincerely do I wish it were otherwise.* You will pardon me for the trouble I give you, and believe me,

my dear Sir, Yours very faithfully
Granville S. Pattison.

PS. Do you suppose Patterson (meaning Dr. Robert M. Patterson) will be elected to fill that chair of Chemistry. (Dr. Patterson was not a candidate).

About two months afterwards, I received the following letter from him, also marked *"private and confidential."*

New Port, R.I. Aug. 1, 1847
My dear Sir,

I was much obliged to you for the kind information contained in your late letters, and I would before this time, have answered them, and thanked you; but I wished before writing, to be enabled to inform you of the choice we had determined to make to supply the vacancy caused by the death of our friend Dr. Revere. When this melancholy event took place, my mind was, at once, directed to the consideration of a suitable person to fill the chair. *As*

I told you, in a former letter, you were of all men living the one I would most desire to have associated with me as a Colleague; but as all hope of this was, for me, unfortunately out of the question, the next most desirable person, which occurred to my mind, was Dr. S. H. Dickson of S.C.[24] In communicating with my colleagues I found the most delightful harmony. Although there were candidates, who had very strong personal claims of friendship to urge on some of the members of the Faculty, they were all disregarded. The single object of every Professor was the interest of the University, and an earnest desire to select that candidate, who, in the opinion of the Faculty, would best promote that object. Where this feeling was unanimous in the electing body, there was no difficulty to be encountered, and although we cannot let our determination be known, until the term specified in the advertisements, has expired, we have finally and absolutely determined to elect Dr. Dickson, and, with the view of enabling him to make his arrangements, we have, in confidence, communicated this to him. Our choice is, therefore, now made, and I hope it is one which you will approve of. Until the election has been actually made, I should, of course, wish nothing to be said about it. The experience derived from the task of filling this chair, has strongly impressed on my mind two things—*First.* The importance of the electing power being placed in the hands of the Medical Faculty, and *Secondly.* The difficulty of finding suitable candidates to fill the chairs of a medical school. Had the electing power been in the hands of the Council, I am satisfied, from what I know of family and personal influences, that the candidate, of all others the most unfit to fulfill the duties of the office, would have, in all probability been the successful one. And although the list of applicants we have had is beyond all calculation, still, numerous as it is, how few are there, who are desirable. Amongst the candidates we have some of the first names in the country, but how many things require to be taken into account, in selecting a professor, a man who is to be associated with you, in the most intimate relations, for the rest of your life. In Dr. Dickson, I believe, we have secured everything most desirable— a gentleman, a scholar, and an able and eloquent teacher. I must say, when I heard that Henry (Professor Joseph Henry)[25] was not to be elected to supply Hare's [25a] chair I regretted on your account. Although he is, without all question, a man of high talents, still, I would consider him a very unfit person to occupy the Chair of Chemistry; and, although I have no doubt that your school will continue to hold its precedence, still I must say, I shall not regret to see a very unsuitable person elected as Professor of Chemistry in the University. The prosperity or the adversity

[23] Elisha Bartlett (1804–1855), M.D. 1826, Brown University, peripatetic professor at Berkshire Medical Institute (1832), Dartmouth (1839), Transylvania (1841), Maryland (1844), Vermont (1844), Louisville (1849), University of the City of New York (1850) and College of Physicians and Surgeons of New York (1852), wrote well on *Fevers* and on *The Philosophy of Medicine.*

[24] Samuel Henry Dickson (1798–1872), M.D. 1819, University of Pennsylvania, founder of the Medical College of South Carolina in 1824, accepted the professorship at the New York University in 1847 but returned to Charleston in 1850. In 1852 he succeeded his deceased friend, John K. Mitchell, at Jefferson Medical College when favored by Dunglison in preference to Austin Flint. He was a fellow of the College of Physicians of Philadelphia and a member of the American Philosophical Society (Samuel X. Radbill, Samuel Henry Dickson, *Annals of Medical History,* 3rd ser., 4 : 382, 1942).

[25] Joseph Henry (1797–1878), physicist, whose basic research produced the electro-magnetic telegraph and systems of electric lighting and power, was Secretary of the Smithsonian Institute (1846–1882).

[25a] Robert Hare (1781–1858), Honorary M.D. 1816, Harvard, brewer, chemist, inventor of the oxy-hydrogen blow pipe, was Professor of Chemistry at the University of Pennsylvania (1819–1847).

of the University of Pennsylvania can have little or no effect on our interest; but they can affect the interests of Jefferson Medical College; and this, as your school, is the one in the prosperity and success of which I feel an interest. I am now at Newport, where I shall remain until I am obliged to go to New York to make the election. I trust you will not pass through the city without seeing me.

Yours very faithfully,
Granville S. Pattison.

This was the last letter I received from him. Dr. Dickson's appointment was an excellent one. He did not, however, remain more than three sessions, after which he returned to Charleston, Dr. Geddings, who had been appointed to his chair, with great liberality, resigning it in his favour. I had an opportunity of seeing Dr. Dickson on his way to New York, and of expressing to him my opinion of Dr. Pattison. I informed him that it was not from many of the physicians in Philadelphia that he must expect to get a proper character of him. The difficulties which had, at one time, occurred between him and Dr. Chapman had tinctured his more immediate friends; and, by tradition, the ill feeling had extended to others who did not even know him;—that he would find him a zealous and popular colleague; a little too urgent, when a measure, perhaps of doubtful expediency, was eagerly embraced by him; but generally willing to yield to the voice of his colleagues if calmly and firmly expressed; full of noble impulses, so that when any act of liberality or charity was proposed, he would not be the last to favour it. In this respect, indeed, he contrasted favourably with Dr. Revere, who was far from being generous; and, although courteous in manner, was suspicious as to motives, and narrow-minded, I always thought. A great fault with Dr. Pattison was, his unscrupulousness if a favourite object had to be carried; and the only way to manage him was to resist him, and if he found the majority were against him he had to yield. Often, indeed, I have known him blamed for measures, which could have been defeated by his colleagues, but who had not moral courage to resist his importunities. All these matters I stated to Dr. Dickson, in the presence of my colleague Dr. Mitchell, at whose house he was staying; and I had the pleasure of learning afterwards, that he had written to say that my portraiture of Dr. Pattison was correct. After Dr. Dickson's withdrawal from the University of New York, Dr. Bartlett, spoken of in a letter to me from Dr. Pattison (p. 96), was appointed to his Chair, who, after one year's service, passed over to the College of Physicians and Surgeons, to fill the Chair of Materia Medica, vacated by the death of Dr. John B. Beck![26] In his place, Dr. Clymer, to whom I had drawn the attention of the Faculty in 1847

[26] John Brodhead Beck (1794–1851), M.D. 1817, College of Physicians and Surgeons of New York, a founder of the New York Academy of Medicine and of the *New York Medical and Physical Journal,* joined the teaching staff of his alma mater in 1826.

(p. 95); and again in 1850 whilst he was absent in Paris, having, before his departure, asked me to watch over his interests in this direction—received the appointment. Prior to this, however, I received the following letter from Dr. Martyn Paine, the Professor of Materia Medica in the Institution:

New York July, 1850.
Prof. Dunglison,
Dear Sir,

I take the liberty of consulting you confidentially about Dr. Clymer, who, in respect to qualifications for teaching medicine, would be an acceptable candidate to myself for our vacant chair of Institutes and Practice of Medicine. But I have heard it alleged, that he is rather impetuous, and might not harmonize with us in the school. I would be glad to know your opinion upon that subject, if you have no objection to give it to me confidentially. Another inquiry relates to his pecuniary means. The Faculty are the proprietors of the Medical Edifice in which they deliver their lectures, and each has invested in it five thousand dollars. Can Dr. Clymer probably advance that amount, should he be so disposed? Is he now in Philadelphia?

I remain very respectfully and truly yours
Martyn Paine.

DEATH OF DR. PATTISON

From Dr. Clymer I was grieved to receive the following letter announcing to me the death of Dr. Pattison, and confirming the favorable views I had entertained of many points of his character.

University Medical College,
14th St. New York, 12 Nov. 1851.
Dear Doctor Dunglison,

As an old associate of Dr. Pattison, I hasten to inform you of his death, which took place this morning. Before leaving for Europe, I attended him in a severe attack of what I supposed to be hepatic colic. I left him quite ill when I sailed, and on my return, found him very much changed. Everyday he seemed to fall off, and, last Wednesday, I was sent for, and found him in great agony, from a similar attack to that of May last. On Monday morning, he began to sink, and died early this morning. On examination, we found a number of gallstones in the duodenum and common duct, one of which, in the latter, had caused ulceration, and had allowed bile to escape into the peritoneal cavity. What was most singular, that although a good deal of recent bile was found there, there was no recent peritonitis.[27] I feel, that as an old colleague of the doctor, you will probably take some interest in these details, and am sure, that, with myself, you will regret the loss, that we have sustained. My intercourse with him, since my connection with the school, has been of the most agreeable kind. I have always found him kind, courteous and conciliatory. Under severe suffering, he was most patient, eventempered, and confiding. He will leave his wife quite comfortable from a life insurance. Though my only object, when I sat down, was simply to announce to you the death of an old associate, still I cannot refrain from mentioning to you what is now uppermost in my mind—who should succeed him? If I were in Philadelphia, I should have a

[27] While the death here described was probably due to a ruptured gall bladder, no thought seems to have been given to pancreatic abscess or myocardial infarct.

talk with you on that subject, before making up my own mind. Can you assist me? Of course, our communications will be strictly confidential.

> Believe me, my dear Dr. Dunglison,
> Much and faithfully yours,
> Meredith Clymer.

This letter reached me in the midst of our session of lectures, and I had the melancholy satisfaction to announce to my class the death of Dr. Pattison, followed by an honest but feeble eulogy [28] on the many excellent traits of character, which, according to my own experience, he possessed, whilst I did not wholly pass over the faults with which he was charged. And now, when I compare him with many of those, who were uncompromising in their hostility toward him, in Philadelphia, I cannot help placing him far above them, not only in intellectual endowments, but in an expansive liberality, which rendered him a most agreeable member of society, but, at the same time in his pecuniary relations was often most injurious to himself.

TRANSYLVANIA UNIVERSITY

Early in the year 1837, in consequence of dissensions which unfortunately arose in relation to the removal of the Medical School of the Transylvania University from Lexington to Louisville, it was deemed expedient, by the Board of Trustees of the University, to remove all the professors, and to go into a fresh election. In the reorganization, it was thought advisable to leave Professor Caldwell out. Professor J. C. Cross [29] was sent as an agent of the school to enlist my services, if practicable. For this purpose, he visited Philadelphia. I asked him if he had any objection to my friend Dr. Robert M. Patterson, then Director of the Mint, being present at our conversation; to which he said no! These two gentlemen, accordingly, dined with me, and after dinner Dr. C. offered to guarantee to me, in the name of the Medical Faculty of the University of Transsylvania with Lexington endorsers—men of wealth and reputation there—the sum of *Five thousand dollars* a session, for three sessions. On declining the offer, he asked me what additional sum I would require. I told him the offer was ample, but I did not think I should be acting honourably to my colleagues of the Jefferson Medical College, if I were to withdraw from them, at the expiration of a single year. On my declining, Dr. Eberle, then of Cincinnati, was appointed. He did not live, however, to officiate one whole session, dying

on the second of February, 1838. On the following day, Dr. Cross wrote me a letter, of which this is a copy; evidently, as the second letter shews, intended as a *feeler;* the main subject being brought in as it were incidentally.

> Lexington, Feb. 3, 1838.

My dear Sir,

I have not written to you since I had the pleasure of seeing you in Philadelphia, but did trust, that I would have been favoured with a line from you before this, inasmuch as the stirring events of your city would much interest me, while we have little here, in our comparatively quiet place, with which I would take the liberty of troubling you. Indeed, I much regret, that the occasion of my writing at this time is one, that must be a source of pain to you, in common with all the friends of medical science, in the United States. On yesterday, I was so unfortunate as to lose one of my most esteemed colleagues. Professor Eberle is no more. He died on yesterday, about 3 o'clock pm, of a lingering illness, of nearly two months duration, which of course, deprived us of his services the whole of the present session, with the exception of about five weeks at its commencement. I have been appointed by my colleagues, to deliver a eulogium on his scientific and intellectual character sometime before the commencement.

Flattering myself, that you take some interest in my success in teaching a department for which you are so peculiarly fitted, and in which Dr. Caldwell stood so conspicuous, permit me to say, that when it was ascertained, that Dr. Eberle would not be able to lecture again this winter, the class determined to request Professor Dudley or myself to discharge the duties of his chair for this winter, and although I publicly announced to the Class, that I did not wish to be selected, and earnestly solicited them not to fix their choice on me, I was still ballotted for, and was beaten only 19 votes. When the great popularity of Prof. Dudley is considered, this must be regarded an achievement, for a young lecturer delivering his first course on the Institutes. So soon as it was understood, that the chair of Theory and Practice was vacant, a memorial was immediately got up, and signed by nearly all the class, to have me transferred to it. So you see, if nothing happens, I am destined to be a rather popular teacher. I should have remarked, that though not elected to the place, I have half of its duties to perform—lecturing nine times a week, six times in my own department, and three times on the Practice of Physic. But I have boasted enough, and I trust you will pardon it. I do not intend to permit the transfer of myself, but to ask you to suggest someone to us, that you may believe would be a suitable successor to Dr. Eberle. What do you think of Griffith [30] of Va.? Will you speak to Gerhard [31] on the subject? We consider our school established, and able to defy any opposition, that can be made in the Western country, and yielding a certain salary of between 3000 and 4000 dollars, must be able to command

---

[28] Delivered November 14, 1851 (E. R. Squibb, *Journal,* p. 237, 1930).

[29] James Conquest Cross (1798–1855), M.D. 1821, Transylvania Medical College, successor for a while to Daniel Drake, practiced in Alabama (1827–1835), joined Eberle on the faculty of the Medical College of Ohio, but the two soon moved to Transylvania. Cross, in the public press and in court, an inveterate foe of Charles Caldwell and Benjamin W. Dudley, was a lifelong friend of Henry Clay, who was kept busy getting him out of trouble, and Dunglison apparently felt need of a witness to any discussions with him.

[30] Robert Eglesfeld Griffith (1798–1850), M.D. 1820, University of Pennsylvania, succeeded Robley Dunglison at the University of Maryland in 1836, then succeeded A. T. Magill at the University of Virginia in 1837 and retired two years later because of ill health. He was a member of the American Philosophical Society.

[31] William Wood Gerhard (1809–1872), M.D. 1832, University of Pennsylvania, was seriously ill with typhoid fever in 1837, having established the pathological distinction of this disease from typhus the year before. He was a fellow of the College of Physicians of Philadelphia and a member of the American Philosophical Society.

the best unengaged talents in the country. I should be pleased to hear from you frequently.

Yours, with sentiments of the highest respect, and feelings of the warmest friendship.

James C. Cross.

A month afterwards, I received the following letter from him.

Lexington, Mar. 3, 1838.

My dear Sir,

Your kind letter has been received, and cannot refrain expressing my gratification at the recollection you seem to cherish of our brief acquaintance. The feelings I have personally, and by letter, expressed to you renders it useless, that I should now say, that I am flattered at the request you make, that I should avail myself of your services in any way that I may consider them advantageous to me. I appreciate the candour, with which you have expressed yourself in regard to the gentlemen whose names I mentioned to you. Be pleased, now, to answer my enquiries about another individual whose services I would like very much to secure and as promptly, if in your power, as you responded to my former letter. The distinguished gentleman to whom I allude is Prof. Robley Dunglison. You are perhaps ready to say, that I should consider your frank and undisguised expressions, when I was in Philadelphia, sufficient to deter me from alluding again to yourself in connection with Transylvania. If such should be your view of the subject, ascribe my pertinacity to my respect for you, and not to your equivocation. Our situation now is different from what it was when I was in Philadelphia. Then, the fate of the school was uncertain—now, we entertain no fears on that subject. Your school, judging from the number of its pupils, as compared with the number in the University, does not seem to me as promising as it was then. You stated to me, that had you been in the Jefferson Institution two years, you would feel less reluctance to leave it. That time you have now been there. With almost any teacher, we shall have as many students next winter as you had last, and with your assistance I doubt not that we shall have at least 300. The high character of our faculty would then be such as to put down all opposition, and we should be able to calculate with much confidence on an annual increase in the size of our class. The only ground, upon which our rivals could injure us, was the supposed insufficiency of our anatomical advantages. The public mind, as you have seen, has been disabused on that subject. Our salary is, at present, nearly equal to your own, and when you reflect, that we can live here for at least half of what it will cost you in Philadelphia, you must come to the conclusion, that it is greater. Look at this matter thoroughly, and tell me as speedily as possible, whether or not you can be induced to leave Philadelphia and on what, if any, terms you can be persuaded to come to Lexington. Excuse me for inclosing to you, instead of your publisher, ten dollars for your Journal. From our class, you will receive, I should think, at least 50 subscribers to your Journal. I have had so many occasions of quoting Stokes,[32] and, indeed, I formally recommended to my class the taking of it, that I think I convinced them of the necessity of having it.

Yours most sincerely
James C. Cross.

The proposition scarcely gave me a thought, and I returned an answer, respectfully declining to be a candidate. Immediately after this, I received the following letter from Dr. Thomas D. Mitchell,[33] the Professor of Chemistry, which sufficiently indicated the want of harmony amongst the Professors, and would have removed my doubts had I been possessed of any.

Lexington Mar. 5, 1838.

"Confidential"
Prof. Dunglison, Sir,

I have just learned, accidentally, that a letter has been addressed to you, touching our vacant chair of Theory and Practice, in the hope of detaching you from Philadelphia. This movement is without a faculty sanction in any form; and is, I believe, prompted by a desire to keep me from that chair. By request of the Trustees, I supplied the vacancy to the extent of 22 lectures, and am now seeking a transfer, having determined to quit my present chair for the Institutes, Theory and Practice, or Materia Medica, whenever a favourable opportunity shall present. You are the best judge of your own interests. Nevertheless, if you conclude to look West, I will not object, but would rather see you in the chair than any other man I can now think of. At the same time, governed by my settled purpose, I should like to succeed you in Jefferson, should you quit that school—I doubt not you will excuse the freedom with which, as a stranger, I take the liberty to address you.

I am very respectfully yours
Thos. D. Mitchell

CHARLES CALDWELL

It was soon reported abroad, that the Faculty of the Medical Department of Transylvania University were desirous of obtaining my services. Dr. Geo. McClellan was written to, on the subject, and it was on this occasion, that he held the singular dialogue with my wife and brother, which I have mentioned elsewhere (p. 89). But one of the strangest communications I received on the occasion was from that erratic genius, Dr. Caldwell, who—it will be recollected—had been left out of the Transylvania Medical School, a short time before, and been received into the Louisville School, which now became the object of his laudation, as the Transylvania School had been previously. His activity in favour of the Louisville Medical Institute, now subjected him to attacks from his old friends in Lexington. He proposed to visit the Atlantic states to obtain what was deemed necessary for it in its inception, and was appointed to even cross the Atlantic for the same purpose. His peculiarities, which were great, were ever and anon hit off in the *Lexington Observer*. One instance I recollect to this purpose.—

Where is the Louisville Medical Institute? When last seen, it was in the streets of Philadelphia, walking about with a large cane in its hand, saying to everybody, "I tell you, Sir"

---

[32] William Stokes, *Treatise on the Diagnosis and Treatment of Diseases of the Chest, Dublin,* 1837; reprinted by Dunglison in *The Medical Intelligencer* (see Appendix, p. 197).

[33] Thomas Duché Mitchell (1791–1865), M.D. 1812, University of Pennsylvania, in private practice at Frankford (Philadelphia), Pa. (1822–1831), taught at Miami Medical College, Medical College of Ohio, Transylvania University, Philadelphia College of Medicine and Jefferson Medical College.

—a favorite expression with the Doctor. In a letter, which I had from Dr. Carpenter, dated, Bristol, June 17, 1842, he remarks

Dr Lee (Dr. Charles A. Lee,[34] of New York) mentions, that Dr. Caldwell is doing great things in Animal Magnetism down in the West. I met that worthy Professor, at Dr. Prichard's [35] table, and was excessively amused at his dogmatism and self reliance, and at the cool way in which he *set down* everybody who presumed to differ from him, even Dr. P. himself.

The following is a copy of Dr. Caldwell's letter, with his own infralineations.

Louisville Feb. 12, 1838.

Dear Sir,

Eberle is dead; and the report is current, that you are to be invited to become his successor. Though I have not even a suspicion, that you will spend on the invitation a *second* thought, nor even a *first* one of a serious nature, still, as no doubt a highly emblazoned account of the prosperity and prospects of Transylvania will be presented to you, I think it but fair, that you should receive so much of the *contra* side of the ledger, as may enable you to strike a balance,—to "look on this picture and on this,"—and thus, see things as they are. Under this impression permit me to lay before you hastily and briefly my view of the matter, or rather to give you facts as they are. And, to remove all suspicion should you entertain any of the correctness of my statement, it contains no secrets. Communicate its contents to whomsoever you please. Lexington is a small healthy inland town; population five to six thousand, and decreasing; no hospital or infirmary, and never can have either of any value; country around very healthy, and inexorably prejudiced against *resurrection business;* living very high, and market very indifferent; school declining every year in numbers; opinion in the west and south becoming universal, that it must go down, because it is improperly located; because, in fact, there exists no reason why there should be a medical school in Lexington, but because there *has been and is one* there. Were the school to be now *first* established, Lexington would be no more fixed on, nor even *thought* of as its seat, than the smallest country town, or any other hamlet in the West. Louisville's present population about 30,000, and rapidly increasing. New Albany and Jeffersonville containing near 15,000, separated from Louisville only by the river; a floating population amounting to at least 20,000 per annum more; an excellent hospital, and an infirmary already; and an United States' hospital about to be erected; the site of it already purchased; anatomical material sufficient for *two* schools; this proved by the experience of the present session; public opinion universal, that, in Louisville, is to be the Great School of the West; a school, which will inevitably swallow up Transylvania and public opinion in this country is *omnipotent.* Our Faculty, to say the least, equal to that of Transylvania; in all points *but one* decidedly *superior:* Dudley's [36] *reputation* in surgery as yet above

that of Flint; [37] but his talents, education and professional science greatly inferior. In two years, Flint will be the *better* teacher (His views of Flint afterwards experienced a great change). Dudley not comparable to Cobb.[37a] In truth, Cobb is one of the ablest and most pleasing lecturers on anatomy I have ever seen, and our anatomical dissector and demonstrator greatly superior to Bush,[38] the adjunct ot Dudley. All the foregoing I would declare on oath, to the best of my knowledge. Our Faculty was not made up and our advertisement of our Lectures sent out, until *September* or the last of August, when the pupils had already made their choice and arrangements, as to the Schools they would attend; and even then our advertisement was very limitedly circulated. In fact, we might almost as *well* have had no announcement out. Yet, *under all* these disadvantages, our class is 80; my class of 20 on Medical Jurisprudence excluded; in all 100; the largest *first* class any medical school on record has ever had. The cornerstone of our new splendid college edifice will be laid with appropriate ceremonies on the 22nd inst. and be in readiness for our next course of lectures. Our Faculty will be full and our announcement *general.* The issue will then be seen. Though I do not defy fate, neither do I dread it, our resolution is *inflexible,* and our hopes high, and our strength will be exerted to the last pennyweight.

Very truly yours
Ch. Caldwell.

PS. It is thought best, that I should labour, for the present year, in the home department of our School. Instead of going to Europe myself, therefore, to purchase a library, apparatus, etc. I have transferred the agency to my colleague Professor Flint. He will pass through Philadelphia on his way. I shall introduce him to you by letter, and will receive as a particular favour your giving him a few letters to Europe.

This well expressed letter was not a little amusing to me, who had heard from Dr. Caldwell most glowing descriptions of the Transylvania Medical School, whilst he was one of its professors, and, when I saw him subsequently, I asked him how he reconciled these descriptions with the views which he then entertained,—in regard to the comparative advantages of a village over a large city in many respects. He replied, that he hoped I did not regard his praises of the Transylvania School as being *his* sentiments; he was only giving utterances to the views of his colleagues!

[34] Charles Alfred Lee (1801–1872), an eminent peacemaker like Dunglison, taught at eight different medical schools and with his colleagues was the first to open the doors of a regular medical college to Miss Elizabeth Blackwell at Geneva, New York (James Bryan, Sketch of Charles A. Lee, *Boston Medical and Surgical Journal,* n.s., 9: 147, 1872).

[35] James Cowles Prichard (1785–1848), M.D. 1808, Edinburgh, anthropologist and psychiatrist noted for his book on the *Natural History of Man.*

[36] Benjamin Winslow Dudley (1785–1870), M.D. 1806, University of Pennsylvania, Professor of Surgery in Transylvania

University, 1815–1850, expert lithotomist, was a very popular teacher (David A. Tucker, Jr. *Transylvania Journal of Medicine,* p. 1, Cincinnati, April, 1936).

[37] Joshua Barker Flint (1801–1863), M.D. 1825, Harvard, joined Caldwell in the Louisville Medical Institute as Professor of Surgery in 1837, was Dean of the Kentucky School of Medicine from 1852–1854, became a great surgeon (D. P. Hall, Joshua Barker Flint, First Professor of Surgery, University of Louisville, *Kentucky Medical Journal* 34: 448–496, 1936).

[37a] Jedediah Cobb (1800–1860) M.D. Bowdoin, 1823, Professor, Theory and Practice of Medicine, Medical College of Ohio (1824 and 1852), of Anatomy, Louisville University (1837–1852).

[38] James Mills Bush (1808–1875), M.D. 1833, Transylvania University, was made Adjunct Professor of Anatomy in 1837 and full professor, 1844–1857 (Robert Peter, Doctor James Mills Bush, *Kentucky Medical Journal,* Historical Number, 15: 76, 1917).

Up to about this period, Dr. Caldwell and myself had been on excellent terms. In the number of May 15, 1837, I had made a favorable notice of Dr. Sewall's Lectures on Phrenology,[39] of which he, at the time, took no notice. It could not, however, have been agreeable to him, uncompromising phrenologist as he was. In the number of the corresponding date for 1838, however, there appeared a notice from my pen, for which he probably never forgave me, and, on various occasions, published strictures upon the productions of my pen, which were far from being free from personality. The notice in question was of a small volume, which consisted of a "vindication" of phrenology, directed especially against the remarks of Dr. Sewall. After enumerating the contents of the volume, and the circumstances under which it was published I stated, that it was "markedly controversial,"—adding—

The author has long been esteemed the apostle of phrenology in this country. . . . Identified as he is, and always has been, with the cause, we are not surprised, that he should endeavour to repel every attack, that may be made upon it; but we confess we have not seen in the work of Dr. Sewall any—to us—adequate reason for his frequent recourse to the *argumentum ad hominem*,[40] an argument not to be employed—if ever admissible—except on extraordinary occasions. The author's own remarks, at the termination of his essay, are indeed, well worthy of being borne in mind by him. "Let the Professor" (Sewall) he says—"or any other writer, call in question the truth of phrenology, and discuss the subject with the candour, calmness and courtesy, which should always characterize a scientific controversy, and if I reply to him at all, my language, matter and manner, shall be marked with a corresponding exemption from passion and reproach, and, as far as I can render it so, from every other exceptionable quality. Fact and plainness, courtesy and argument, shall be alone employed. But they shall be employed with whatever of force and efficiency I can bring to the contest." Such are the sentiments, that ought always to be entertained on such occasions. How difficult it is for us, in our controversies, to attain that happy desideratum, which Dryden has so well depicted. "How easy is it"—he observes—"to call rogue and villain, and that wittily! But how hard to make a man appear a fool, a blockhead or a knave, without using any of those opprobrious terms! To spare the grossness of the names, and to do the thing yet more severely, is to draw a full face, and to make the nose and cheeks stand out, and yet not to employ any depth of shadowing. This is the mystery of that noble trade, which yet no master can teach to his apprentice; he may give the rules, but the scholar is never the nearer in his practice; neither is it true, that this fineness of raillery is offensive. A witty man is tickled, while

[39] Thomas Sewall (1787-1845), M.D. 1820, Harvard, Professor of Anatomy, Medical Department of Columbia College, Washington, D. C., published *Examination of Phrenology*, 1837 (republished 1839), which he sent to many parts of the country for review. Caldwell had deep convictions about the scientific importance of phrenology, delivered a course of lectures on the subject to the medical class of the Summer Institute of Philadelphia, published *Phrenology Vindicated and Antiphrenology Unmasked* and many other writings relating to it, while Dunglison was allied to the antiphrenologists (Emmet Field Horine, *Biographical Sketch and Guide to the Writings of Charles Caldwell, M.D.*, Brooks, Ky., 1960).

[40] Argument directed against the person.

he is hurt, in this manner, and a fool feels it not; the occasion of an offence may possibly be given, but he cannot take it. If it be granted, that, in effect, this way does more mischief—that a man is secretly wounded, and though he be not sensible himself, yet the malicious world will find it out for him; yet there is still a vast difference betwixt the slovenly butchering of a man, and the fineness of a stroke that separates the head from the body, and leaves it standing in its place. A man may be capable, as Jack Ketch's wife said of his servant, of a plain piece of work, a bare hanging; but to make a malefactor die sweetly, was only belonging to her husband."

I added that the essays exhibited "great vigor and independence of thought"; but that I was quite convinced their controversial spirit would detract greatly from "the effect they were intended by their zealous author to induce."

That this notice was far from being satisfactory to Dr. Caldwell is shown by the following letter.

Louisville, May 25, 1838

Dear Sir,

In this letter my purpose is neither to complain, remonstrate nor request, but simply to represent, narrate, and possibly reason, leaving the sequel to yourself. Having previously noticed Dr. Sewall's Lectures in the Intelligencer for May 1837, you have noticed my reply in the same work for May 1838; The two notices, however, are altogether different. I do not mean in *tone* and *sentiment* (for on them I do not feel myself privileged to remark) but in *material* and *manner*. Your notice of Dr. Sewall embraces facts, and, therefore, involves knowledge and judgment. Your notice of myself consists of *opinions*, and involves almost exclusively *taste* and *fancy* with something of feeling. From your notice of Dr. Sewall's lectures, your readers can form some idea of his matter and substance. From your notice of my "vindication," they can judge only of my *manner* as you *conceive* it to be, and very imperfectly, even of that. They merely learn the title of my reply, and that in its character it is "controversial," and there the drama closes, and the curtain drops. Neither fact, argument nor illustration from it is allowed to tread the stage. To explain myself more fully.

After a few introductory clauses of a commendatory nature (on which were I inclined to be "controversial" I could pass strictures not easily thrown off), you extract Dr. Sewall's five or six general propositions, with the whole of his remarks under the third of them, respecting the comparative size of the head and the brain, and from your previous approval of the work, you not only *allow*—you virtually *compel* your readers to believe, that you concur on the soundness of your author's remarks. I mean his remarks respecting the immense difference in the size of brains, contained in skulls of the same dimensions. In truth, you tacitly concur in a position, which pronounces Phrenology to be not only fallacious, but *arrantly preposterous*. For, were Dr. Sewall's representation a *general truth*—as he intends it to be so considered—instead of a very striking *exception* from such truth, Phrenology would be one of the silliest *impostures* of the day—or of any day. But the representation, viewed *as a general fact or principle*, is *not true*. On the contrary, it is conceived and prepared, not "in the true spirit of science" (your own words) but in a spirit of *deceptiveness*, as I, in my reply, have made *fully appear*. This you may say is *confident* language. Be it so. It involves as much of *truth* as it does of *impudence* or self estimation. And for a decision on this point, which time shall neither reverse nor annul,

I fearlessly appeal to the judgment of that portion of the public that is competent to decide. The Intelligencer, however, furnishes the public with no ground on which to form a judgment. Why? Because your notice contains not the slightest reference to the substance of my reply. Yet, without that or some other similar or better source of information, you must be fully aware, that not one person in a million is prepared to decide correctly on the subject. To save you the trouble of research on this point, permit me to refer you to my "vindication" from page 83 to page 90, where you will find my exposition of the unsound and deceptive character of Dr. Sewall's representation, or rather gross and *culpable misrepresentation* of the *average* thickness of the skull, and concomitant size of the brain. After carefully reading those pages, and comparing their contents with your quotation from Sewall's Lectures, I leave *for the present*, the sequel to yourself. As you had already praised Dr. Sewall's lectures, I could not, and did not, expect you to arraign your own decision, by also praising my condemnatory reply to them. But I did presume you would allow the reply to speak briefly (however feebly) for itself. In plain terms, I expected you to quote from me on the same topic on which *you* had quoted from the pages of my antagonist. This would have bespoken *impartiality*. In this way alone, could you have enabled the public to judge between us. And of such judgment only, built on fair and solid ground, am I desirous. And that I shall endeavour, in some shape, to attain. Nor shall I make the attempt from personal motives. For my own reputation in the matter (however presuming and vain-glorious the declaration may be deemed) I have no apprehension. If *it* cannot pass the ordeal unscathed, it is not worth possessing. But for the interest of truth, I feel intensely; nor am I without some feeling for the interest of my *publisher*, who holds the work on his own risk. And as far as its influence may extend your notice is adverse to both those interests. Such, at least, is my opinion. You will hardly, therefore, feel surprised at my earnest desire, that my "vindication" should be made more fully and *fairly* known to the public for what it is *really worth*—be that of *approval* or of *condemnation*. Let it be looked at *as it is*, and treated accordingly. You think, that Dr. Sewall has given no "adequate reason for my frequent recourse to the *argumentum ad hominen*." I, on the contrary, think he has. If he be, as I have pronounced, and I think *proved*, him a deliberate *falsifier*, a plagiarist, a garbler, an intentional misquoter and perverter of other writers' meaning, and a malicious slanderer—if he be, as I conscientiously believe him to be, all this, it is scarcely possible in my estimation, to rebuke and denounce him with too much severity. Truth and literature are sacred things, and should not be wantonly and sacrilegiously sported with by the reckless and the unprincipled. Receive this as my *old fashioned* creed, and believe me,

truly and respectfully yrs
Ch. Caldwell.

PS. Your implied approval of one point in Dr. Sewall's Lectures, is to me as singular as it is unexpected. "Dr. Sewall"—you say—"enquires: I. How far Phrenology is sustained by the structure and organization of the brain." The Doctor's very *pretension* to an inquiry into this indicates either blank ignorance *in himself*, or a design to deceive *others*. That we cannot, from the "structure" of a part, deduce its *function* is a settled maxim in physiology.—But I must close this letter of such merciless length."

It was obviously impossible to publish a letter of so libellous a character in the shape of a reclamation; and, had it been practicable, I should have declined, owing to the hint, that such publication might do *for*

the present; and consequently, that I should be exposed to fresh demands from him. I took no notice of it therefore; in consequence of which he wrote weekly for sometime, a series of denunciatory communications in a newspaper, published in New York—the *Weekly Whig*, I think was its name, and the caption "The Editor of the *American Medical Intelligencer* weighed in his own balance and found wanting." Of these, also, I took no notice. Dr. Caldwell was evidently desirous of drawing me into a controversy, and I was equally determined he should not succeed in so doing. An unmerciful critic himself, he was in the highest degree sensitive at the gentlest criticisms of his own productions. His was, indeed, an excellent instance to be added to those alluded to by Dr. Forbes in the admirable "Postscript" to the *British and Foreign Medical Review*, to which I have referred elsewhere. "More than once," he remarks—

in the course of the publication of the Review, I have had an amusing illustration of the different way in which the same mind will view a criticism when directed at the production of another and at its own. I refer to the case in which a contributor has himself afterwards published a book, when this book has been reviewed by one of his colleagues, and undergone gentle criticism—it might possibly be somewhat more gentle on account of the previous relationship of the author to the Review—it has been amusing to see the quondam contributor exhibit the same sensitiveness to reproof as if he had never bestowed any on others; entirely forgetting how he himself had formerly laughed at (in others) the very feelings he now indulged as legitimate (*British & Foreign Med. Review*, No. 48, p. 594.)

I doubt not, that his vanity, which was excessive, was wounded, in the first place, by the notices I gave of Dr. Sewall's book and his own; and, in the second, by my silence in regard to his letter to me. His self respect was really inordinate. It has been told of him, that in speaking to some one of phrenological heads, he stated, that there were three, which were preeminent; one of these of Cuvier, and the other of Webster. The third modesty forbade him to mention. Although there could be no doubt of the cause of difficulty between him and me,—that it arose in consequence of my published opinions of Dr. Sewall's book and his own, he assigned another cause for it to Dr. Harlan,[41] of Philadelphia, which had in reality not the slightest foundation. He correctly narrated to the Doctor a conversation between us, in which he remarked to me, that my head had the organs of perceptivity largely developed, and that he had no doubt that I could beat him in writing off any composition on the spur of the moment. "In me, however," he added, "the organs of intellectuality are larger than in you; and, give me time, I think I could compete with you." I need scarcely say,

---

[41] Richard Harlan (1796–1843), teacher of anatomy, surgeon to the Philadelphia General Hospital, member of many learned societies, published, in 1825, *Fauna Americana*, a work of great erudition.

that the conversation amused me exceedingly, and I have often mentioned it as characteristic of Dr. Caldwell; but it never excited, in the slightest possible degree any other feeling in me. Moreover, the attacks were altogether on his part, and every one of them remained unanswered by me. To one of them I shall have to refer hereafter.

THE AMERICAN MEDICAL LIBRARY AND INTELLIGENCER

Soon after my removal to Philadelphia, Dr. Pattison proposed to me to become associated with him in the editorship of a Journal, similar to one, which he had conducted previously, and which had been published in Washington by Genl. Duff Green. The plan was certainly a most useful one. It was to be published semi-monthly, in numbers of 128 octavo pages, 112 of which were to consist of a reprint of a standard work,— the remainder of original matter: the whole so arranged, that each work could be bound separately. The subscription price 10 dollars per annum. The publisher was Adam Waldie, who had become well known as the publisher of *Waldie's Circulating Library* or *Waldie* as it was often called, under the able editorship of Mr. John J. Smith, the Librarian to the "Library Company of Philadelphia," which was the means of diffusing a vast amount of information amongst the public, by reprinting the best literary works, as they appeared. I consented to be associated with Dr. Pattison; but, in the spring of 1837, he determined to go to Europe, and, therefore, proposed to leave the editorship solely in my hands, which I agreed to.

The title chosen, was *The American Medical Library and Intelligencer; a concentrated Record of Medical Science and Literature,* and the first number appeared on the first of April, 1837. The "Notice by the Editor" will indicate the proposed nature and objects of the work.

The Editorship of the *American Medical Library and Intelligencer* having devolved wholly on the undersigned, his utmost zeal and assiduity will be devoted to fulfill the design of the work, as detailed in the prospectus, extensively issued some time ago. No effort shall be wanting to place before the subscribers every medical fact and observation of importance, which may appear, from time to time, at home or abroad. The desire of the undersigned will be, to make the work cosmopolite, as regards the sources whence its information is derived, and the reflections to which such information may give rise; whilst it shall be truly American in its character,—its pages being open to appropriate communications from every intelligent individual, and respecting every honourable association of individuals, who may be labouring to promote the great interests of the republic of science. As one of the objects of the work is to publish short original communications, the undersigned solicits contributions on any topic, which may be esteemed, by the writers, of interest to their professional brethren; and as such communications must, from the nature of the work, be brief, many, who would pause before they commenced a long essay for a quarterly publication, may readily decide upon furnishing a less elaborate account for a

Journal, which appears more frequently, and which must necessarily be less formal and stately, although, it is hoped, equally dignified and useful. In the selection of works for the *Library,* the undersigned, it need scarcely be said, will exert his best judgment. The correspondence established with Great Britain, France, Germany and Italy, will enable him to command the best productions of those favoured countries, and although the *Library* will be chiefly made up from the most valuable works that issue from the press of Great Britain, he will eagerly embrace the opportunity, afforded by the publication of a superior work in those other countries, to place an English version of it before his readers. At this moment, he has before him many late French, German and Italian periodicals, from which, as well as from the Medical Journals of this country, the best materials shall be transferred, in some form or other, into the pages of the *Intelligencer.* In conclusion the Undersigned may repeat, in the language of the prospectus, that no effort will be wanting on his part to give interest and value to the publication, and to render it what it purports to be, "a concentrated record of medical Science and Literature."

Robley Dunglison.

The Journal and Library were certainly calculated to be extensively useful, and at the end of the first year, I was enabled to state "with no little satisfaction" on the authority of the publisher, that not more than a dozen complete sets of the original edition remained unsold. It was well printed, on good paper, and its appearance, in every respect, creditable. "That the Editor's attempts have not been futile"—I remarked—

has been sufficiently shewn by the favour with which the work has been received both at home and abroad. Whilst many of the articles of the *Intelligencer* have been transferred to other Journals of this country, it has been gratifying to the Editor to observe, that its pages have furnished, and will doubtless continue to furnish, valuable extracts for one of the best—if not itself the very best—of medical periodicals—the British and Foreign Medical Review, edited by Drs. Forbes and Conolly. The correspondents of the "Intelligencer" may consequently expect, that their various communications will meet with that attention abroad, which they may merit: at home, alas! in all countries, it frequently happens, that they are either totally and undeservedly neglected, or experience ungenerous and uncandid treatment, too often suggested by unworthy and illiberal motives, by which those at a distance can scarcely be swayed. *Preface to Vol. 1 of Intelligencer.*

The reprints with the articles of the *Intelligencer,* for the year, exceeded the enormous amount of *three thousand closely printed pages,* and all for *ten dollars!* One of the works, translated wholly by myself, was *Medical Clinics of the Hospital Necker, or Researches and Observations on the Nature, Treatment and Physical Causes of Diseases, By I. Bricheteau, Physician to the Hospital etc.* which amounted to 150 pages.[42]

Unfortunately, the affairs of the publisher became more involved about the appearance of the second year of the *Library & Intelligencer,* and from that time forward until the stoppage of the concern at the end

---

[42] Isidore Bricheteau (1789–1862), associate of Philippe Pinel, appointed to the Necker Hospital in 1830, published his *Clinique Médicale de l'Hôpital Necker* in 1835.

of March 1841 each successive number appeared with greater and greater effort on his part. All the new works, to be reprinted in the *Library*, had to be procured by myself; the paper and printing for the goodness of which he had formerly been celebrated became execrable; and I should have withdrawn from the concern, long before it was finally arrested, but for my respect for Mr. Waldie, and my distress at the condition of his intellect, which was manifestly tottering, and ultimately became so much affected, that, to preserve the little property he was in possession of,[43] and to restore him to the state of sound mind again if practicable, I deemed it necessary to advise, that he should be taken to the Pennsylvania Hospital for the Insane at Blockley;[44] not long after which he died. He was an excellent hearted, liberal man, who, from being a journeyman printer, had raised himself to some eminence in his vocation, but was little acquainted with the management of money matters, or, indeed, of any kind of business, which required much art or knowledge of men.

The work was transferred to Mr. J. J. Haswell, a bookseller, who had retired from the firm of Haswell, Barrington & Haswell, publishers and booksellers, and, after much entreaty, I consented to continue as Editor for another year, from July 1841 to June 1842 inclusive. It was, during the period, published monthly, at the price of *five dollars*, after which it was finally discontinued on the removal of the bookseller—who was not successful—from the city. During this last portion of its existence an American edition of "Hope on Diseases of the Heart"[45] with notes etc. etc. by Dr. C. W. Pennock was published in it. My work on "New Remedies," of which I shall make mention hereafter, also appeared first in the "Library," in the numbers from July the first 1839 to October the first inclusive.

DR. SAMUEL FORRY

It was in consequence of what ought to have been deemed a favourable notice, in the *Intelligencer*, of Dr. Forry's work, entitled *The Climate of the United States and its Endemic Influences: based chiefly* on the *Records of the Medical Department and Adjutant General's office, United States Army*, that a most wanton attack of plagiarism was brought against me in an ephemeral sheet, published in New York weekly, under the title of the *New York Lancet* and edited by a so called *Dr.* Houston, who was a reporter to the *Herald,* and subsequently to Congress. The *Lancet*

was, indeed, an offset from the establishment of the *New York Herald,* and had all the recklessness and insolence of its prototype.

In the article in question, I spoke of Dr. Forry's work as follows.

This is an excellent contribution towards the medical statistics of this country—based chiefly—as the author states in the title—on the records of the medical officers of the army. The work is divided into two parts,—the *first* comprising "Researches in elucidation of the Laws of climate in general, and especially The Climatic Features peculiar to the Region of the United States" and the *second* embracing "Researches elucidating the Endemic Influences peculiar to the Systems of Climate developed in part 1st." With most of the author's views we entirely accord; indeed, the work affords evidence, that our own inquiries have received a portion of Dr. Forry's attention. On the subject, however, of the vegetable origin of malaria, we are compelled to differ from him; his facts and arguments on this subject are not, indeed, more conclusive than those of his predecessors, whilst he adduces, we think, sufficient testimony against the soundness of his own views. We do not, however, on this account, the less recommend Dr. Forry's labours to the notice of our professional brethren. His work is unquestionably one of the most interesting productions, that have appeared on this interesting subject. *American Medical Intelligencer* for March 1842.

The remark, that Dr. Forry's work afforded evidence that my own inquiries had received a portion of his attention, doubtless made him apprehensive, that I might point out the numerous cases in which he had *used* me, without the slightest acknowledgment, and determined him to forestall me through the pages of the *Lancet*, by charging me with plagiarism from his own work, which actually was not at the time in existence, in my description of "Dengue," in the first edition of my *Practice of Medicine.* In that article, I had referred to the *Vital Statistics of the Army,* published under the direction of Surgeon-general Lawson; but, I was informed afterwards, drawn up by Dr. Forry.[46] I could not, consequently, credit Dr. Forry with what purported to proceed from the Surgeon-general himself. Yet the alleged plagiarism consisted in my having done an impossibility! I took no further notice of the scurrilous article in the *Lancet* than by writing to Surgeon-general Lawson. The character of my communication to him will be understood by his reply of which the following is a copy.

---

[43] A receipt signed in 1839 by Dunglison on behalf of Adam Waldie is in possession of the Historical Society of Pennsylvania.

[44] Blockley Township is now part of West Philadelphia. This hospital was opened for admission of patients on January 1, 1841 (Thomas G. Morton and Frank Woodbury, *History of the Pennsylvania Hospital,* p. 122, 1895).

[45] James Hope (1801–1841), whose *Diseases of the Heart and Great Vessels* (1831) is a classic.

[46] Samuel Forry (1811–1844), Assistant Surgeon in the U. S. Army, whose *Statistical Report* has Surgeon-General Lawson's name on the title page and Forry's name effectively hidden in a note on the reverse of the title page, which was apparently missed by Dunglison when he quoted from the work in an article on Dengue in his *Practice of Medicine* (2: 656). He was said to have a singularly mild and amiable disposition and had many friends in New York (J. S., The Late Samuel Forry, M.D. *New York Journal of Medicine* 4: 7–9, 1845; Lee, Charles A., Epilepsy Terminating Fatally . . . Being a History of the Case of the Late Samuel Forry, M.D., *ibid.*, pp. 9–16).

City of Washington,
May 30th 1852

Dear Sir,

Your letter of the 22nd inst. calling my attention to the publication in the New York Lancet, charging you with "plagiarism in not having given Dr. Forry the merit of affording you the information upon which your article Dengue was based" has been received. In reply to your interrogatory, "whether, in the event of another edition of your work being called for—there is any reason, why the credit should not be still given to the original public document," I beg leave to make the following remarks.—Doctor Forry, late Assistant Surgeon of the Army, was employed in the Surgeon General's office, and, while there, collated and condensed, under my orders and superintendence, the matter contained in the various Sick Reports etc. on file in the office. In compiling the materials from the vast number of documents put in his possession and giving a chain of connection to the facts and observations selected, preparatory to their publication, Dr. Forry must, necessarily, have employed much of his own language, and put forth many original ideas. To what extent, however, he may have contributed original matter to the work, you and others can judge as well as myself, and may, with greater propriety perhaps, speak on the subject than I can. As relates to your article on Dengue it would seem, that you had, in giving the *Vital Statistics of the Army* as the authority for some of your remarks, not only not claimed originality for yourself, but had virtually given Dr. Forry, the compiler of the work, the necessary credit. Doctor Forry is not now in the Army. He resigned his commission in 1840. Touching the attacks, made upon you in the *Lancet* of the 14th inst. I am free to say that the article is written in bad taste, and with an asperity of language altogether uncalled for. And I regret the publication for the reason, if for none other, that it can be productive of no good to Dr. Forry now, while it may have the effect of creating a feeling of hostility on the part of those, who might otherwise be disposed to advance his literary fame.

I have the honour to be, very respectfully,
Your obedt. servant,
Ths. Lawson

Robley Dunglison MD
Philadelphia.

But although I took no farther notice of the article in the *Lancet*, my friend and colleague Dr. Huston, struck with its injustice, wrote to the Editor of the *Lancet* pointing this out to him; and giving passages from the work of Dr. Forry, which he had taken *verbatim* from my *Elements of Hygiene*: but, instead of inserting the article, he merely remarked, that he had received a communication from Philadelphia complaining of the strictures on me, and exhibiting a similarity of passages, which he accounted for by Dr. Forry and myself having drawn our information from the same source! At this time, the Editor did not anticipate, that his moral delinquency in the matter would meet with its reward thereafter. Dr. Houston—the Editor—was the son of an Irish clergyman, who was known to Dr. Wylie of Philadelphia; and as it appeared that Houston had never graduated, Dr. Wylie applied to Dr. Ely—one of the Trustees of Jefferson Medical College—to have a degree conferred upon him, without examination. Dr. Ely mentioned the application to Dr. Huston, who detailed to him the conduct of Dr. Houston, whilst editor of the Lancet, in my case; and the affair was abandoned. I shall place here in juxtaposition a few passages from Dr. Forry's work, published in 1842; and from my *Elements of Hygiene*, published in 1835,—seven years previously.

Climate of the U. States (Forry)	Elements of Hygiene (Dunglison)
Although the thermometer may be 15° or 20° higher here than in England, during the heats of summer, yet we suffer but little more from its effects: for as the air of the latter country is more loaded with humidity, causing a diminution of the cutaneous and pulmonary transpiration—the evaporation of which constitutes a cooling process—a languor and a listlessness, with an indisposition to mental and corporeal exertion are induced. p. 78	If the air be greatly charged with moisture, especially during the heat of summer, owing to a diminution of the cutaneous and pulmonary transpiration, the evaporation of which constitutes a cooling process—we feel languid and listless, with an indisposition to every mental or corporeal exertion. This is the cause why we suffer little more during the hot summers of this country, than in those of Great Britain, where the air is always more loaded with humidity, although the thermometer may be fifteen or twenty degrees higher here than there. p. 64
Although the morbific cause may be general, and widely diffused, yet it is mostly modified by local influence, constituting an *endemico-epedemic*. An example in point is presented in the history of epidemic cholera, the visitations of which in the United States at least, were much favoured by the high temperature of summer, and by the peculiar atmosphere of towns situated on seas and rivers. We sometimes see a district signalized for its salubrity, desolated by a malignant fever, the production of which required a combination of certain local and general causes; but as this precise concatenation of causes may never re-occur, so the inhabitants may remain exempt from a similar scourge. p. 306.	For instance, there may be something in the locality, connected with a favouring state of the atmosphere, which may occasion one place to be insalubrious, whilst others in the immediate vicinity, are entirely healthy. Of this we have had a striking example in the case of malignant cholera, which has attacked several of our towns in the most virulent manner, whilst others, and some of these to all appearance similarly circumstanced, have wholly escaped. The locality, which has seemed to favour its visitations, has been the confined air of towns, and these towns on the seas or rivers. This complaint, then, required a combination of atmospheric and local causes to induce it; in other words, the causes were of an *endemico-epidemic* character. Perhaps the requisite union of local and atmospheric causes may never again meet in some of those places, and the scourge may not reappear. We frequently see the most salubrious districts desolated by malignant fever, by a complaint, perhaps not previously known there, and possibly destined never again to return, because the precise endemico-epidemic influence may be wanting. p. 93.

These extracts are sufficient to exhibit the correctness of my remark on Dr. Forry's book in the *Intelligencer*, that "my own inquiries had received a portion of his attention";—and many other examples might be given of the same fact. The impunity, however, which he met with at my hands rendered him bolder; until ultimately, he extracted whole paragraphs from my *Elements of Hygiene*, and even gave them as editorial comments in the *Democratic Review*, and in the *New York Journal of Medicine*, of which he was Editor. It was by mere accident, that I took up the former Journal, and read the following extract purporting to be from the pen of the Editor; and not much more than a year after he had falsely charged me with plagiarism from him. Immediately after reading it, I wrote the following ironical article, which I never, however, presented for publication. It refers to, certainly, one of the most barefaced cases of literary piracy that has ever fallen under my notice; and one which was liable to such ready detection, as the Journals in question were published in New York, whilst I was living in Philadelphia; and my *Elements of Hygiene* had been published eight years before, and had been extensively distributed—the whole edition of 1500 copies being exhausted the year after. I made, however, no reclamation for the outrage.—

"COINCIDENCES OF SENTIMENT" ETC.

It must be not a little gratifying to those, who have closely investigated a subject, to find their sentiments corroborated by subsequent inquirers; and still more convinced must they be of their accuracy, when they discover a coincidence not only in sentiment, but in language. We have been struck with one of those strange coincidences in an article headed "The Medical Philosophy of Travelling" in the *Democratic Review* for July 1843, which purports to be a notice of the works on climate and change of air of Dr. James Johnson, Sir James Clark,[47] Dr. Forry, and Dr. Drake. With the general tenor of the article we have no fault to find. It conveys, indeed, the views of some of the best *hygienists* of the day. It is to place on record the signal similarity of sentiment, and even of language, between the reviewer and previous and contemporary writers, that we draw attention to it. For this purpose, we place in parallel columns some of the editorial remarks, and the observations on the same subject by Dr. Dunglison in his *Elements of Hygiene*, which, if the Reviewer had ever seen, he has omitted to notice directly or indirectly, and which, if he has not seen, indicate one of those singular cases of identity of views, and almost of language, which is sometimes, yet rarely, met with!

Democratic Review July 1843

Elements of Hygiene (Dunglison)

When we consider the multitude of valetudinarians, who annually visit the watering places of this country and of Europe, and who return to their homes renovated in health and inspired with confidence in the virtues of the waters near which they may have resided, the inference is obvious, that the salutary effects are attributable more to the change of air, and other extraneous circumstances than to the various waters. This is well illustrated in the circumstance, that many a valetudinarian in leaving an Atlantic town for the interior mountainous region as for example the White Sulphur in Virginia, finds himself, during the journey, fatiguing as it is, almost restored. Many springs, which are inert, as the Bath and Matlock waters of England have thus acquired a high reputation for their medicinal qualities. These agreeable watering places are constantly crowded during the season of visiting; the latter in consequence of the surrounding beauties of nature, and the former for the ceaseless round of amusements, which, keeping the mind agreeably and lightly engaged, produce a beneficial reaction on the mental or corporeal disorder. Were such waters bottled and transported to a distance, it is obvious, that no beneficial effects could follow their use by an invalid. It was proposed, at one time, to carry sea water by means of pipes to London, to place within the reach of all of its inhabitants, the advantages of sea bathing at home; but had the scheme been carried into execution, it is much more than probable, that the usual effects of sea bathing would have been no longer realized. p. 58	The multitudes of valetudinarians who annually leave their habitation to visit the watering places of this country and of Europe, and who return to their homes in the enjoyment of health, and full of confidence in the waters near which they may have resided, and of which they may, or may not have partaken, furnish satisfactory replies to these musings. Long before the citizen of our Atlantic towns reaches the Alleghany Springs of Virginia, he has an earnest of the advantages he is about to derive from change of air; and many a valetudinarian finds himself almost restored during the journey, fatiguing as it is, through the mountain regions, which have to be crossed before he reaches the White Sulphur in Green-Brier county. . . . We can thus understand the reputation acquired by the inert Bath and Matlock waters of England, the latter of which has scarcely any solid ingredient, and yet what crowds flock to those agreeable watering places, to the former for the perpetual amusements, that keep the mind engaged and cause it to react beneficially on the corporeal or mental malady; to the latter, for the enjoyment of the beauties of nature, for which Derbyshire is so celebrated. It is obvious, that were such waters bottled, and sent to a distance, so that the invalid might drink them at his own habitation the charm would be dissolved. The garnitures, more important in this case than the dish, would be wanting, and the banquet would be vapid, and without enjoyment or benefit. Not many years ago, amidst the bubbles, that were engaging the minds and money of the English public, it was proposed to carry sea water by pipes to London, in order, that the citizens might have the advantage of sea bathing without the inconvenience of going many miles after it. Had the scheme been carried into effect the benefits from metropolitan sea-bathing would not have exhibited themselves, in any respect, comparable to the same agent employed at Brighton or Margate. p. 151.
(This article is repeated without the slightest alteration in the *New York Journal of Medicine* for the same month, July 1843, edited by Dr. Forry, as part of a review of Dr. Drake's *Northern Lakes*.[48])	

---

[47] James Clark (1788–1870), M.D. 1817, Edinburgh, author of *The Influence of Climate in the Prevention and Cure of Chronic Diseases*, 1829, high in favor with Queen Victoria and the Prince Consort (William Munk, *Roll of the Royal College of Physicians* 3: 222, 1878).

[48] Daniel Drake, *The Northern Lakes a Summer Residence of the South*, 29 pp., Maxwell, Louisville, Ky., 1842, reprinted by the Torch Press, Cedar Rapids, Ia., 1954, originally appeared in the *Western Journal of Medicine and Surgery* 6: 401–426, 1842.

The above is one of many specimens of "coincidence" in sentiment and language between Dr. Dunglison and the writer of the Review! One other we may mention. It concerns Dr. Forry and the Reviewer. Dr. Forry has employed himself profitably in researches on climate, and has, doubtless, added to our stock of knowledge on some points. Yet he seems to us to overestimate many of his own deductions. Thus, certain of them are contained *verbatim;* first of all, in Surgeon General Lawson's *Statistical Report on the Sickness and Mortality in the Army of the United States,* which, we are informed, was compiled by Dr. Forry; afterwards, in Dr. Forry's own work on the climate of the United States; and, again, in the pages of the *American Journal of the Medical Sciences* for April 1842, p. 307; and a singular repetition occurs, in which the *Democratic Review,* in the article cited, is likewise concerned. In the *American Journal of the Medical Sciences* for October 1841, p. 295, there is the following sentence— "As in the corporeal structure, different effects result from the dry and restless air of the mountain, compared with those evidenced in the moist and sluggish atmosphere of the valley; so, as regards the mental manifestations, the observation of the poet is founded in nature.—

> An iron race the mountain cliffs maintain,
> Foes to the gentler manners of the plain.—Gray

In Dr. Forry's work on the "Climate of the United States," the same sentence and quotation appear, with the modification of the words "in nature" being omitted. In his *Meteorology* it appears like the last without any modification and finally we have it for the third time, without the slightest alteration, at the commencement of the article in the *Democratic Review.* Yet there is a wonderful coincidence between this and the following from the article "Endemic Diseases" in the *Cyclopaedia of Practical Medicine,* to which, however, no reference is made. "As in the body, different effects result from the dry and bracing wind of the mountain, compared with those from the moist and sluggish air of the valleys; so, as regards the mind, the observation of the poet is philosophically true.—

> An iron race the mountain cliffs maintain,
> Foes to the gentler manners of the plain.
>   Gray.

The late Dr. Samuel G. Morton [49] complained to me bitterly of Dr. Forry's plagiarisms on him, but, like myself, took no public notice of them. Dr. Forry did not live long after the publication of those on which I have animadverted; and, after his death, he received the most laboured eulogies in the medical journals of New York, and, if I mistake not, had a monument erected to his memory by the subscriptions of his medical brethren of that city. He undoubtedly stood out in relief amongst them. [50] The subject of "Vital Statistics" was a favourite with him; and his opportunities, whilst in the Surgeon-general's office, were numerous for cultivating it. Of his work on the Climate of the United States, I expressed publicly the honest feeling I entertained of its merits; and deeply regretted, that by the obliquity on which I have animadverted, it was necessary to qualify the good opinion I should otherwise have formed of him as an able and honorable member of the profession. [51]

ATHENIAN INSTITUTE

In the year 1837, it was proposed, I believe, first of all by Mr. Nathan Sargent, [52] then a resident of Philadelphia, that a course of lectures should be delivered gratuitously by literary gentlemen of the city, once a fortnight in turn; the subjects to be selected by themselves. The proposition met with favour, and the "Athenian Institute" was incorporated, of which Judge Hopkinson was made President; Mr. John J. Smith, Treasurer, and Mr. Sill, Secretary. A Board of Counsellors was also chosen, of which I was one. The lectures were delivered, first of all in 1838 in the Masonic Hall in Chestnut St. and subsequently, in the Musical Fund Hall in Locust St. They certainly were eminently successful for the first two years; crowds of the most respectable persons attending; so that when they were ultimately discontinued, upwards of three thousand dollars remained in the Treasurer's hands; the disposition of which was left to a Committee, consisting of Mr. Wm. B. Reed, Mr. Fisher Leaming and myself; and we recommended that a portion should be lent, without interest, to the American Philosophical Society, and a portion to the Historical Society of Pennsylvania. [53] I gave the third lecture of the course on "Instinct," which was subsequently published, at the request of the publisher, Mr. Morton Macmichael, in the *Young People's Book.* A second lecture, in a subsequent year, I gave on "Popular Superstitions." There was, about this period, a kind of mania for popular lectures; and frequent calls were made upon me by different literary institutions, which I was compelled to decline; and, in order not to give offence, I made the

---

[49] Samuel George Morton (1799–1851), M.D. 1820, University of Pennsylvania, M.D. 1823, Edinburgh, was a member of the American Philosophical Society, the College of Physicians of Philadelphia and a score of other learned bodies which threw him into Dunglison's social swirl (Charles D. Meigs, *Memoir of Samuel George Morton, M.D.,* Phila., 1851).

[50] Dunglison must have received pretty rough treatment at the hands of his New York colleagues, among whom were honorable men like John W. Francis, John H. Griscom, and Alexander Hosack.

[51] Dr. Squibb, in his *Journal* (p. 287), notes this lecture by Dunglison, saying: "a clear case of stealing of thunder. . . . How should Forey [*sic!*] have known whose original he was taking!"

[52] Nathan Sargent (1794–1875), Philadelphia lawyer, widely known under the pen name of "Oliver Oldschool."

[53] The minutes of the Athenian Institute from 1847 to 1864 are preserved in the Historical Society of Pennsylvania.

rule absolute, and, consequently, declined at an after period, lecturing before the Smithsonian Institution; the Gettysburg College; the Medico-Chirurgical Faculty of Maryland; the Mechanics Institution of New York, and other similar bodies, as I had refused my services to smaller and yet highly useful institutions.

A great amount of information was certainly communicated in those lectures, but, after a while, the novelty passed away; it became necessary to obtain the service of less practised lecturers and writers, and the course was not resumed.

INSANE POOR OF PENNSYLVANIA

The condition of the Insane poor in Pennsylvania had long been a topic of great interest with the philanthropist; and at length, in the year of 1838, a meeting of citizens was held in Philadelphia, which was presided over by Thomas P. Cope, Esqr., with Thomas Bradford, Esqr. and Judge King as Vice Presidents, and Frederick A. Packard and James J. Barclay, Esquires as Secretaries. At this meeting, the object of which was to take into consideration the propriety of adopting measures to establish, at public expense, an Asylum for the Insane Poor of Pennsylvania, the following resolutions, appended to an appropriate preamble, were brought forward by Joseph R. Ingersoll Esqr. and adopted by the meeting.

Resolved, that it is expedient to make application to the Legislature for the passage of an act to authorize the purchase of extensive grounds, and the construction of a state asylum for the relief of the Insane Poor of Pennsylvania.
Resolved, that a Committee be appointed to prepare, print and circulate memorials to the Senate and House of Representatives, and cause them to be presented—to procure and publish information (statistical and otherwise) on this interesting subject, and to adopt such other measures as may, in their opinion, contribute to the success of the undertaking.
Resolved, that J. R. Ingersoll, Dr. Robley Dunglison, J. M. Read, Rev. Dr. Tyng, Saml. R. Wood, J. K. Kane, S. B. Morris, I. Collins, Edward Yarnel, Rev. Dr. Demme, T. Earp, Dr. B. H. Coates, J. R. Tyson, Dr. Charles Evans, W. S. Hansell, Dr. Joseph Parrish, Townsend Sharpless, Dr. Caspar Morris, Dr. J. R. Burden, Ambrose White, Dr. I. N. Marsellis, John Goodman, J. J. Smith Jr., Rev. Dr. Mayer, John Farnum, G. N. Baker, together with the officers of the meeting, be the said Committee.

By the general Committee named in the last Resolution, a subcommittee, consisting of myself, Frederick A. Packard Esqr. and Dr. Caspar Morris was appointed to prepare

a summary account of institutions for the safe keeping and treatment of the insane poor, and especially the number and condition of such within this state, accompanied with such arguments as the committee think will promote the establishment of a state asylum for the insane paupers of this commonwealth, and to forward the same to the members of the legislature, and also to circulate them among the citizens generally,

and at a meeting of the general committee, held on the 17th of December 1838, the subcommittee

presented, through their chairman, Dr. Dunglison, a report, which was approved, ordered to be printed, to be signed by the Chairman and Secretary of the general committee, and to be published under the direction of the subcommittee in such form as to secure its most extensive circulation.

The Report was entitled *An Appeal to the People of Pennsylvania on the subject of an Asylum for the Insane Poor of the Commonwealth.*[54] Several thousands of it were printed, and it was most industriously and extensively disseminated, especially through the exertions of Mr. Isaac Collins, and Mr. Packard. In this appeal, I gave many statistical details in regard to the ratio of the insane to the general population in different countries of Europe, and in various portions of this country, and, so far as could be estimated, in Pennsylvania; the comparative number of the insane amongst the poor and those of easy circumstances; the ratio of lunatics to idiots; the comparative curability of recent and of chronic cases; the importance of proper classification of the sufferers; the striking amelioration in the treatment of the insane in modern times; a brief description of the public lunatic asylums in the different States of the Union; the production labour in which so many of the insane could be profitably employed; and the Report concluded with the following "appeal" proper:

Such being the facts in regard to the condition of the insane in this commonwealth, can farther arguments be needed to point out the necessity of an establishment of the kind that is contemplated? Shall we be content with inglorious inactivity, whilst our brethren elsewhere are sedulously employed in their endeavours to restore to mental existence those, who are afflicted with the most awful of dispensations, and generally from no fault of their own? Can we remain satisfied with their condition at home in their own miserable hovels, or with banishing them from our sight to be immured in institutions, where but imperfect attempts at restoration are practicable, and where they are merely kept from inflicting injury upon themselves or others, with the moral certainty, in too many of the cases, that hallucinations, which, under other management, might have been wholly removed, must become more and more firmly implanted, until, ultimately, the wretched maniac sinks, prematurely under his excitement, or subsides into a state of incurable melancholy or fatuity? Or can we hesitate to exert all our energies to diminish evils of heart-rending extent, and to adopt measures—so eminently within our reach—for restoring the miserable lunatic to his relatives and to his country, or of ameliorating and softening his condition when perfect recovery is impracticable?

Satisfied, that only one feeling can prevail upon this deeply interesting and momentous subject, it is but necessary, perhaps, to urge the importance of *speedy* action,— if not on the ground, that, already, much precious time has been suffered to pass by unimproved, for the overwhelming reason, that every year's delay removes the chance of restoration from hundreds of our fellow creatures, whose

---

[54] Printed for the Committee, Philadelphia, 1838, by A. Waldie, printer.

reason is, as it were, in our keeping, and lays the foundation of evils, which may descend to all future ages.[55]

Subsequently, memorials were sent to the Senate and the House of Representatives, before whom a bill was introduced, accompanied by a report from Mr. Konigmacher [56]—the chairman of the Committee to whom the subject had been referred. The bill passed the house of representatives with slight opposition; but on account of the exhausted condition of the treasury, it did not receive the sanction of the executive. Governor Porter, however, deplored in feeling language, "the stern injunctions of duty by which he was governed," and expressed his belief, that "at no distant day, the Commonwealth would be so far extricated from her embarrassments as to be able, without inconvenience, to accomplish the laudable undertaking." Participating in the benevolent sentiments expressed by Governor Porter; impressed with the irreparable evils, which must result from delay, and believing, that the period had arrived for farther action, and that certain objections, which were made against the former bill could be entirely obviated, the committee of citizens, at a meeting held on the 25th of September, 1840, appointed a subcommittee, consisting of myself, Mr. Isaac Collins and the Rev. Dr. Demme, to prepare a "second appeal" to the people of Pennsylvania, which should embrace such portions of the former appeal, and of the report made to the legislature, and such other information as the subcommittee might think proper. "At a meeting of the Committee, held on the 9th of October 1840, the subcommittee, through their chairman—Dr. Dunglison" —reported an appeal, "which was approved; ordered to be printed, to be signed by the Chairman and Secretary of the Committee, and to be published under the direction of the subcommittee, in such form as to secure its most extensive circulation." In this "Second Appeal" the same statistical information was given as in the first. It contained, however, a large amount of matter in regard to the wretched condition of the insane in Pennsylvania, obtained in the interval between the publication of the two "Appeals"; pointed out the evils resulting from the condition of the indigent insane of the commonwealth and drew a bright picture of the advantages, that must accrue from adopting a course like that which it was the object of the Appeal to inculcate. The application to the Legislature was again successful; and it did not fail a second time with the executive. The most strenuous exertions were made

by the friends of the measure and especially, perhaps, by Mr. Packard, and Mr. Isaac Collins; the latter of whom remained some time in Harrisburg; and, through the newspapers there, acted upon the members of the Legislature, whilst I did the same here (in Philadelphia) under his suggestions. In the number of the *American Medical Intelligencer* for Feb. 1, 1841, I had the satisfaction to insert the following paragraph:

Pennsylvania Asylum for the Insane Poor. It affords us heartfelt pleasure in being able to state, that the bill for the establishment of an institution for the insane poor has passed the legislature by an overwhelming majority, and has received the sanction of the governor; so that it is now a law. It is a glad triumph to those whose philanthropic exertions have been instrumental in the result; and a bright spot in Governor Porter's administration.

The appointment of commissioners for the erection of the building, and of trustees for the management of the Institution was vested in the governor, who strangely overlooked many of the most prominent philanthropists in the undertaking, without whose active exertions it would have been difficult to carry the measure through. The venerable Thos. P. Cope, Mr. Packard and others ought assuredly to have had the preference over others that were appointed. Of all these, indeed, who had been energetic from the commencement, Mr. Isaac Collins, and myself, were alone named; —he as a Trustee for two years, and myself for three years. The Commissioners were—John K. Kane, George Rundle and John W. Ashmead. They were not harmonious, however. A site was purchased for the building under circumstances, which were not approved of by my friend Mr. Kane, now Judge Kane, and he withdrew. Proceedings were arrested; and in the pecuniary condition of the commonwealth the act was suffered to remain dormant. It, doubtless, however, paved the way for the after movement, in which I had no part, and which resulted in the establishment of the present Asylum for the State at Harrisburg.[57]

VACCINE VIRUS

In the year 1838, I was favoured by Mr. Estlin,[58] and by Dr. W. B. Carpenter, then of Bristol, England, with some vaccine matter procured immediately from the cow. At that period, the same confidence in the effects of vaccination, as a preventive of smallpox, was not entertained by many as formerly; this diminished confidence was encouraged by an epidemic, which had affected many persons seriously and even fatally, who had been deemed secure. The following is a copy of Mr. Estlin's letter to me.

---

[55] Among the responsibilities of the attending physicians at the Philadelphia General Hospital was the charge of the lunatics, so that Dunglison had a practical knowledge of this problem. Besides, he had studied in Paris soon after Pinel first removed the chains from the insane in their treatment, and in this *Appeal* a full account of the work of Pinel is presented.

[56] *Report in Relation to an Asylum for the Insane Poor,* Harrisburg, 1839, read in the House of Representatives, March 11, 1839.

[57] In 1844 Dorothea L. Dix prevailed upon the state legislature, and the Pennsylvania State Lunatic Hospital finally opened its doors to patients October 6, 1851, in Harrisburg, Pa., with Dr. John Curwen as the first superintendent (James Clark Fifield, *American and Canadian Hospitals,* p. 1064, 1933).

[58] John Bishop Estlin (1785–1855) was the leading ophthalmologist of Bristol, where he established a dispensary for diseases of the eye in 1812.

Bristol, Oct. 25, 1838

Sir,

As I am sending to New York and to Boston some vaccine virus, which I have lately procured from the cow, Dr. Prichard (my brother-in-law) thought if I were to send some to you, it might be acceptable to yourself or to some medical man of your acquaintance engaged in vaccination.

But I am in great ignorance as to the acceptableness of such a thing in America? *Here*, in consequence of the frequent failures of vaccination in protecting from Smallpox, an idea is very prevalent that the virus is *worn out;* certainly its effect on the arm and on the constitution, both immediately after inoculation and in its antivariolous power, is not what it was twenty years ago, and recourse to the Cow, for a fresh supply, has long been a desideratum. *Why* there has been a difficulty about it, I know not, but such is the case. I have, at length, succeeded in getting some, and applications for it are coming to me from medical men from all parts of the Kingdom. Should the London Medical Gazette of September the 15th, and October the 20th, fall in your way, you will there see all I have to say on the subject. I have sent copies to Dr. Jackson of Boston,[59] and to Dr. A. Brigham [60] of New York. I inclose you eight points, which will do for four children, two places of insertion in each child, and one point for each insertion. I find the best way to vaccinate with points is, to make a cluster of small scratches so as just to bring blood of this [#]size, and to rub the point well upon it.

To authenticate the descent of this virus from its original source, I will give you its correct *genealogy* since it has been in my possession.

The Cow diseased about August 11, 1836.

Miss A. infected by milking.

Jane inoculated from Miss A. Aug. 11 at the Farm near Berkley.

Stott, vaccinated in Bristol Aug. 23:—

Stiff—do. Sept. 1: W. Norris, Sep. 12: Frankham, Sept. 19: W. Webb, Sept. 26: W. W. Holden, Oct. 3: Hatton, Oct. 10: Geo. Chalk, Oct. 17; from whom the lymph sent was taken on the 8th day: viz. Oct. 24, 1838. Thus, the supply I forward is removed from the Cow only 10 degrees. Should you know anyone, who would like to use the lymph, and were I to be informed at any future time of the result, I should be much gratified. If I have erred in my belief of this communication being in anyway acceptable to you, you must ascribe it to the blunder of my imagining, that my American brethren participate with us in our wants and notions in regard to vaccination.

I am Sir, yours respectfully,
J. B. Estlin.

In a letter from Dr. W. B. Carpenter, then of Bristol, dated, Oct. 26th 1838, he remarks—

I do not know, whether there has been, in America, the same complaint of the inefficacy of the protecting power of the vaccine virus, as has been growing in this country for some time, and which a very severe epidemic of Smallpox seems to have justified, many having suffered greatly who were supposed to be quite secure. My former instructor, Mr. Estlin, surgeon in this city, has recently been fortunate enough to procure some fresh virus from the cow, the effects of which seem to resemble those produced by Jenner's vaccination, which is well remembered here, more closely than do those of the virus commonly used. I have been pretty extensively employed in propagating it; and the enclosed points I charged yesterday from the most beautifully regular vesicles I have ever seen. If you are disposed to try its effects it would afford me much pleasure to hear your opinion of the comparative powers with those of the old lymph. If it should be favourable, I would suggest your dispersing the virus through the States, as we are now doing through this country.

In another letter dated Bristol, June 26, 1839 he remarks:

—I shall hope to hear from you continued good accounts of the success of our new vaccine lymph. It has certainly quite fulfilled the expectations, which were at first, excited, regarding its general external characters and its increased effect on the constitution; but the actual value of its protective powers still remains to be tested. Mr. Estlin has employed it with complete success for the removal of a naevus over which the vesicles spread, and in a very remarkable manner.[61]

I subjected the new virus to an extensive trial; and it was much used by my friends and (now my) colleagues Drs. Meigs and Huston; and by Dr. Bridges,[62] at the time one of the city vaccinators,[63] by Dr. Kirkbride [64] and others; and in the *American Medical Intelligencer* Dec. 15, 1838, I was enabled to make the following remarks.—

Vaccine Matter fresh from the Cow. The supply of this virus, which was received by us has proved inadequate for any extensive dissemination. In the arms, which we have seen, the vesicle has gone on beautifully until the eighth day; but on the ninth, it has become sunken, and the areola

---

[59] James Jackson (1777–1867), M.D. 1796, Harvard, spent three years in England where he was able to study Dr. Woodville's early vaccination cases. When he returned to Boston in 1800, he shared with Waterhouse the reputation of being a pioneer authority on vaccination in this country (Henry B. Viets, *Brief History of Medicine in Massachusetts*, pp. 129, 142, 1930).

[60] Amariah Brigham (1798–1849), superintendent of the New York State Lunatic Asylum, Utica, N. Y., 1842–1849.

[61] Although Dr. William Woodville, of London, was able to supply Jenner's vaccine to all the world, after some years, the virus became attenuated and new sources were constantly sought. When Estlin succeeded in propagating his vaccine matter through cows, as well as successive groups of children, he was able to obtain a more liberal supply, which he offered to "any medical gentlemen connected with a public institution for gratuitous vaccination."

[62] Robert Bridges (1806–1882), M.D. 1828, University of Pennsylvania, assistant and successor to Dunglison's friend, Franklin Bache, was a Councillor of the American Philosophical Society and Librarian of the College of Physicians of Philadelphia (obituary, *Medical News* 40: 256, 1882).

[63] The Philadelphia Vaccine Society, established in 1809, was supplanted in 1816 in Philadelphia city limits by city vaccinators, who not only had the duty of vaccinating the poor but were required to keep on hand a constant supply of fresh vaccine for distribution among the physicians of the city, free of charge.

[64] Thomas Story Kirkbride (1809–1883), M.D. 1832, University of Pennsylvania, Superintendent of the Pennsylvania Hospital for the Insane from its inception until his death, a founder and President of the Association of Medical Superintendents of Insane Asylums of America (now the American Psychiatric Association), was a member of the American Philosophical Society and fellow of the College of Physicians of Philadelphia.

has not been as regularly defined as in the cases originally depicted by Jenner, and in those, which we are in the habit of seeing, but the attending indisposition on the eighth day is marked. Although the character of the disease varies in these respects, we have little doubt, that it will afford the due protection, for it is altogether unlike the spurious forms described by Jenner and others. Still, the matter remains to be tested and to accomplish this satisfactorily, we have written to Bristol for a further supply, which will, doubtless, be forwarded to us by the earliest opportunity. It is to be expected, that the characters of the vaccinia will vary somewhat from those induced by virus, which has passed, as it were, through thousands of individuals.

The pages of the *Intelligencer* contain favourable testimony in regard to the new virus from Dr. Bridges and others. With some, however, extensive sloughing was observed occasionally, which led them to hesitate in employing it farther. Smallpox and varioloid having been unusually prevalent in Philadelphia, during the spring of 1840, Dr. Kirkbride, who was, at the time, physician to the House of Refuge, and to the Pennsylvania Institution for the Instruction of the Blind, was induced to revaccinate all the inmates of those institutions; and for this purpose, he used the new virus, and he remarked, that although some members of the profession appeared disposed to reject it from the severity of the symptoms which it induced, yet, except in three cases, he never witnessed sloughing or other unpleasant effects. His own observations induced him to put more confidence in its prophylactic powers than in the old virus, although, as he properly remarked, "this point can only be settled by time, and an enlarged experience by the profession generally." After this, I lost sight of the new virus as it passed from individual to individual; and it fortunately happened, that confidence became again restored in regard to the preservative powers of the old virus; so that little has been said on the subject since that period.

VI. PHILADELPHIA DAYS

William B. Carpenter. Friedrich Oppenheim. William Stokes and Robert Graves. The Philadelphia Hospital. Harlan-Gibson feud. Revolt of the resident physicians. Brandy and delirium tremens. The insane department. Mr. J. P. Kennedy. The pharmacopoeia of the U. S. Mr. John Vaughan. Mr. Du Ponceau. Mr. William B. Wood. *The Medical Student.* New remedies. Roget's "Physiology." Traill's *Medical Jurisprudence.* Practice of medicine. Charles Caldwell again. *Cyclopedia of Practical Medicine.* Dunglison's industry. Club of Five. Baron von Raumer. Judge Story. Scientific medicine. Sir James Clark. Sir Henry Holland. Lead poisoning. Dr. Pereira. Sydenham Society.

WILLIAM B. CARPENTER

The first communication I ever had from Dr. Carpenter was the letter from which I took the first extract in regard to the new vaccine virus.

22 Park Street, Bristol
Oct. 26, 1838.

Dear Sir,

Although I have not the pleasure of being personally known to you, and probably not even by name, I find, that I may claim acquaintance with you by my written productions, as I am the author of most of the later physiological articles in the *British and Foreign Medical Review*, amongst which was the critique of your work on Physiology. I am happy to find, from the third edition of that work, which has just come into my hands, that you feel satisfied with the notice there taken of it; and I am also very glad to perceive, that you have introduced into this Edition so much that will render it of increased value to the Student, who is desirous of pursuing the subject for himself.

Having, for several years, been endeavouring to bring together the principal facts relating to the vital actions of living beings of *all* classes, and to render them subservient to general principles of *universal* application, I have been encouraged, by the rapidly extending taste of Physiology in this country, to put them forward in a form, which may be interesting to the general reader, and valuable to the Medical Student under the title of "Principles of General and Comparative Physiology."[1] My friends Dr. Forbes, Professor Alison,[2] Sir James Clark and Dr. Prichard have led me to believe, that such a work would be useful and acceptable, and Dr. A. has promised to do what he can to promote my views, by recommending my work to the Students of his class in the University of Edinburgh. You will see from the accompanying sheets, which constitute about two thirds of the volume, that the plan of the work is quite distinct from any which has yet appeared; and as your own is confined to *human* physiology, I am induced to think, that you may view this as an appropriate addition to it. It will be illustrated by about 220 figures in copper and wood. One of the plates which I send is nearly complete;—the other is a proof I have this morning got of the etching only. The remainder I shall send you by the next trip of the Great Western. I have been induced by Dr. Forbes's opinion to believe, that there might be a probability of my work being reprinted in America; as, from the success of your own work, it is evident, that the taste for the science of Physiology must be on the increase. I take the liberty, therefore, of requesting your mediation in the affair, as, although I have many friends in the United States, I have none among the medical profession; and if your opinion of the work is favourable, it might probably weigh much with a publisher. The volume consisting of about 480 to 500 pages, and with six plates, will here sell for 14/ or 14/6—a price, which, if the whole edition sells, will barely remunerate me for the expences of printing and engraving. I hope for my profit in future editions, which will, of course, require a much less outlay in proportion to their extent. If the work were reprinted in America, I should propose an arrangement of this kind,—that I should retain the copyright, but that the publisher should print an edition of what size he thought proper,— that I should send him the necessary number of impressions of the plates, and that he should give me a certain sum as compensation. The *engraving* of the plates will cost me nearly £60; and an American publisher might, therefore, well afford to give me £50 for the edition, paying, of course, the expenses of the printing and paper of the plates. Of course, if I could get more, I should not be sorry, as guineas are not yet sufficiently plentiful with

[1] 1st ed., London, 1839.
[2] William Pulteney Alison (1790–1859), Professor of Medicine, Edinburgh University.

me for me to neglect picking up what I can; but I should not stand upon this. I believe, that to secure copyright in America, there must be something more than a *reprint* of the English edition. If this is the case, I should add a few pages near the close, amplifying the chapter on the nervous system. I am afraid that there will not be time for you to come to any decision previously to the return of the Great Western. Should you be able, however, to make any bargain, which you would consider fair towards me, there would be no objection to beginning the printing at once, and, with this view, I have made one or two corrections on the sheets, in which there are, I believe, but very few errata besides these. Dr. Forbes has mentioned to me the house of Broaden and Otis (he thinks) in Boston, as one which might probably undertake the work; but I should much prefer its coming out in conjunction with *yours* if it appears to you, as it does to me, that the plan and scope of the works are at once *distinct* and *harmonious*. You will, I hope, excuse all the trouble I am now giving you, from the consideration, that it is in the cause of science; and from the more personal one, that I should be most happy to do the same for anyone of your country so far as any opportunities admit. As yet, my influence is more with Reviews than with publishers, and having a reputation to make, I have found it a hard matter even to get a publisher to divide the risk of this Edition with me. . . . With the sheets of my work I enclose a copy of the Centenary Address, which it fell to my lot to deliver some time ago, and which if you were an Edinburgh student, may interest you.

Believe me to remain, Dear Sir,
respectfully and sincerely yours,
William B. Carpenter.

Professor Dunglison.

My publishers—Messrs. Lea & Blanchard—were unwilling to lay hold of the "Principles of General and Comparative Physiology," either on its first or second edition, but of the third edition published in 1851, and immensely enlarged, after the Author's reputation had become so completely established by the republication of his *Human Physiology*, his *Manual of Physiology* and other works, they purchased a considerable number of the English copies.

Sometime after the date of the above letter, I received the following, complaining of an attack, which had been made upon him in one of the most prominent medical Journals of Great Britain.

Bristol April 14, 1840.

Dear Sir,

If you have seen the January number of the Edinburgh Medical and Surgical Journal, you will, I dare say, have noticed the violent attack, which is there made upon me, and my Principles of Physiology. To me this was most unexpected, as I had fully calculated from the general reputation I had left behind me in Edinburgh, and from the friendly footing I was on with Dr. Craigie,[3] upon a favorable notice. The violence of it, however, overshot its mark; and forcing me, as it did, onto the field, it has enabled me to make what is, I believe, universally (with the exception of the Editor and his clique) regarded as a triumphant reply. The names of some whose testimony I have adduced in my behalf are probably better known on

this side of the Atlantic than on yours. Dr. Pye Smith has a very high rank in Britain as an orthodox theologian; and he has lately shown himself well versed in Geology and its connected sciences, by a series of lectures, which he delivered and published on the relation between Scripture & Geology, for which he has received from the Royal Society the unsolicited honour of a Fellowship. Professor Powell, of Oxford, is well known in Britain as one of our ablest Mathematicians and *Physicists* (most barbarous word) and is a fearless advocate of liberal principles on religious subjects. If you have not seen his "Connection between Natural and Revealed Truth," I think you would be pleased with it. He has lately published a very able pamphlet upon "State Education." I did not put these testimonials into circulation, until I had used all the influence I could exert to get my own Reply admitted into the Journal itself; but this I could not effect, the Editor not answering my own application, and giving a point-blank refusal to Dr. Alison, who was kind enough to go to him in my behalf. I scarcely know why I trouble you with these personal details; but I cannot help thinking, that you may be interested in them. I believe, that the "head and front of my offending" is my being an acknowledged contributor to the B. & F. M. R., which has already a circulation by one half larger than the E.M.J. I should much like to know your opinion of my Review of the *Bell* Controversy in the January No., and of the article on Education in the last No. if that reach you before you write. If you have not seen the original accounts of Schleiden's and Schwann's researches, you will, I doubt not, be interested in the analysis of them, which is a joint concern between Dr. Baly[4] (the translator of Müller[5]) and myself. Owen[6] has just published the first part of the Odontography,—one of the most splendid and remarkable contributions to Comparative Anatomy that has ever appeared. To think, that the microscopic structure of a fragment of a tooth should afford a better indication of the affinities of the animal than the bones themselves! I am now collecting testimonials in readiness for any physiological vacancy that may occur. If you would favour me with one, founded upon my Principles of Physiology, Thesis, and my papers in the B. & F. M. R. I should feel much obliged. I should also be glad of any notices of my book in American periodicals. I think you mentioned *three*. Of these, only the Philad. Medical Examiner has reached me. If I am not asking too much in requesting copies of the others, or extracts from them if they are long, I should feel much obliged by them, as such testimonials are very useful with bodies, that are governed by documentary evidence. Believe me to remain, dear Sir, respectfully and sincerely yours,

William B. Carpenter.

The next topic of correspondence between Dr. Carpenter and myself was the unfounded and unjustifiable attack made upon him by Dr. Martyn Paine, of New York, to which I referred at page 81, and on which he wrote to me the following letter:

---

[3] David C. Craigie (1793–1866), editor of the *Edinburgh Medical and Surgical Journal*.

[4] William Baly (1814–1861), after acquiring his M.D. in Berlin, spent the next four years translating Johannes Müller's *Physiology*, adding copious notes of his own.

[5] Johannes Müller (1801–1858), pioneer of modern physiology, teacher of Schwann, Helmholtz, Virchow, Köllicker, Dubois-Reymond and Henle, who were also famous medical educators.

[6] Sir Richard Owen (1804–1892), Hunterian Professor at the Royal College of Surgeons, published his *Odontography* in two volumes, 1840 and 1843.

Bristol Feb. 3, 1841.

My dear Sir,

I feel extremely obliged by your very kind letter of Dec. 29; and for the trouble you have so kindly taken to set me right with the American Public, even at the risk of annoyance to yourself from my pertinaceous opponent. I received from Dr. Jackson of Boston an account of Dr. Paine very closely corresponding with that which you gave; and I certainly feel glad, that the estimation in which he is held is not such as to prejudice me in the minds of those of your countrymen, whose good opinion I should most value. In England it is very different. Any person, who comes much before the public is liable to be attacked in a manner that may give him great annoyance, even though there be not the least foundation for the charge, by anonymous scribblers; and from the small size of our country, such calumnies are circulated very rapidly, and spread through the whole community. If such charges come with a name, even though that have no weight in itself, they are all the more believed, and if the name be that of a person, who is himself in any degree a public character, it is at once presumed, that the charge is true, and the unlucky person attacked finds it difficult enough to get rid of the stigma. The charge brought against me by Dr. Paine had so much *vraisemblance* [7] in it, as to be readily caught at; and people did not suspend their judgments, until they knew who and what Dr. Paine was; but jumped at once to the belief of my delinquency, without waiting to hear what I had to say for myself. I except, of course, my own circle of acquaintances among whom I flatter myself, that my character stood sufficiently well to induce a general discredit of the charge, or, at any rate, a suspension of judgment, until I should have time to clear myself. I had no idea, that *Universities* were so easily got up among you, as they would seem to be, by the account I hear of the school with which Dr. Paine is associated. I wish his colleagues joy of Pattison. He did all but ruin our so-called London University—now sunk into the comparative insignificance of London University College—during his brief connection with it. I shall be much obliged by your sending me any numbers of their Lancet, that may contain an attempt at reply to my letter. The pages cut out and inclosed in a letter, will reach me more speedily than if sent through the bookseller. I have had no reply from him to my communication; nor do I much expect one. I learn recently from Dr. Holland,[8] that he has written Dr. P. a severe letter on his own account, charging Dr. P. with dishonesty in his quotations from his "Medical Notes." I believe that the matter is now so completely set at rest in this country, that it will be unwise to disturb it again, but it may be necessary for me to take some notice of Dr. P's defence should he make one. I know, that there are one or two of my kind *friends* here (Dr. Marshall Hall,[9] for instance) who believe, and express their belief, that Dr. Forbes and I are the *receiver* and *thief*. Should any insinuation of this kind be publicly made, I shall urge Dr. F. to give up the name of the real delinquent, which he may do with the less hesitation, as the man has left Britain under not very favourable circumstances. I have now to express to you my most grateful thanks for your kind interest in regard to my new work on Physiology, the republication of which in America would be a high

---

[7] Appearance of truth.

[8] Sir Henry Holland (1788–1873), physician-extraordinary to Queen Victoria and Prince Albert, visited America eight times, published *Medical Notes and Reflections,* 1st ed., 1839.

[9] Marshall Hall (1790–1857), English practitioner and physiologist, discovered reflex action of the nervous system.

---

gratification to me, even though I gain nothing by it in a pecuniary view. I had hoped to have been able to send you the complete work by this packet; but its appearance has been delayed, in consequence of the large amount of new matter I have had to introduce at a very late period, in consequence of the publication of some important papers in the Phil. Trans. and of Gerber's General Anatomy, including some important researches by Mr. Gulliver.[10] I have not attempted to do more than bring together, in a concise and systematic form, the results of the latest and most satisfactory inquiries in different departments of Physiology; there is little that can be called *original* in the work, but much that will be *new* to most readers, even to those well acquainted with the elaborate Treatise of Müller. I have endeavoured to make it a Student's book; and have dwelt much on the *practical applications.* Dr. Paine will doubtless find much in it to criticise severely; as I am a strong upholder of the Humoral Pathology in its present modified form; and I have indicated many important paths of observation and enquiry. I have made enquiry of Churchill respecting Casts of the Woodcuts; but these he does not seem very willing to supply. There are about 100, many of them executed quite in first rate style; and are very well printed (I cannot compliment *the States* on any productions in that line that have fallen under my notice). As all the expenses of the publication are undertaken by him, I know nothing of the cost of these, except that he has grumbled not a little at some of them for which he has had to pay between 3 and 4 guineas. I should think the cost of the whole has not been less than £150. He says, "I have so frequently had applications of this sort, that I now never entertain them, for I never once brought the negotiation to bear. If you offer blocks at 10/ each, which have cost 40s/ or 60s/, they will offer 4/ or 5/; consequently, it is better not to injure the books by the process of stereotyping." I scarcely suppose, that the sale of the work in America is likely to be such as to make it worth while to reengrave the cuts; and if your publishers think so, they must either offer Churchill a fair price for a set of casts (say a *quarter* of the original cost of the blocks) or give up the scheme. As I know, that there is nothing to prevent an American publisher from reprinting an English work, I do not see how Lea and Blanchard can be expected to make me any compensation for the use of mine. But if they should still desire to reprint it, I would suggest this plan. I believe, that, even now, I could considerably improve the work, so rapid has been the advance of Physiology, whilst it has been passing through the press, and from some investigations on which I am now entering, I think it likely, that I shall have still better means of doing so a few months hence. Now, if Messrs Lea and Blanchard thought it worth while to make me a sufficient compensation, I would supply them with a copy carefully revised and corrected, and with such additions as I may think desirable,—in fact, a new Edition in time for publication at the beginning of the next Session, which is, I presume, with you as with us, the great time for the publication of Medical Books. Will you be so good as to inform them of this proposal, and request them to state their intentions to me. I should add, that there is a steel plate besides the woodcuts, embracing subjects, that could not be well expressed in wood. If they should make any proposal regarding the blocks, they might negotiate also for impressions of the

---

[10] George Gulliver (1804–1882) supplied Notes and Additions to the English translation of Fr. Gerber's *Elements of the General and Minute Anatomy of Man and the Mammalia,* London, 1842.

plate. With many thanks for your most kind and disinterested conduct in these concerns, believe me to remain,

dear sir, most sincerely yours,
William B. Carpenter.

About the latter end of the same year, I received the following letter from him inclosing the communications of himself and Dr. Forbes on the subject of the plagiarism laid to his account by Dr. Paine (p. 536). It will be seen how much he continued to be disturbed by the conduct of that gentleman.

Bristol Dec. 1, 1841

My dear Sir,

I cannot but anticipate, that you will peruse the enclosed letter containing my denial, backed by Dr. Forbes's statement of the serious charges brought against me by Dr. Martyn Paine, of New York, (which are quite new to me, although his pamphlet is dated *April,* and which will have become almost a dead letter with you by the time you receive this) with some satisfaction. I trust, that I am not taking too great a liberty in the request, which concludes mine, that you will procure the publication of them in the leading medical Journals of the United States. I do not know in what estimation Dr. P. is held amongst you, but I cannot help remarking, that his conduct in this business gives me a very low opinion of his character. He has absolutely taken the trouble to send 1000 copies (so I learn from New York through a private source of information) by post into this country; and has thus circulated his calumnies against me into quarters in which my denial can never follow them; and all this because he *fancies,* that I am the author of an unfavourable Review of his book, with which the crime I am charged with has *nothing whatever to do.* It is curious, that his opinion of the merits of the Review, and of my contributions to it should have undergone such a complete change within the last eighteen months. Had the charge been true, I could never have held up my head again among men of Science; and as it is, I cannot but believe, that the diffusion of it in this country will have done me irreparable injury among those good opinion may be of great importance to me in future whose good opinion may be of great importance to me in future life. The only way in which the evil can be remedied, one who has received his pamphlet, with a request, that this may be mentioned, wherever the pamphlet has been spoken of. This I shall call upon him to do; and, if he refuse, I think I shall have a right to place him in a very disadvantageous condition with regard to the public. In order to save you trouble, I have forwarded several printed copies of the enclosed, by a parcel to the care of Messrs. Cary: with a copy of the new edition of my Physiology, in which I think that you will find much to interest you, and of which I have much pleasure in requesting your acceptance. Believe me to be, dear Sir, respectfully and sincerely yours,

W. B. Carpenter.

The following letter is on the same subject.

Bristol June 17, 1842.

My dear Sir,

I am much obliged by your letter of April 28, and am glad to find, that you had received my volume on Human Physiology. I shall hope to obtain from you some *critical* remarks upon it, either as a whole, or as a collection of distinct physiological doctrines. Hitherto it has met with nothing but commendation in this country; and I am really vexed with some of my friends for not giving me the opportunity of benefiting by the suggestions, which must, I

doubt not, occur to them. I anticipated a complete demolition from my *friend* Paine in the New York Lancet; but from what you tell me of the position of that Journal in regard to the school, it seems possible that I may escape. I shall esteem it a favour if you will forward to me any Reviews of it worth reading that may appear in American periodicals. I am very glad to find, that my credit is sufficiently good amongst *you,* to bear up against Dr. Paine's repeated and vehement onslaughts. In *this* country, the whole matter has been settled from the time that my Reply was in circulation; and not the least notice has been taken of his subsequent effusions except a request, by some of the correspondents of the *Lancet,* that Dr. P. would pay the postage of his pamphlets. It is amusing enough, that even my old antagonist, Dr. Marshall Hall, who is evidently the "distinguished philosopher," quoted with such gratulation by Dr. Paine, is now on the best possible terms with me. I believe, that he is at last persuaded, that I never meant anything but justice to him; and has been smoothed down by my presenting him with a copy of my *Human Physiology,* which I sent to him as a just tribute to the value of his labours, and as an acknowledgment of the use I had made of them. To finish with Dr. Paine. I am much obliged to you for your caution respecting Dr. Lee, with whom, however, I have had no direct communication. The letters of his to Dr. F., which I have seen, however, appear written openly enough; and he does not disguise his leaning to Dr. Paine. At the commencement of the affair, he used to Dr. F. the expression, that I should find Dr. P. a formidable opponent on account of his stores of literary knowledge, his retentive memory. Still, he expresses himself as fully satisfied of my innocence of the plagiary; and gives a very amusing description of the mode in which Dr. P. received the intelligence of the *real authorship* of it. My friend, Dr. John Reid (now Professor of Medicine in Saint Andrews),[11] who has been visiting me lately, told me, that having put Paine's last letter into the hands of a person, who had previously heard nothing of the matter, he found him possessed with the notion, that the plagiary was contained in my review of Paine's book, so completely are the different questions jumbled together. Dr. Lee mentions that Dr. Caldwell is doing great things in Animal Magnetism down in the West. I met that worthy Professor at Dr. Prichard's table, and was excessively amused at his dogmatism and self reliance; and at the cool way in which he *set down* everybody, who presumed to differ from him, even Dr. P. himself. I have desired Churchill to forward a copy of my two volumes to you for the American Philosophical Society, which he has done through Messrs. Wiley and Putnam. When Dr. Gibson (of the University of Pennsylvania) was in England, he told me, that he thought it not improbable, that I might be elected a Member, if I would send these as my credentials. Perhaps you will be so good as to remind him of this. (Dr. Gibson took no action whatever in the matter. Some time after this, I brought him forward, and he was elected.) In regard to the reprinting of my *Human Physiology* in America, if the thing should be determined on, I should be glad to send over some corrections and improvements, though I may get nothing for them, as I should wish the edition to be as perfect as possible. These I will send forthwith, if desired by the Publishers. Dr. Lee spoke of having seen the work announced, and said, "it will have an extensive sale." I have never received the work you mention on *Practice* (my own *Practice of Medicine*) nor do I find, that Dr. F. has seen it. There is a likelihood of a vacancy in the Chair

---

[11] John Reid (1809–1849), Chandos Professor of Medicine, University of St. Andrews, Scotland, noted for his investigations upon the glossopharyngeal and vagus nerves.

of Physiology at Edinburgh before long, by the transference of Dr. Alison to the Chair of Practical Medicine. I am not sure, whether I shall start for it, as the School has very much declined; and as I may not improbably be excluded on the ground of my religious opinions (He was a Unitarian.) I think it as well, however, to be prepared; and I shall feel much obliged if you will transmit to me, at your earliest convenience, such a testimonial as your knowledge of my writings may enable you to give me. With many thanks for your kind attention to my wishes in past concerns, believe me to remain, Dear Sir,

respectfully and sincerely yours,
William B. Carpenter.

Prof. Dunglison

In spite of Dr. Paine's denunciation of the Microscope, I am, at present, employing it to great advantage in the examination of the supposed unorganized skeleton of the Invertebrata; and I have made some very interesting discoveries proving their high organization.

I sent him such a testimonial as I could conscientiously give him; but his application was fruitless, as is shewn in the following communication.

Bristol, Oct. 22, 1842.
My dear Sir,

I am much obliged by your kind letter, and am very sorry to learn, that you have been suffering severely from illness.[12] I could wish to send you a long reply on various topics; but I have been so much engaged, that I am driven up to the last hour, that the Post remains open. In regard to the Edinburgh Professorship, I very early found, that my unfortunate heresy would prevent my success, rendering it improbable, that I should have even a single vote, and, therefore, making the expense and trouble of a personal canvass quite useless. I was led, a few years ago, to a belief, that this would not be the case; but parties run very high in Scotland at the present time, on religious matters, and both sides join in antipathy to the unfortunate Unitarians. There is a law requiring subscription to an orthodox test on the part of a newly elected Professor; but this has become quite obsolete in Edinburgh.—Nevertheless, I am assured, that it would have been revived by the clergy in my case, had any strange concurrence caused my election. I thought it right, however, to collect and print my Testimonials, which I enclose. I was sorry, that having mislaid the one, which you were so good as to send me two years ago, I could not insert it. I could have got many more, had time permitted, and had I been disposed to increase the bulk of my collection. Of the influence of this decision upon my ultimate prospects I shall presently speak. In regard to my *Human Physiology*, I am very glad to be able to tell you, that, although it was unfortunately delayed, until the end of our medical publishing season its sale has been very good—500 copies in the first three months. My publisher reckons, that a new edition will be required at the end of next year (1843); this will necessarily contain a great deal that is new—so rapid is the progress of physiology—and will be improved in various other respects. Now, it seems to me a great pity, that the first edition should be reprinted in America, when the second will be close on its heels in England; and I should, therefore, strongly advise, either, that the republication should be delayed, or that your publishers should make some agreement with me for the early supply of the new matter. I do not intend to increase the bulk of the volume, but to leave out the most unimportant matter, replacing it by what

[12] See p. 95 where this illness is also mentioned.

is of more value. There will be a dozen or so of additional illustrations.

I quite agree with you, that Liebig [18] has been much overpraised as to many points, but with regard to some of the subjects he has least touched upon, I think his hints very important. You will see my general estimate of this work in the Oct. No. of the B. & F.M.R.[14] I did not review the former treatise. The most intelligent men in this country are coming fast towards the establishment of a *rational humoral pathology,* in spite of the opposition of Dr. Paine. I do not trouble myself anymore about that worthy. I have not seen his last Reply, save a paragraph extracted in the Lancet, which contains one of his usual blunders. Dr. F. never *could* have said, that the plagiarist and I wrote the whole. It is not improbable, that Dr. F. *did* let out, that I wrote part of the Review on *Inflammation,* though not a syllable of that on *Hunter* which was all done by the plagiarist, Morgan, then of Aberdeen. I reviewed Macartney and Carswell; and Rasori was done by another hand. That is the whole case, and I wish it had been so stated at first; but Dr. F. was not willing to let the public so much into the secrets of his Review. I have been latterly turning my thoughts a good deal towards America; and I should be glad of your opinion as to the probability of my being able to make an income of £300 or £400 a year (which with a little from other sources, will suffice for my very moderate wants) by Lecturing, Private Tuition on medical and scientific subjects etc. etc. all which is much more to my taste than practice. I have been informed, that if I could get invited over by the Lowell Institute in Boston, I should acquire an advantageous *status.* Do you think this so? I have a good many friends in that town, who would, I think, be glad to serve me for my own, and my father's sake. It is not for me to speak of myself, but I think you will see, from some of the inclosed, that my personal qualifications are not unfavorable to my success. I could do very well on this plan in England were it not for my heresy. I should of course, hope to gain a footing in some of your chief medical schools, when a vacancy might offer. I do not know, whether you are aware, that the little book on Vegetable Physiology, which I see advertised for republication is mine. It has had a very fair success, and I shall probably continue the series. I shall hope to hear from you soon in reference to the last part of this letter. Pray understand, that I ask only for your *individual opinion,* and do not wish you to consider yourself as incurring any responsibility in advising me. Believe me to remain, most sincerely yours—

W. B. Carpenter.

I did not encourage Dr. Carpenter to come to this country under all the prospects that presented themselves to him in his own; and, more especially, as I had learned from a gentleman, who had heard him speak, that he was by no means happy in his delivery. I stated farther to him, that whilst the delivering of a course of lectures in the Lowell Institute might be honorable, itinerant lecturing was considered somewhat derogatory, in consequence of the field being occupied by so many pretenders, whilst men of true science, in the United States, were exceedingly shy of appearing before the public as popular lecturers. To

[18] Justus Liebig (1803-1873) in *Die organische Chemie in ihre Anwendung auf Physiologie und Pathologie,* 1842, first classified organic food stuffs and the processes of nutrition.
[14] 14: 492-521, 1842.

my letter on that subject, I received the following reply.

Bristol, Apl. 3, 1843.

My dear Sir,

I ought to have thanked you ere this for your kind reply to my inquiries respecting my prospects of success in America; but I have been partly waiting for something else to say, and have also delayed in consequence of various circumstances, which have very closely occupied my time. Your letter, together with the results of other inquires satisfied me, that I shall be more prudent in waiting for sometime longer to see what may turn up for me in this country. My kind friend, Dr. Forbes, is sanguine of my ultimate success in London; and if I can obtain the situation, which will be vacant next year, of Fullerian professor in the Royal Institution (which is now held by Mr. Rymer Jones)[15] I shall probably remove there at once. Everytime that I visit London I feel an additional longing to be a resident there and to be able to avail myself of the numerous opportunities of advancing in my objects of pursuit, which will there be opened to me. I am much pleased to find, that Messrs. Lea & Blanchard have determined upon reprinting my *Human Physiology;* though I could have much wished, that they would wait for the new edition, which will almost certainly be required by the end of this year, and which I shall take the greatest pains to render as complete as possible. It is now but little more than twelve months since it first came out; and I understand from Churchill that nearly 1000 out of 1250 copies have been sold, a measure of success with which I have reason to feel very well satisfied, especially as it has to contend not only with Müller, but with Wagner,[16] which is being translated by Dr. Willis.[17] Had I been a little less closely occupied than I am at present, I would have drawn out a list of the most important alterations, which I should desire to see made. But I must confine myself to requesting you to draw the attention of Dr. Clymer (whom I have not the pleasure of knowing) to the following points: 1. I should wish the Addenda and Corrigenda at the end of the volume to be inserted in their proper places, where consistent with the still later expressions of my views which I shall presently indicate. There is a stupid mistake also in p. 399, line 3, where *tricuspid* should be substituted for *mitral,* which I should wish corrected. I have not noticed any other errata of this kind: 2. I should wish as much as possible of my Report on the Physiology of cells, or at least of that part of it, which relates to the connections of cell life with the organic functions—to be incorporated (see Dr. Forbes's Review Jan. 1843).[18] The *general doctrine* should be introduced somewhere about §§88 (I have not time to look for the appropriate spot); and the other

portions under their respective heads of Absorption, Assimilation and Secretion. 3. I should be glad to have all those passages expunged, in which there is anything like a favourable opinion expressed as to the conversion of blood corpuscles into tissue cells, as I now believe this entire doctrine to be a mistake of Barry's.[19] My general reasons are stated in my Report, which, as expressing my more matured opinions, I should be glad to have made the standard for all that portion of my book, which relates to the Blood and the act of Nutrition. 4. On the general anatomy of the tissues a good deal of additional and corrective information may be introduced from Mr. Paget's Report (in Dr. Forbes's *Review,* since reprinted separately with some important additions[20]) and from Dr. Todd and Mr. Bowman's *Physiological Anatomy* of which the first part is just published. This last contains the best account of the Formation of Bone (by Mr. Tomes[21]) that I have yet seen. 5. I should like some notice taken of the very interesting confirmation, afforded by Mr. Owen, to my views on the Nervous System, which you will find in Dr. Forbes's *Review* Jan. 1843, p. 212. 6. In regard to the questions treated of by Liebig I should wish my Review of his work in Dr. F's *Review* for October, and a short article on Dr. Bence Jones's book on the Urine etc.[22] in the present April No. to be made the guide in any alterations. I find intelligent men in London very *unfavourable* to Liebig, though a few cry him up as infallible. Of course I do not wish to tie Dr. Clymer's hands as to additions or annotations, but any *alteration* in the *text* should be made in accordance with my present views, as elsewhere published. I do not think that many omissions can be made without injury. Pray accept my thanks for the very kind interest you have shewn in this affair and believe me to be most sincerely yours

W. B. Carpenter.

The following letter was mainly in acknowledgment of the aid I had rendered him in the affair of Dr. Paine, and in the republication in America of his *Human Physiology.*

Bristol Dec. 30, 1843.

My dear Sir,

I have too long delayed replying to your last kind letter, which ought to have drawn an immediate acknowledgment from me of my strong feeling of gratitude for your generous conduct in defending my reputation at the expense of trouble and annoyance to yourself, and in recommending the republication of my *Human Physiology,* which must, to a certain extent, affect the sale of your own excellent work. But I was much occupied just at that time, with family matters, in consequence of the death of a near relative, and various other causes have kept me almost a slave to my work ever since. I cannot let the year slip by, however, without writing to you, which I am the more moved to do, in consequence of having lately received from Dr. Clymer a copy of the Reprint of my *Physiology,* which he was good enough to forward by a private hand, so that I have had no trouble about it. The book is got out in very creditable style; and I am very much pleased with the manner in which it has been edited. I think it just as well,

---

[15] In 1833 Mr. John Fuller established two chairs in the Royal Institution of Great Britain (founded in 1799 and sometimes confused with the Royal Society), one of chemistry, given to Michael Faraday for life, the other of physiology, the incumbent to be elected every third year, both professors to be called Fullerian Professor. Thomas Rymer Jones (1810–1880), M.R.C.S., a zoologist, held the Chair of Physiology 1840–1842.

[16] Rudolph Wagner (1805–1864), Professor at Göttingen, *Lehrbuch der speziellen Physiologie,* [1838]–1842.

[17] Robert Willis (1798–1878), a great translator of books, among them Rudolph Wagner's *Lehrbuch der Physiologie* bearing the English title *Elements of Physiology.*

[18] Report on the results obtained by the use of the microscope in the study of anatomy and physiology. Part II. On the origin and functions of cells, *British and Foreign Medical Review* **15**: 259–281, 1843.

[19] Of some thirteen Barrys, I assume this is Martin Barry (1802–1855), English physiologist.

[20] Sir James Paget (1814–1899): *Report on the Chief Results Obtained by the Use of the Microscope in the Study of Human Anatomy,* London, 1842.

[21] Sir John Tomes (1815–1895), pioneer in English dentistry.

[22] Henry Bence Jones (1814–1895), eponym of Bence Jones protein in urine.

that Dr. C. has abstained from Liebigism: [23] for it is difficult, without entering *critically* upon the various questions, which are now afloat, to give anything like a satisfactory account of them; and I am myself much puzzled as to the extent to which it will be wise to go, in the new Edition I am now preparing. His estimate of the merits of the book in his own short Preface is just what I like. I have not been able to do anything in regard to your work on the Practice of Medicine, in consequence of not having received the concluding sheets of it, which has prevented my sending it to Dr. Forbes for review. My copy stops at p. 460 vol. 2. The public appreciation of American Medical Literature is not, I think, so great in this country as to induce a publisher to undertake a reprint of a bulky work until attention has been drawn to it by previous favourable reviews in the leading periodicals. We certainly have wanted a good book on Practice, more than almost anything else, but I am inclined to think, that the recent publication of Dr. Watson's Lectures,[24] which were very highly thought of at the time they appeared in the *Medical Gazette*, will have. supplied the deficiency. He is not a man of brilliant talents, but of high mental cultivation and of most excellent judgment. I know him well, having been his pupil whilst in London. Allow me to call your attention to the recently published work of Dr. C. J. Williams on General Pathology.[25] It is extremely well up with the physiology of the present time; and is, to my thinking, a great advance upon any previous work of the kind that I have seen. I am not disposed to agree with him in his views upon Inflammation, which seem to me too mechanical; but I have not met with anything else that I should decidedly object to. I send you, with this, a copy of a short account, which I have lately published of my microscopic researches on Shell etc. The field is so extensive, and my power of prosecuting the inquiry (from limited time and pecuniary means) is so small, that at the present rate, some years must elapse before I can work out the subject completely. But I have been encouraged to do so by the British Association, and I hope to communicate to them at their next meeting, a detailed account of some of the families in which my results promise to be of most interest. Through a mistake of the printer, I have received but a very small number of spare copies of the paper, and the one I send you is the only one I have left. I should be much obliged to you to draw the attention of Prof. Silliman to it, if it could be reprinted in his Journal it would do me much good. He may probably have seen it in the *Annals of Natural History*, which I presume to be forwarded to him. If it should incite your American Geologists to send me a box of fossils for examination, I can only say, that I shall most gladly receive them. You will, I am sure, be glad to hear, that better prospects are opening to me in this country; and that I have, at present, no thoughts of being obliged to seek a livelihood elsewhere. I cherish the hope, however, of paying a short visit to the U.S. before many years are over, for the sake of becoming personally acquainted with the many kind friends I have amongst its scientific and literary men, and this might, perhaps, be arranged by my obtaining an engagement at the Lowell Institute in Boston, which would pay my ex-

penses. I shall have the pleasure, shortly of sending you a copy of the new edition of my Human Physiology, now in the press. It will be, in many respects, much improved, but I shall avoid enlarging it more than I can help; thinking that, for the use of the student, it is quite bulky enough already. It has kept its ground very well in spite of its competitors, which are now becoming numerous—Wagner (translated by Willis)—Todd & Bowman,—and Valentin (very likely I should think, to be translated)—in addition to Müller. But I know, that it can only *remain* a standard book by being kept well up with the advance of science. Believe me to remain, dear Sir, yours respectively and sincerely Prof. Dunglison

William B. Carpenter.

In the year 1845, I wrote to Dr. Carpenter by my friend and colleague Dr. Meigs. He had then removed to London, where, along with his other pursuits he was engaged as a private tutor in Lord Lovelace's family. In answer to my letter he sent me the following.

24 Regent Street, London
June 2, 1845

My dear Sir,

By a note, which I have lately received from you, through Dr. Meigs, I fear, that you have not had a letter, which I wrote to you on receipt of the last edition of your Physiology, thanking you for that kind present, and adverting to a few other matters. I have had much pleasure in the limited amount of intercourse, which Dr. Meigs's short stay, and numerous objects in London permitted me to have with him, but he has promised to spare me a day, if possible, at Repley, after his return from the continent. Dr. Forbes has placed in my hands the commission, which you sent to him respecting a Microscope; (see p. 78) and I have done my best to get it executed in the best manner, that the price you named will admit of. . . . Whenever I am able to despatch it, I shall do myself the pleasure of sending you a series of preparations of my own making, illustrative of the structure of shell, bone etc. together with a few samples of the perfection at which the art of minute injection has arrived in this country, with which I shall shortly be furnished by my friend Mr. John Quekett, who has charge of the Museum of the College of Surgeons, holding the situation formerly occupied by Owen, who has now succeeded old Clift in the higher office. . . . I have been rather struck with finding from Dr. Meigs, how little attention seems to be paid to microscopy in the United States. I believe, from his account, that Bristol alone contains more first class instruments than the whole of your country could furnish, and the number of those, who are really working at minute anatomy, is very considerable. Mr. Simon's recent prize essay on the Thymus Gland is a very good example of the success which attends this new method of inquiry, in unravelling the intricacies of Physiological Anatomy. . . . My own plans, after which you are good enough to enquire, have been lately quite unsettled again, by a change of residence on the part of Lord Lovelace (whose children I was educating) which necessitates the dissolution of the engagement. I am, at present, on the lookout for something else, and have two or three different things in prospect. I asked Dr. Meigs, whether he thought, that I could make a popular course of Lectures answer in some of your principal towns, as I have long had a wish to visit America, and might, in this way, do something more than pay my expenses. He told me, that if I could get up an attractive course on Physiology or Microscopy, I might be secure of a very large class in Philadelphia, and of a good one in

---

[23] Dunglison opposed Liebig and his school in regard to theories of "fat and gelatin being non-nutritious" (E. R. Squibb, *Journal*, p. 194, privately printed, 1930).

[24] Sir Thomas Watson (1792-1882): *Lectures on the Principles and Practice of Physic*, London, 1843, first published in the *Medical Times and Gazette*, 1840-1842, was highly regarded for almost three decades.

[25] Charles J. B. Williams (1805-1889) delivered the Gulstonian lecture on inflammation in 1841, which apparently formed the basis of his *Principles of Medicine*.

6 or 8 other towns in the Northern States, my reputation being now pretty widely diffused. But he told me, that he thought, that you would be a better judge on the subject. I am thinking of getting up a course on the Microscope, which should embrace its modern applications to Animal and Vegetable Physiology, to Geology & to Crystallography, —with a series of large drawings,—a first rate oxy-hydrogen microscope,—and a number of achromatic microscopes for the exhibition, after the lecture, of objects too refined for the oxyhydrogen. "Though I say it, that should not say it," I believe, that I have some *gift* as a Lecturer; (such had not been the character given of him to me) and, independently of any pecuniary advantage, I should really be very glad to be instrumental in diffusing a knowledge of my favourite branches of Science. Perhaps you will be good enough to turn this over in your mind, and to tell me the result of your cogitations. Would it be advisable for me, in forming such a plan, to seek an engagement at the Lowell Institute, Boston, as a certainty to begin upon? I shall have the pleasure of forwarding to you, by Dr. M. a copy of the first part of my detailed Report on Shell Structure, from the Transactions of the British Association just published. I am, dear Sir, Yours most sincerely,

W. B. Carpenter.

I, again, did not encourage him for the reasons I have elsewhere given (p. 115), to come to the United States as a peripatetic teacher; and I presume he soon abandoned the idea. His next letters to me dated 1 Clarence Terrace, Stoke Newington near London Sept. 29, 1845; and June 17, 1847; relate to the admirable instrument he had sent me, made by Smith & Beck, under his kind superintendence. In the latter communication, he says—

I received, a few months since, an invitation to lecture at the Lowell Institute next season; but I have been obliged to decline it, on account of previous engagements, though with much regret; and I am now becoming so fixed in London, by various appointments, which I have lately been fortunate enough to obtain, that I see no prospect of carrying into effect my *vision* of visiting the United States for some years to come. I expect, after the present year, to be able to devote a great deal more time to original investigations than I have yet been able to afford, as a large part of my income, during the last few years has depended upon the delivery of popular lectures, of which I shall now be able to relinquish all such as would take me away from London, and withdraw me from scientific pursuits. I am especially anxious to work out some of the problems, which connect Physiology and Metaphysics; and more especially to study the comparative anatomy of the Encephalon with express reference to the distinction between the Sensory Ganglia and the Cerebellar Hemispheric Ganglia. You doubtless saw my hand in the article in the B. & F. M. R. in October last, in which this subject was dwelt upon. The ire of the Combe school of phrenologists has been greatly raised by my animadversions; but as not one of them knows anything of comparative Anatomy, they have not yet ventured upon a reply. My friend Noble (whose book I reviewed) though still holding by the observations of Gall, as furnishing an adequate foundation for the subdivision of the Cerebrum, now fully assents to my views of the fundamental importance of the Sensory Ganglia, and many other unprejudiced phrenologists have taken my views into serious consideration. In the next edition of my Human Physiology, I think I shall be able to put the subject in a still clearer and more perfect form. I am now commencing

the preparation of a new Edition of my General and Comparative Physiology, which will be almost a new work (so rapid is the progress of discovery) and yet not one of my original fundamental principles has been shaken, but everything rather confirmed.

The last letter, which I have received from Dr. Carpenter, was the following, written to acknowledge the receipt of a copy of the last edition of my *Human Physiology* (1850) which I sent him by the hands of Dr. S. W. Mitchell,[26] the son of my colleague Professor Mitchell. I took this means of introducing Dr. Mitchell to him.

6 Regents Park Terrace
London June 27, 1851.

My dear Sir,

I fear, that you must have thought it very strange, that I have allowed so long a time to elapse without writing to you, and that, in particular, I have omitted to thank you for the kind present of your new edition of your excellent Physiology. When, however, you see my recently-published volume (His *Principles of Physiology, general and comparative*) and form some idea, from its contents, of the range of study requisite to have produced it, I think you will be kindly disposed to pardon my shortcomings. I have forwarded a copy for your acceptance, through Mr. Miller, and I trust, that you will receive it by the time this reaches you. For the last two years, I have been obliged to put aside *everything that could* be postponed, in order to proceed with this work; and it will be some time before I shall be able to do much in the way of original investigation of any subjects but those which I can *think* out. For I have now to prepare a new edition of my Manual, which has been out of print for sometime; and as soon as this is through the press, the *Human* must be taken up, this having been out of print since last October. I am going to make considerable changes in this last work, eliminating all, or nearly all, the *Comparative* Physiology that it contains, and extending several parts that have more particular reference to the *Human* subject, and especially the psychological portion, which has occupied my mind a good deal since the last edition was published. We have been considerably astonished by some of the phenomena exhibited by two Americans, Messrs. Darling and Stone, under the designation of Electro-Biology. These being shewn upon individuals of whose good faith the public is well assured, have made a strong impression; and medical men have shewn themselves more disposed to admit their validity, and to inquire into their rationale, than they have done in regard to Mesmerism. Nobody here attaches the least weight to the so-called electrical part of the affair. It is evident, that the effect may be produced by gazing at any *fixed* object, and that the subsequent actions are entirely determined by the *suggestions* (occupying wholly the mind of the patient; and consequently incapable of being corrected by ordinary experience) which are put into it from without. I consider these phenomena as of great value in throwing light upon some of the Mesmeric wonders, which we could scarcely disbelieve as *facts* but which we scarcely knew how to explain. I am come to the conclusion, that [there] is nothing [that] is too marvellous for a *sensitive subject* to

[26] Silas Weir Mitchell (1829–1914), M.D. 1850, Jefferson Medical College, whose interest in medicine was greatly stimulated by Dunglison, became a member of the American Philosophical Society and President of the College of Physicians of Philadelphia. For an account of this visit to England with his sister in 1851, see: Anna Robeson Burr, *Weir Mitchell*, p. 63, 1929.

be made to do, if she can be only made to *expect* its occurrence. The various results of Reichenbach's experiments all fall under the same category, being, I believe, purely *subjective.* The view, which these phenomena open up as to the influence of suggestion even upon minds in their normal state, is extremely extensive, and I believe, that Psychology will be much the better for the direction of the minds of *thinking* people to this inquiry. I know two cases at this present time, in which from the habitual practice of biologizing, (as it is absurdly called), the minds of the parties are so open to the influence of suggestions from a particular individual, that they cannot resist doing what he assures them, that they must, and will, do. Of course, this is a most undesirable state to be in; and I think it very wrong, that it should be induced. Of its reality, however, my knowledge of the parties forbids me to doubt. The subject has been pretty fully worked in Edinburgh, and you will find a paper on it, very sensibly written, by Dr. Alex. Wood of Edinburgh,[27] in the Edinburgh Monthly Journal of Medical Science for May last. . . . Shall we not see you over here at our World's Fair? I assure you I honestly think it worth the voyage. It will give me great pleasure to see you, and to do *what I can* to make London agreeable to any of my Philadelphia friends. What I have told you of my literary engagements, however, will have shown you, that *time* is my great desideratum. Believe me to be, dear Sir,

yours very faithfully,
William B. Carpenter.

In my reply to Dr. Carpenter, I expressed my accordance with the views he had embraced on the subject of the so-called Electro-biology; and referred him to the last two editions of my *Physiology* for an analogous explanation I had given of the so called mesmeric phenomena, and especially of those manifested in phrenomesmerism [28] (See *Human Physiology,* 7th edit. ii, 623, Philad. 1850) Dr. Carpenter's estimate of the "worlds fair" was sanctioned by Dr. Pereira, who, in a letter to me, dated Finsbury Square, London, July 3, 1851 asks—"Shall you visit Old England this summer? I think the Exhibition is really worth the cost, the labour and the danger of a voyage from America here."

FRIEDRICH OPPENHEIM

The Editorship of the *American Medical Library and Intelligencer* threw me into correspondence with Dr. Oppenheim [29] of Hamburg, one of the able Editors of the *Zeitschrift für die gesammte Medicin;* his co-editors being the distinguished Dieffenbach of Berlin, and Fricke of Hamburg. With this Journal I exchanged; and to its editors I was in the habit of sending my own works more especially. It was clearly from this source, that Callisen obtained a great part of his information in regard to many of the more recent medical writers,

[27] Alexander Wood (secundus) (1817-1884), lecturer on medicine in Edinburgh, first in Great Britain to use the hypodermic syringe.

[28] Friederich Anton Mesmer (1733-1815), denounced as a charlatan, actually stimulated wide and continued interest in hypnotism and related psychic phenomena.

[29] Friedrich Wilhelm Oppenheim (1799-1852), M.D. 1821, Heidelberg, medical teacher at Hamburg, edited the *Zeitschrift* with Dieffenbach and Fricke from 1836 to 1842, then alone.

the titles of whose works are given in his elaborate *Medicinisches Schriftstellers Lexicon*—a Dictionary of the then living Authors. The notice of books at the end of the first edition of my *Medical Student* is, indeed, constantly referred to; and in my own case, an "Annual Announcement of the Jefferson Medical College" is ascribed to me merely because it was announced in the *Zeitschrift* as having been received from me. Callisen's work is one of immense labour, consisting of *thirty three volumes* octavo, ending, however, with the year 1845; and it is not probable, that it will ever be resumed. It is impossible—it appears to me—that it could ever have sold to the extent of paying for the paper and printing. In this country, most assuredly, it would have proved a ruinous speculation. The price at which it is now offered, in the Antiquarian Catalogues of Germany, sufficiently exhibits, that it is not extensively sought after there or elsewhere.

During the first year of my editorial labours in the "Intelligencer," I had the gratification to receive the following letter from Dr. Oppenheim:—

Hamburg May 6, 1837.

Sir,

In handing you this fourth volume of their publication—which they have taken the liberty of dedicating to you—the editors of the *Journal of General Medicine* feel persuaded they could not have better ornamented their work than by prefacing it with the name of him whose incessant labours in the field of medicine and physiology have so essentially contributed towards the advancement of our science. They have, therefore, not only hastened, in this part of their journal, to lay copious extracts of your *Elements of Hygiene* before the German public, but they intend giving to their learned readers further extracts of your last publication—"General Therapeutics or Principles of Medical Practice"—and from all such works as they yet hope to see from so able a pen. Trusting that you will deign to accept this volume of their Journal, they have the honour to sign,

Your most obedient,
The Editors of the Journal of General Medicine
Dr. Oppenheim, Redacteur en chef.

Dr. Dunglison, Prof., Philad.

The dedication of the fourth volume of the 'Zeitschrift' was couched in the following complimentary language.

Seiner Wohlgeboren
dem Herrn
ROBLEY DUNGLISON MD

Professor der Pathologie und der gerichtlichen Medicin am Jefferson College (Philadelphia), ehemaligen Professor der Therapie, Materia Medica, Hygiene und gerichtlichen Medicin an der Universität von Virginia (Charlottesville), Mitgliede der amerikanischen philosophischen Gesellschaft etc. etc. etc.

widmen

diesen vierten Band ihrer Zeitschrift
als einen geringen Beweis ihrer Anerkennung seiner verdienste um die innere Heilkunde-die HERAUSGEBER.

In the year 1841, I received a letter fom Dr. Oppenheim, dated April the first, informing me, that, at a general meeting of the Medical Society of Hamburg, held on the second of January, he "had the honour" of proposing me a Coresponding Member thereof, and that "he had the pleasure of seeing me unanimously elected."

WILLIAM STOKES AND ROBERT GRAVES

The publication of the Lectures on the Practice of Physic by Dr. Stokes of Dublin; and of the Clinical Lectures of Dr. Graves [30] of the same city led to a short correspondence, and inter communication. The following letter I received from the latter gentleman.

9 Harcourt St. Dublin
1st October 37

My dear Sir,

I, this day, received your letter, and am very glad indeed to find, that you approve of my Lectures. The Lectures, published this year in the London Medical Gazette, are very correctly printed, and were sent by myself to that journal. I have detected only one important typographical error; viz. at p. 258—in the 13th Lecture on Paralysis; 21st line of right column, read health instead of breath—I shall feel much flattered by your reprinting these twenty Lectures. I shall willingly purchase a dozen copies from you, and pay the money to Longman & Co. Booksellers, Paternoster Row, London, from whom your correspondent there can get the money. (They were furnished him gratuitously by the Publisher, Mr. Waldie). Doctor Stokes will take the same number of copies of his Lectures—Send both to Messrs. Longman. In the year 1835, I published upwards of twenty clinical lectures in Renshaw's London Medical & Surgical Journal. Some of these on Paralysis etc. are important, and, perhaps, you would do well, if you have that Journal, to publish a few of the best of these with those of the Gazette. You will perceive, that I have long adopted the opinion, lately put forward with great ability in the American Journal by Dr. Gerhard, that the Maculated fever of Ireland is different from the so named Typhus of Paris. It is necessary to specify this. It is a great satisfaction to us here, to observe the progress our brethren in the States are making in all the Sciences, and especially the Medical. I always feel much pleasure in paying attention to physicians and surgeons of your country, who visit us, and, lately, have had the pleasure of seeing two eminent men, Dr. Warren, of Boston, and Dr. Ludlow, of New York.[31] When any of your friends visit Dublin I shall be glad to have an introduction from you by them.

Yours very truly,
Robt. J. Graves.

I did not, however, give many letters of introduction to him, especially after the visit of Dr. Gibson, of the University of Pennsylvania, who gave great offence to Dr. and Mrs. Graves, and more especially to the latter —as I have been informed—by the notice which he gave

of them and their establishment in the Rambles in Europe, which he published on his return.[32] Soon after this, I introduced Dr. Moreton Stillé [33] of this city to Dr. Graves; who informed me, on his return, that the observations of Dr. Gibson had given so much offence, that a friend of his in Dublin advised him not to present his letter of intoduction. He did so, however, and, although nothing could be more polite to him than Dr. Graves; yet he could not help observing, at table, the constrained manner of Mrs. Graves towards him. How much is it to be deplored, that the rashness and impropriety of one individual should mar the enjoyments of so many right minded persons, who may come after him. And this was not the solitary case of offence given in the Rambles. Dr. Randolph informed me, that Dr. Traill, of Edinburgh, and his friends there spoke in the most unmistakeable manner of the impropriety of themselves and their modes of life being held up to public notice by one, who had been kindly admitted into what ought to be regarded as a sanctuary.

THE PHILADELPHIA HOSPITAL

In June 1838, I was elected one of the attending physicians to the Pennsylvania Institution, Blockley,[34] in the place of Dr. Stewardson, who had resigned on his being appointed one of the attending physicians to the Pennsylvania Hospital. Dr. Harlan, who had no affection for the University of Pennsylvania, and especially for Dr. Gibson—both of whom engaged in the discreditable and undignified course of vituperation before the class at the Hospital, first informed me of the vacancy, and urged upon me to proceed to an active canvass of the Members of the Board of Guardians; stating, that where such a canvass had not been made, the Candidates had, in all instances, failed. I told him I could not enter into a personal canvass, and did not. I mentioned it, however, to my friends, stating, that, on account of the College to which I was attached, such an appointment would be agreeable to me; and they, doubtless, exerted themselves; as I was chosen. Dr. Pepper,[35] whose father was a wealthy and influential citizen, was the next prominent candidate.

[32] Gibson characterized Graves as a man of "humour, high spirits, and quizzical propensities," "peeping and prying into every hole and corner . . . cracking jokes with the patients or pupils, or old women . . . too fond of analogy and drawing conclusions from solitary facts" (William Gibson, Rambles in Europe in 1839, pp. 213–216, Phila., 1841).

[33] Moreton Stillé (1822–1855), M.D. 1844, University of Pennsylvania, Fellow of the College of Physicians of Philadelphia.

[34] The Philadelphia Hospital, now the Philadelphia General Hospital, is meant here. Beginning as the Almshouse in 1731, it had a hospital department for treatment of the aged, sick, crippled and infirm from the start, but was first officially called the Philadelphia Hospital in 1835. Medical students were admitted for instruction at least as early as 1772.

[35] William Pepper (1810–1864), M.D. 1832, University of Pennsylvania, the first of a Philadelphia medical dynasty, Professor of Medicine at his alma mater, was a member of the

[30] Robert Graves (1796–1853), eponym of Graves' disease (exophthalmic goitre), and William Stokes (1804–1878), eponym of Stokes-Adams syndrome and of Cheyne-Stokes respiration, Regius Professor of Medicine at Trinity College, Dublin, early exponent of Laennec's stethoscope, whose Diseases of the Heart and Aorta (1854) is a classic.

[31] Which Dr. Ludlow has not been identified.

Soon after my appointment, it was urged, by the Board of Guardians, that I should give a course of Clinical Lectures at the Hospital, in order, that the Institution might reap the advantage which would be likely to ensue from an increased number of the Students of the Jefferson Medical College taking out the hospital ticket. I declined, however, assuming this duty alone, as I should have to follow the two lecturers of the University of Pennsylvania, and I knew it would be impossible to detain students, already fatigued, to hear a third lecturer, whatever might be his merits. As soon, however, as my friend Dr. Pancoast, already one of the attending Surgeons, was appointed a Professor in the College, the difficulty was removed, and we agreed at once, to lecture together at the Hospital on one day of the week, the Professors of the University of Pennsylvania occupying another.[36] In the meantime, from 1838 to 1841, I attended, regularly, every other day, to my duties in the hospital department; and reports of many of the interesting cases in the clinic were published in the American Medical Intelligencer, from records taken by the resident physicians in attendance,—Drs. Edwin A. Anderson,[37] of Wilmington, N. C., Alexander M. Vedder, A.M., of Schenectady;[38] Joseph B. Cottman of Maryland;[39] Dr. W. H. McKee, of Raleigh, N. C.;[40] Wm. B. Page[41] and by Drs. Squibb[42] & White, and others. From 1839, after the reorganization of the Jefferson Medical College consequent on the removal of the Drs. McClellan and Colhoun, I continued my service at the Hospital and my regular clinical lectures there, up to the time of the action of the Board of Managers, which excluded the schools from every opportunity of teaching practical medicine in the hospital. The causes that led to this action of the Board were numerous. Selected, as they were, from every calling, it was not to be expected, that the members could be possessed of absolute wisdom, especially on matters, that concerned a liberal profession; whilst at the same time, the desire of exhibiting their power induced them perhaps to pass regulations, that were occasionally injudicious and annoying;—this more, however, as regards the resident than the attending physicians. A pseudo-philanthropy was likewise entertained by some of the members, which led them to object to post mortem examinations in the dead-house; and to patients being exposed in the Clinic, and subjected to auscultation and percussion; or to operations before the class;[43] and I regret to say, that this feeling was encouraged; and the practice openly condemned in published lectures by a Professor—Dr. Henry S. Patterson[44]—of a college—the Pennsylvania, which was not represented in the Medical Board.

HARLAN-GIBSON FEUD

They were likewise subjected to annoyances at times from the indiscreet and improper remarks of one of the Members of the Faculty—Dr. Gibson, who, in one of his clinical lectures, did not hesitate to reflect on the Board and its members; and, also, from the dissatisfaction which often prevailed amongst the residents, who were in reality, however, appointed by the Board itself and certainly in no case by the Colleges. On one occasion, moreover, a discreditable paper warfare occurred between two of the attending surgeons—Drs. Harlan and Gibson—which was well calculated to be laid hold of by dissatisfied members of the Board of Guardians as exhibiting a mode of treatment, in one case, which to the *laity* must have appeared—to say the least of it—cruel and unnecessary. An operation was performed at the Hospital before the class; which was animadverted upon severely by Dr. Harlan in his next lecture; on this a rejoinder was made by Dr. Gibson, both the animadversion and the rejoinder being published in the pages of the *Medical Examiner,* and both being exceedingly censurable. The following letter, however, which is an answer by Dr. Harlan to Dr. Gibson's rejoinder was published in letter form, and

American Philosophical Society, the College of Physcians of Philadelphia, and many other scientific organizations.

[36] The Philadelphia Hospital (or Almshouse-Blockley, as it was still apt to be called) was open to all of the students of medicine in Philadelphia for a fee of ten dollars. Dunglison lectured there at 11:30 A.M. every Saturday from November first to March first. Because the medical schools were in the City, several miles away, they had to be conveyed back and forth by horse-drawn omnibuses, each school providing its own bus in order to preserve order among the students (Heber Chase, *Medical Students Guide,* p. 17, 63, 1841).

[37] Edwin Alexander Anderson (1816–1894?), senior resident physician in charge of the asylum of the Philadelphia Hospital in 1838, specialized in ophthalmology in Wilmington, N. C., where he was on the roll of members of the North Carolina Medical Society up to 1894–1895.

[38] Alexander Marselis Vedder (1814–1878), M.D. 1839, University of Pennsylvania, was resident physician in the Philadelphia Hospital in 1838, returned to his home town in New York to practice and become Mayor.

[39] Joseph B. Cottman, M.D. 1838, Jefferson Medical College.

[40] William H. McKee (1814–1877), M.D. 1839, University of Pennsylvania.

[41] William Byrd Page (1817–1877), M.D. 1839, University of Pennsylvania, Professor of Surgery, Pennsylvania Medical College and Fellow of the College of Physicians of Philadelphia.

[42] Edward Robinson Squibb (1819–1900), M.D. 1845, Jefferson Medical College, founded the E. R. Squibb Pharmaceutical Company, member of the American Philosophical Society, left a *Journal* which was published in 1930, giving detailed accounts of lectures in 1851 at Jefferson Medical College, including those of Dunglison.

[43] The medical staff was often at odds with the Board of Guardians of the Philadelphia Hospital, who recognized the values of the clinical material to the medical schools, but finding that sick patients were taken from the wards to the lecture room for the purpose of furnishing subjects for the lectures, none being exempt from thus being exposed, felt that "there are rights possessed even by recipients of charity which should be guarded, and feelings which should be respected."

[44] Henry Stuart Patterson (1815–1854), M.D. 1836, University of Pennsylvania, chief resident-physician to the Almshouse for one month in 1845, resigned when it was complained that he was at the same time serving as professor at the Pennsylvania Medical College (Simpson, *ibid.,* p. 762).

extensively distributed. It is not characterized by much delicacy, but, in this respect, there was no great difference between the contending parties, in the opinion of most of their professional brethren.

SECOND EDITION

To the Editors of the Medical Examiner.

Gentlemen,

The fifth number of your Journal contains the report of my lecture at the Philadelphia Hospital, delivered on the 3 rd of February, in which it became my serious duty to animadvert upon the danger and inutility of surgical operations in certain cases, as satisfactorily illustrated in a cruel experiment by my predecessor, in the attempted removal of a tumour involving nearly the entire mouth and nares. On the morning of the operation, Professor Gibson detailed to his class such arguments as he considered sufficient to authorize his surgical interference in so desperate a case. My arguments in opposition to the operation were not alluded to; he merely stated, that his colleagues agreed with him as to the propriety of an operation, with the exception of Dr. Harlan. These arguments, I took the liberty of laying before the students, at the subsequent lecture, the first of my course, disclaiming all personality. I endeavoured to impress upon them the reasons for my unqualified disapprobation of such operations as the one in question. "My learned and amiable friend," the Professor of Surgery, I am informed, has taken umbrage at the expression of an honest difference of opinion on this important point. He is indignant that anyone out of the pale of the University should be bold enough to doubt the infallibility of a professor before his pupils: it is too true, that University opinions are already nearly a unit in the Medical board of the Philadelphia Hospital, and the Professor is anxious to make them entirely so. At the close of my lecture on Saturday last, Dr. Gibson made his appearance in the amphitheatre, *professedly* for the purpose of vindicating his conduct before the pupils, but his remarks, in reality, consisted in gross personalities and puerile witticisms towards myself to the total exclusion of any argument tending to his vindication. I very much mistake the character of the majority of the Medical Class there convened, if they could find themselves in the least degree edified by the indelicate abuse in which he indulged on that occasion, so unworthy the dignity of a Professor of Surgery in the University of Pennsylvania. The principal facts in my statements, and to the refutation of which his answer should have been confined, consisted mainly in this:—That the patient, who had suffered so severely from the hammer, chisel, and red hot pokers, has received no *permanent* benefit from those inflictions; and from the very nature of the tumour could not have been permanently benefited by any operation—that the patient, in fact, was as likely to die now from the tumour as he was previously to the operation. This statement Dr. G. has not attempted to invalidate, notwithstanding his previous assertion, that the operation afforded the patient the only chance of life. Such being the nature of the observations of Dr. G. in his lecture of the 24th inst. I observed, with some regret and considerable mortification for the dignity of our profession, that you considered them worthy to be appended to the report of my lecture in your last number, in which I never alluded disrespectfully to the operator, but confined my observations to the operation. The nature of Dr. G's observations, destitute, as they are, of fact and argument, preclude the possibility of any notice on my part, of a nature suited to the pages of your Journal, which ought always to be devoted to the dissemination of medical truths. Yet, were I disposed to indulge in invective, or

answer Dr. G. in his own strain—the theme is by no means a meagre one;—but I shall confine my remarks, at present, to a few hints, arising from his learned and witty discourse; which, if he be not incorrigible, may prove serviceable to him hereafter. I sincerely hope, that the gentleman will not take offense at my candour—"for, knowing, as I well do, his extreme modesty, meekness, and humility, his tender sensibilities could not fail, I am sure, to be shocked by praise of this public description." In the first place, Dr. G. entirely overlooked the fact, that his criticism of my voice would apply "a fortiori," to some of his colleagues, in whose success, as lecturers, his pecuniary interest is more involved. "Could it be otherwise," continues Dr. G., "that such a voice, loud and sonorous, peculiarly musical in its tone, clear and delightful in its cadence, should fall upon your ear with all the harmony of sympathetic association." What possible bearing could such a personal allusion have upon the point at issue? We admit, that Dr. G. possesses a very *loud* voice (to say nothing of its music) but he is well aware, that it is much easier to make a noise than to make sense in a lecture room. He should, also, know, that a simple truth, modestly told, is of far more importance than all the flowers of eloquence, or the pompous diction with which falsehood and error are occasionally clothed to make them palatable. Dr. G. is peculiarly unfortunate, for one in his position, in his next attempt at a cut. "We feel peculiarly happy, gentlemen, in bearing testimony to the merits of our colleague, and have no doubt, that the managers have retained his important services, in this great establishment, for the purpose of attracting to it full classes, such as I have the honour to be surrounded with at this' moment." It has been the subject of notorious remark, that the medical class attending these clinical lectures this season does not amount to more than one half of the usual number, a falling off due to the insolent and ungentlemanly personality, used by Dr. G. at the commencement of the Course, towards the students of the Jefferson Medical College, who almost to a man refused to honour him subsequently with their presence, preferring to take the Pennsylvania Hospital ticket—thus, by his illtempered and illtimed personalities, defrauding the funds of the institution of at least one thousand dollars annually. The learned Professor next takes much unnecessary pains to expose his own ignorance to his pupils, by pretending to despise those collateral sciences, the study of which has so much tended to elevate our profession. "Not only, indeed, is he (Dr. H.) distinguished in his particular vocation, but he has decided advantages over his less fortunate brethren, in being acknowledged, both in this country and in Europe, as profoundly acquainted with natural philosophy, natural history in all its branches" etc. Very different are the notions in such matters, of the learned Editors of the British and Foreign Medical Review, as is shown by the following extract from one of their recent numbers. "The ungenerous notion, formerly so successfully disseminated, that no medical man should attempt to be a man of science, exists no longer; *a medical man cannot preserve his intellectual rank in society without some scientific acquirement;* and subjects, which were once confined almost to the closets of philosophers are now discussed at dinnertables, declaimed upon in drawing-rooms, and settled at conversazione." The Professor would, consequently, find himself without intellectual rank in European Society. Professor G's own work on Surgery would have been more extensively useful and a less fruitful source of error, had he paid some attention to the study of comparative physiology, comparative anatomy and even zoology. The success of the following attempted depreciation of my recent work the *Medical & Physical Researches* is at least doubtful.

"His large and splendid volumes too, on newts and frogs; toads and beetles, lizards and snakes, intermingled most naturally with aneurisms and tumours; fistulas and strictures etc. affords additional testimony of his industry and talents." Such unqualified praise from such a pure source, might have been considered complimentary, had not numerous encomiastic criticisms alluded to, both at home and abroad, rendered any observations of Dr. G. on the subject entirely supererogatory, and we further regret our inability to return the intended compliment, having searched in vain for any published opinion on his own work on Surgery beyond the Atlantic. But I am in possession of a written opinion near at home, which I copied from the walls within the University, whilst his second volume was yet in press, the justness of which the modest Professor will not fail to acknowledge. It appears, that the medical students, who subscribed to his work, were obliged to pay for both volumes on the delivery of the first: the second volume was delayed to an unreasonable period, the subscribers became restive, which produced reiterated apologies from the Professor—under the impression, of course that the acknowledged merits of number one, created the anxiety for the appearance of number two. The patience of the subscribers became finally, exhausted. An Esculapian neophyte embodied the general sentiments of the class in the following couplet, written in mammoth letters on the wall near the entrance of the lecture room, with a piece of charcoal:—

"A student advises whatever betides,
To leave the next volume alone—
The first is enough for our b---s,
The second pray keep for your own!"

But for the terseness and truth of these lines, the fastidious reader might object to them on the score of delicacy: as it is, I think them worthy to descend to posterity, like the "fly in amber," embalmed by the surrounding medium. They are, besides, in strict unison with the Professor's familiar style with his pupils. Dr. G. dwells on his great *surgical experience* as in contrast with my own. As regards our public opportunities, we commenced our tour in the Philadelphia Hospital, the principal theatre of them, about the same period—say fifteen years ago. As respects the public hospitals of Europe, he visited them only as a student, to think with the brains of others: I enjoyed the advantage of examining those institutions with the freedom of a practical surgeon receiving the honored attentions of the veterans of science, and, as far as private practice is concerned, since he came among us, the least successful practitioner of equal age, would be loath to exchange with him. Dr. G. endeavours to cast an imputation on me for visiting privately, his patient, previously to the operation, when he knows it was my duty, as one of the surgeons in chief, to give my opinion on all cases involving capital operations, especially when requested so to do by the house-surgeon, by whom I was informed of the intention of Dr. G. to operate, in open violation of the articles governing such matters—and he was only enforced to consult his colleague by an order from the managers. This very important lecture on "Tumor of the Antrum" was further elucidated by Dr. G's informing the class of my "unrivalled ornithological" standing; and amused the class by dwelling on the facts that the celebrated Audubon had named a new species of American bird, *"Falco Harlani"* or *Black Warrior!* and concludes his transcendently witty and valuable *clinical* remarks, by a humble petition that our *Warrior Eagle* in all the nobleness of his nature, will "still suffer little birds to sing." The "little birds" are too well paid for singing, and too "snugly ensconced" behind the fortified walls of an invulnerable aristocracy

to be easily disturbed, even by a "war eagle's scream." It is unreasonable, however, that they should desire to monopolize all the singing. As the Professor appears interested in ornithology I take the present occasion to communicate a fact in that science, from the knowledge of which he may draw consolation. That classical ornithologist, Illiger, has classified certain birds of ignoble habits, under the denomination of *Cathartes* (including the turkey-buzzard:) these the "Warrior Eagle" holds in utter detestation, and never, *voluntarily*, associates with them—secure in their filthiness, they revel in their own excretions! I might extend this letter to an indefinite length, were I disposed to recriminate. It would be easy to amplify upon the well known *candour* and *love of truth* for which the learned Professor is famous—the amiable, honourable and magnanimous disposition he has ever displayed in his intercourse with men, and which has secured to him the respect and esteem of so large a portion of his professional brethren without the sphere of the University of Pennsylvania, since he became a citizen of Philadelphia. The numerous foreign complimentary titles which have been conferred on him in acknowledgment of his industry, talents, and improvements in surgery! The great bravery, which he displayed, some years since, in boldly attacking a professor in a rival school, and his dexterity, wisdom and discretion in obtaining a friend to fight his battle for him! —but *"verbum sat,"* and, for the present, I forbear. This splendid specimen of the "Clinical" Lectures, delivered in the Philadelphia Hospital, by the very popular Professor of Surgery in the University of Pennsylvania, taken in its literal sense, is no very great exaggeration of his everyday style of discourse, both public and private, hence his obstreperous volubility on the present occasion.

Your much obliged, and very obedient etc.
R. Harlan.

Philad. Feb. 28, 1838.

"Cathartes Gibsoni is up again!" Since the distribution of the first impressions of this letter, very numerous and urgent demands for copies, both in and out of the profession, render a second edition expedient. I learn, that the honorable Professor introduced the subject of this controversy, brought on by his own unprovoked and ungentlemanly attack, before his class in the University of Pennsylvania, on Saturday last!—Surely it is a bird of ignoble habits, that "befouls its own nest." He commenced his lecture with a military flourish, and the ominous announcement, that the "Warrior Eagle is up again"—not doubting, it appears, of the success of his ineffectual attempt to *put me down.* The Professor, at least, is no judge in this matter. In reference to my letter, before his surgical class, the courageous Professor attempted no vindication of himself, nor any refutation of the grave imputations therein promulgated; but, in lieu of which he regaled his hearers with false misrepresentations and Billingsgate slang. His object is obviously to secure to himself an immunity by thus adopting a course of conduct which places himself beyond the pale of personal responsibility. Abundant testimony exists, to prove that the assertions he made to his pupils, on the 3d. inst. are utterly untrue having no other foundation than in his own distempered imagination. The imputations contained in my letter are not to be avoided by such contemptible subterfuges—these will adhere like "the shirt of Nessus,"[45] to this modern Hercules—this tunic can be exchanged only for the shroud.

---

[45] The Centaur, Nessus, was shot with a poisoned arrow by Hercules; when dying, he gave his infected cloak to Deianira, who sent it to Hercules, her husband, and when Hercules put it on, he died in fearful agony.

REVOLT OF THE RESIDENT PHYSICIANS

It need scarcely be said, that nothing could be calculated to do more injury than such a correspondence, which, if not laid hold of at the time, by the Board of Guardians as an incentive to changes in the hospital department, was likely to be treasured up for future action. They, too, who could be guilty of such an imprudence, would be sure to repeat it; and, thus, the medical service, as connected with the Colleges, became gradually less and less popular with the Board; and they were prepared for ulterior changes as soon as a fitting occasion occurred. It was not, however, until the year 1845, that a decisive step was taken by them. The resident physicians, who, as before remarked, were not appointed by the attending physicians, but by the Board,[46] feeling themselves aggrieved by a physician holding no office (a Dr. Curran) being permitted to board and sit with them at table, notwithstanding that his company was highly offensive to them "and chiefly on account of his voluntarily prescribing and otherwise intermeddling in the cases of patients with whom he had no concern," a "firm but respectful application" was made to the chief resident officer of the house, either to remove him or to provide another table for the Physicians."[47] This application was ultimately refused, when the matter was referred to the Board of Guardians by the physicians, who required them to continue to board where they were. On this, the whole body sent in the following resignation:

Blockley Hospital June 30, 1845.

Gentlemen,—Circumstances having compelled us to dissolve all farther connection with the Institution under your care, we hereby tender our resignations as Resident Physicians to Blockley Hospital. Signed, William V. Keating,[48] E. G. Higginbotham,[49] R. J. Farquharson,[50] John B. Sherrerd,[51]

[46] The Board of Guardians, on the other hand, finding that the resident physicians assumed too much authority, questioned the wisdom of placing the care of the patients in charge of such "mere novices." (Charles Lawrence, History of Philadelphia Almshouses and Hospitals, p. 156, 1905).

[47] This occurred June 30, 1845. As Dr. Agnew, years later, told the story (with tongue in cheek, I suspect), "in consequence of the want of due formality and decorum in the destruction of an unfortunate cockroach, which had rashly taken a short cut across the table instead of going around, these gentlemen became indignant and demanded of the managers to be transferred to the table of the matron."

[48] William Valentine Keating (1823–1894), M.D. 1842, University of Pennsylvania, successor in the Chair of Obstetrics to C. D. Meigs at Jefferson Medical College; member of the American Philosophical Society and of the College of Physicians of Philadelphia.

[49] Edward Garrigues Higginbotham (1824–1901), M.D. 1845, University of Pennsylvania, Confederate surgeon, practiced in Richmond, Virginia.

[50] Robert James Farquharson (1824–1884), M.D. 1844, University of Pennsylvania, naval surgeon, frequently mentioned by E. R. Squibb in his Journal, settled in Iowa (History of Medicine in Polk County, Iowa, p. 33, 1951).

[51] John Browne Sherrerd (d. 1852), M.D. 1845, University of Pennsylvania, died in Scranton, Pa.

A. Porter,[52] E. B. Jones,[53] R. F. Mason,[54] Job Haines,[55] and A. Lee Brent.[56]

The same evening I was called upon by a respectable Member of the Board of Guardians, who informed me of the act of the young gentlemen. It met with my decided disapprobation, and I remarked to him, and soon afterward to one of the young gentlemen in question, that I could find no apology for their having quitted their post, and left so many sick in the hospital uncared for.[57] From others, they received the same frank expressions of opinion,[58] and on this account, as well as from their own subsequent reflections probably, they addressed a letter to the Medical Board, of which the following is an extract.

Philadelphia Hospital
Blockley July 1, 1845.

Gentlemen,

We wish it distinctly understood by the Medical Board, that it is not, and never has been our intention to abandon our patients in the Philadelphia Hospital, Blockley until a suitable provision is made by the managers for the performance of our duties. To shew the sincerity of this declaration, we are willing, at your suggestion even now to return to our posts until other appointments, more satisfactory to the Board of Guardians, are made. We however, ask a redress of grievances in regard to the Steward etc. etc.

The Board refused to accept their resignations,[59] and, in a summary manner, having too much the appearance of vengeance for a dignified body, proceeded to dismiss every one of them. A notice appeared in the public prints animadverting on the conduct of the young gentlemen, to which they replied, in a card "to the public," exhibiting more indignation than good taste. In reference to their first letter they say, "The undersigned are, also, happy in being able to state, that they subsequently consulted seven of the attending physicians of the hospital, in reference to the propriety of their course, and that it was throughout fully approved by them"—I was not one of the seven; for, when called upon, I frankly stated my objections; and the views of some of the seven must have been misunderstood, judging from

[52] Andrew Porter (d. 1859), M.D. 1846, University of Pennsylvania, died at Cape May, N. J.

[53] Edwin B. Jones, M.D. 1845, Jefferson Medical College, was from Virginia.

[54] Not identified.

[55] Job Haines, M.D. 1845, Jefferson Medical College, was from New Jersey.

[56] Arthur Lee Brent (d. 1872), M.D. 1845, University of Pennsylvania.

[57] Drs. Horner and Clymer went to the hospital to take care of the patients that evening.

[58] At a special meeting of the medical staff held the next day, July 1, 1845.

[59] The steward, residents, and nurses were ordered to appear before the Board of Guardians to present their versions of the difficulties. The medical staff, meeting on July 2, determined to send a representative to urge the managers to permit the residents to remain.

their expressed sentiments at a meeting of the Medical Board summoned soon afterwards.

The Board of Guardians [60] embraced this occasion for remodelling the whole of the medical department of the Philadelphia Hospital; deciding, that the existing medical officers should terminate their labours by a certain day; [61] that a permanent resident physician should be appointed, who should have the whole direction of the Medical Service of the Hospital, at a fixed salary,—hitherto no compensation whatever having been allowed to the attending physicians,—that a certain number of consulting physicians should be appointed; [62] and of junior or resident physicians, who should be wholly under the control of the Chief Resident. A notice of the determinations of the Board of Guardians was sent to the Secretary of the Medical Board; and a meeting summoned at the house of Professor Saml. Jackson, [63] the senior attending Physician, to determine as to the course the Medical Board should pursue under the circumstances. It was fully attended. It appeared to me, that there was but one course to be adopted,—to instruct the Secretary of the Medical Board to acknowledge the receipt of the communication of the Board of Guardians, and to state, that the Medical Board would be ready to vacate their position as soon as their successors were appointed. Drs. Jackson and Horner had prepared written statements in the way of reclamation, in regard to the services which had been gratuitously rendered for so long a period and the pecuniary advantages the clinical lectures had been to the Institution and the public, which appeared to me injudicious, and not altogether dignified, for the occasion.

I, consequently, after some discussion wrote, at a side table, a resolution in accordance with the view stated above; but it appeared to the meeting to be so tame, that it was received with anything but approbation. I, therefore, folded it up, and put it in my pocket. It was then proposed, that the papers of Drs. Jackson and Horner should be referred to a special committee, to report to the Medical Board to be convened on that day week (Tuesday) what steps should be taken in the matter. I was named as chairman of the committee; but I objected, in consequence of my being opposed to *any* paper of the nature proposed, or any letter of reclamation or recrimination being sent to the Board of Guardians. It was urged, however, that I was the proper person; and I consented; and the Committee consisted of myself, Dr. Jackson, and Dr. Horner. I

confess I considered it a difficult task to know how to reconcile the matter,—two of the gentlemen having propositions of their own to which it might naturally be supposed, that they would be wedded. On the Sunday following, we met; and I was gratified to find that Dr. Horner, on reflection, had changed his view and was not disposed to press the reception of his paper, nor did he express himself favorable to that of Dr. Jackson. I now drew out of my pocket the identical resolution, which I had offered on the Tuesday before, and stated, that it would, perhaps meet their views. Both assented. We recommended it for adoption at the meeting on the Tuesday following and it was carried without a dissentient voice—a striking instance of the good effects of delay, whenever feelings are greatly enlisted. I had little or no feeling, however, about the affair. I knew, that all power rested with the Board of Guardians and that, as they had appointed, so could they remove us; and, accordingly, all that we had to do was to submit. [64]

Subsequently, when the consulting physicians were to be appointed—the Hospital Committee consisting of Mr. John Price Wetherill, Mr. Abbot, and Mr. Williams, of the Northern Liberties, sent Mr. Robbins, the Secretary of the Board to me, to know, whether, if I were appointed one of the Consulting Physicians, I would serve. I asked, if he was empowered to offer me the situation. He said no, but that the Committee had no doubt of my being appointed, if I would say, that I would accept the office. I replied, "Tell my friends Messrs. Wetherill, Abbott, and Williams, that I am greatly obliged to them for their kind wishes; but that I have made it a rule, never to accept or refuse an office until it has been offered to me." I had made up my mind, that I ought not to accept it, but still, notwithstanding my answer, my name was brought forward, and I received several votes, although, when a question was asked, whether I would take the office if chosen, no one felt authorized to say, that I would. I have some reason, however, to think, that the fact of my name having been brought forward was believed by some to be with my consent; and that the inoffensive course I had recommended the Medical Board to pursue was dictated by motives of policy. The fact I have just stated will sufficiently show, that if I had wanted the situation I could have readily obtained it, for, of the different members of the Medical Board, I was—Mr. Robbins informed me—the only one who had been asked to serve by the Hospital Committee. [65] That Committee was, indeed, well aware of the extent and nature of my services. I certainly was a faithful officer; and was the cause of some modifications in the Hospital department, calcu-

[60] On July 21.

[61] October 1, 1845.

[62] One consulting surgeon, one consulting physician and one consulting accoucheur.

[63] Samuel Jackson (1787–1872), M.D. 1808, friend of Nathaniel Chapman, was Professor of the Institutes of Medicine at the University of Pennsylvania (1826–1863), fellow of the College of Physicians of Philadelphia, member of the American Philosophical Society and President of the Philadelphia Board of Health (William S. Middleton, Samuel Jackson, *Annals of Medical History*, n.s., 7: 538–549, 1935).

[64] D. Hayes Agnew (Philadelphia Hospital Reports, 1: 18, 1890), in a humorous account of this affair, names only Jackson, Horner, Clymer, Gillingham and Pancoast, but the account here given by Dunglison is corroborated by the records still preserved at the Hospital. Dunglison was there; Agnew was not.

[65] The medical schools were excluded for nine years (Agnew, *ibid.*, p. 19).

lated to be of advantage to both the patients, and the community.

BRANDY AND DELIRIUM TREMENS

It was the universal custom, when I took charge of the wards, to administer but one article in delirium tremens and that article was alcohol.[66] The quantity employed was consequently astounding, and excited the attention of the Board of Guardians, especially when they saw, that, under my eclectic treatment of the disease, scarcely any was employed, and yet the success of the course was, at least, equal. The pages of the *American Medical Intelligencer* (May 1842 p. 225) and of my *Practice of Medicine* contain the results at which I arrived, and subsequent observation had not led me to modify them. It was asserted, indeed, that my directions were not followed by the resident physicians, and that they administered brandy to the females affected with delirium tremens,—the males being under the charge of one of my colleagues; but I believe this was mere assertion, in as much as it was positively denied by some of the residents, whilst the spirit books of the Hospital exhibited, that in one case from November 1, 1841 to May 1, 1842 "not a drop of alcohol liquor was used in the treatment of the disease in the Women's asylum, although some severe cases, in the third stage, had occurred, which, notwithstanding, terminated most satisfactorily."

THE INSANE DEPARTMENT

The most defective part of the Philadelphia Hospital was allotted to the Insane. There were no accommodations for classification, and hence the ferocious maniac was put along aside the most tranquil. At an early period of my attendance I brought this subject before the Hospital Committee; and in this I was powerfully aided by Dr. Pennock,[67] who had been previously one of the attending physicians to the hospital; and who was, withal, a skilful physician, and an exemplary philanthropist. In the public prints he repeatedly drew the attention of the public and of the Grand Jury to the faulty arrangements in existence,[68] but it was not until a

thorough revolution had occurred, consequent on the removal of the whole corps of attending Physicians, that they felt compelled to adopt improvements, which the increased knowledge of the community called for. It was not long before my services ceased in the Hospital that I wrote the following letter on the subject to the Board of Guardians.—

The undersigned—as one of the attending Physicians to the Philadelphia Hospital, respectfully submits the following suggestions to the Board of Guardians.

Several years ago—it may be recollected by some of them—a Committee of the Medical Board presented sundry propositions to the Board of Guardians in relation to the Hospital and especially to that portion of it which is appropriated to the insane poor, urging the necessity of a proper classification of the inmates, and of additional means for exercise and amusement. So far as the undersigned is aware, no action of any consequence was taken on these propositions. An additional ward was, indeed, opened on the female side, which was so far judicious. The undersigned being, at the time—as he has since been—under great hopes, that some salutary enactment would be passed by the Legislature in regard to these unhappy objects of our commiseration and philanthropy, did not press the Board of Guardians for any farther action.

In the years 1838 & 40, as Chairman of a Committee for procuring an Act to be passed by the Legislature for the amelioration of the condition of the Insane poor of this Commonwealth, he prepared two appeals to the people in which he assigned to the Board of Guardians every credit for their anxiety and endeavours to subject the insane in the Philadelphia Hospital to such treatments of a physical and moral nature, as might be deemed desirable; but, at the same time, he did not conceal the difficulties that existed in an Almshouse in having such cares and attention bestowed on them as their condition demanded. The last argument he urged, indeed, in his "Appeals," with the view of inducing the Legislature to establish a large institution for the insane poor, to which the insane of this hospital might be transferred; and in this course he was encouraged by Members of the Board of Guardians themselves, who philanthropically felt, that the insane would be more likely to be restored to reason in an establishment erected for the very purpose, than in the insane wards of their own hospital, under any modifications of which they might be susceptible. By dint of great exertions on the part of some judicious and philanthropic individuals, a Bill was passed for the erection of an Hospital for the Insane poor under certain fixed regulations; and Commissioners were appointed, who actually purchased a site for it. Owing, however, to circumstances which it is unnecessary to detail to the Board, the whole matter was suspended by the Legislature. Still, the undersigned indulged a hope, that the Legislature might remove the suspension, and that an asylum, worthy of this great Commonwealth, might be erected which would enable the Board of Guardians, if they saw fit, to free themselves from the care of a class of patients, for which their existing accommodations were by no means adequate. The inefficient and unsatisfactory legislation of the present session on this subject has shewn that these hopes were illusive; for, although an insane hospital for the poor has been directed by the Assembly it is so restricted in every respect as to render it improbable, that its salutary agency can extend hither. Had the former bill been resuscitated, and proper Commissioners appointed, the undersigned had reason to believe, and to affirm, that the whole of the contemplated loan of 120,000 dollars could have been raised without much difficulty. Impressed, then,

[66] Alfred Stillé credited this to William W. Gerhard as a "radical reform in the treatment of mania-a-potu and delirium tremens," good food and alcohol replacing "the murderous rise of opium" (Reminiscences, *Philadelphia Hospital Reports* 1: 60, 1890).

[67] Caspar Wistar Pennock (1799-1867), M.D. 1828, University of Pennsylvania, pioneer American cardiologist, fellow of the College of Physicians of Philadelphia, was attached to the Philadelphia Hospital 1835-1845.

[68] Frequent grand jury inquests were held, with much public clamor and political fanfare, concerning the plight of Blockley lunatics, but real or lasting results were seldom achieved. 120 cells, with posts and rings to shackle the maniacs, were provided when the almshouse moved to Blockley in 1834, and a young doctor, two years out of medical school, was appointed to care for them and the small-pox patients. These cells were finally abandoned in 1860.

with the entire conviction, that nothing can be expected from the State, the undersigned reluctantly abandons a long cherished hope; and now turns to the condition of the Insane Asylum in this establishment with the view of respectfully requesting the attention of the Board of Guardians to such ameliorations as it may require. Already, our large cities—New York and Boston for example, have their own asylums for the Insane poor, arranged according to the best lights afforded by the improved treatment of the Insane, within the last fifty years more especially. The undersigned is well aware, that the Hospital Committee have instituteed modifications in the department for the insane, which is under his care for six months in the year, and he cheerfully testifies to the anxiety, which they have exhibited to improve the condition of the unhappy inmates. Yet there is room for farther improvements, which should harmonize with each other; and all of which should tend to the same great object. Classification may be adopted in a convenient or in an inconvenient form; and after changes have been made, without a full consideration of the most satisfactory plan, they may be found to work inefficiently and, hence, subsequent alterations may be demanded, which cannot fail to occasion great expense and embarrassment. The most important error in the present arrangement is, that there is no separate and distinct building in which the furious and noisy maniac can be widely separated from the quiet and convalescent,—nay, according to the existing practice of the establishment, persons of sound mind, who may have been guilty of offences in the Almshouse are actually sent to the insane wards for punishment; and, in order to give vent to their rage and mortification, and to excite still farther annoyance, they at times, occasion so much noise and disturbance throughout the night as to endanger or undo what the physicians of the asylum have accomplished by days and weeks of anxious and philanthropic exertion. To inquire into the best course that can be adopted for the restoration or for the amelioration of the condition of these unfortunates is the object of this communication to the Board of Guardians. By laying down a well conceived plan for the improvement and extension of the accommodations and grounds, so that proper classification, exercise and amusement may be adopted; by so modifying the order of medical attendance, that the attention of a resident may be devoted exclusively to the insane, and by other suggestions, that may be made after a full examination of the subject, the undersigned feels satisfied, that the insane department of the Philadelphia Hospital may be placed upon as good a footing as the nature of the Institution will admit of. Already—as the undersigned has remarked—the subject has been brought before the Guardians, in the report of a Committee of the Medical Board, and he confesses, that it would be more agreeable to him, as it doubtless would be to the Board of Guardians, that suggestions for improvement, proceeding from the Board themselves conjointly with the Medical Board—who, it must be presumed, are fully informed on the whole subject—should be carried into execution rather than those which emanate from casual and irresponsible visitors, no matter how worthy, in all respects, such visitors may be.

Robley Dunglison.

Soon after this was written, in the year 1845, my connection with the Philadelphia Hospital terminated. I had the pleasure to learn, however, that alterations were made in the spirit of the above letter, which could not fail to have a salutary influence.[69] The new system

did not, however, on the whole, work as well as was anticipated. The chief resident, instead of being "permanent" was often changed: and one of the Board of Guardians—Mr. John Price Wetherill—informed me, that he often regretted the action of the Board in dissolving all connection with the former physicians and surgeons, and would be glad to have the ancient system restored; and I have since had reason to believe that he was not solitary in this view of the subject. It became necessary however, for the professors of the Jefferson Medical College to make unusual exertions to provide clinical instruction for their students. To require all to take the ticket of the Pennsylvania Hospital would be to compel them to pay 10 dollars without an equivalent inasmuch as the amphitheatre couldn't possibly accommodate the students of all the schools. It was, therefore, determined not to require the ticket of the Hospital to be taken; and to establish an energetic clinic at the college, at which patients should be received gratuitously, be furnished with medicines, prescribed for, and be operated upon, in surgical cases, before the class, and thus arose the clinic [70] at which during the year ending April 2, 1852, *two thousand and twenty nine cases, nine hundred and twenty six being medical, and eleven hundred and three surgical,* presented themselves, and *three hundred and eight operations* were performed. It would have been desirable to have the Philadelphia Hospital as a school for clinical instruction and it was discreditable to the city, that it was not continued as such, but it was undoubtedly, a great relief to me to have to take my share of the clinical duties at the college rather than at the Philadelphia Hospital, the access to which was, at times, during the winter, almost impracticable, and the amount of time, required in the discharge of the necessary duties, more than I could well afford. The expenses of the college to the Faculty were, however, greatly increased by the change, amounting, in the clinical year from April 1, 1851 to April 1, 1852, to nearly *seventeen hundred dollars.*

MR. J. P. KENNEDY

In the latter part of the year 1838, my friend Mr. John P. Kennedy of Baltimore, of whom I have already made mention (p. 71), being then in Congress, came on to Philadelphia to consult me, in conjunction with Dr. Jackson, in regard to his health, which had been far from good; and one evidence of which was an annoying herpetic eruption about the lips. His friends were exceedingly anxious in regard to him and his father-in-law—Mr. Gray—had serious apprehensions that he was not long for this world. He had been kept on a very restricted diet, with abstinence from all exhilirating drinks; and being naturally of a nervous temperament, and lax fibre; he was, when I saw him, under the most

---

[69] Alluding to the appointment of Dr. Patterson; see footnote 44 above.

[70] Squibb described the practice at this clinic in his *Journal* (2: 369) and alluded to Dunglison's "economic expectantism" in prescribing.

favorable circumstances for a thorough change of system. Dr. Jackson agreed with me in permitting him the use of champagne; and we prescribed for him a gently stimulating salve for the lips. It was determined by me, that the champagne experiment should be tried at my table, and its effects were so satisfactory, that Mr. Kennedy was directed to persevere with it. He was instructed by Mrs. Kennedy, by all means, to avoid the night air. We went, however, to McAran's garden [71] together, and after a sojourn of a few days, he returned to Baltimore wonderfully renovated. Soon afterwards, I received the following letter from him, which, with those given afterwards, will indicate his lively mind and social disposition.

Balt. Nov. 27, 1838.
My dear Dunglison,

I went to Washington after my return from Philadelphia, and then to Mr. Gray's, only getting back in time to prepare for the Club (mentioned at pp. 71–72) at my house, last night. These wanderings will account for my not immediately sitting down to return you thanks for your kind letter. Your prescription works à merveille. I am the wonder of the town, and drank your health in whiskey-punch last night in a circle of your old friends, whose merriment only wanted the addition of your hand at a laugh. I regretted very much, that my occasions compelled me to leave Philadelphia so soon. I had just begun to find the profit to which I might turn my enfranchisement from my late restrictions, in the cultivation of the society of the many good fellows you have about you. I mean, however, to indemnify myself in future for the loss, by more frequent visits. I am about making a most decisive change in my menage. At present I am somewhat overhoused— have too much roof and lodgement. I do not spend more than six months in Baltimore, and, hereafter, shall spend still less, being determined, in the Spring, to bid adieu to the law. My purpose, therefore, is now to build up a very neat gem of a place on Elk Ridge within a few miles of Mr. Gray's, where I shall move my books, wines etc. and always have a few spare chambers for friends. Here, when I wish to be at home, I will repair; and at all other seasons, I shall travel as it suits me, spending a month in Philadelphia, New York, or where I like it best. I am already in contract, or near it, for 100 acres on the very highest point of the Ridge in an exceedingly beautiful region, and hope, next summer, to be at work upon my cottage. I think you did right to sustain Ingersoll. God knows we have political acerbity enough to turn our whole social system sour, and unless we can sweeten it with a dash of the "simple syrup" of letters (which form of purification I shall hereafter prescribe, upon the authority of my own experience, for the feculencies of the blood) there will be nothing but bitterness left. Tell Dr. Patterson (the then Director of the Mint, whom he had met at my house) that when I get into my cottage, I mean to set up a rival mint concern, and hope to show him how much more passable my coinage will be than his. I shall challenge him to a personal inspection, and as he may require a medical adviser I shall exact of him not to travel without you.

Remember me very kindly to Dr. Jackson. I have written to him by this mail to remind him of a duty, which you,

on your part, seem inclined prepensely to forget. Pray treat me as a patient, by charging me a fee. I speak from my own professional experience when I say, that a fee is essential to the due preservation of one's professional repute. Make my respects to Mrs. D. and tell her, that I recommend her to employ you in her case, as a most excellent and satisfactory prescriber of agreeable medicines.

Very truly yours
John P. Kennedy.

In my reply to this in the frank and unreserved intercommunication which existed, I drew his attention to the following case of strange tenure of lands, which I copied from the History of Court Baron and Court Leet, (History of the High Court of Parliament vol. 2. p. 586) to which my attention was first directed by Mr. McIlhenny, the Librarian to the Athenaeum:—[72]

In Edward 1st's time, Rowland de Sarcere held 110 acres of Land in Serjeanty in Hemingston in Suffolk, to play Christmas games before the King: viz. to dance, puff up his cheeks, making therewith a comical sound, and to let a fart; and to burn a fart with a candle.

To this letter I received the following answer.

House of Representatives
Dec. 28, 1838.
My dear Dunglison,

This is the first moment, that some half dozen letters of business a day, and the multitude of little cares belonging to my station, have afforded me to reply to your kind note. The idleness of this place, with its insignificant but perpetual calls upon one's time, is altogether too intense for any profitable occupation, even that of friendly correspondence. We get up late; we sit in the house late, dine a great deal in company, and go to bed late. The consequence is, that in summing up each day's account, I find myself perpetually in arrear, not only on the score of friendship, but of those more reluctant duties, which my position casts upon me. I had some hope—a faint one— of seeing you here during these holidays. I understand, that Bonnycastle (Professor at the University of Virginia, see p. 62) was to be here from the University, and founded thereon a surmise, that you perhaps might be aware of it, and would come to meet him. He is an old chum of yours I think? I find, that he married a cousin of mine, whose sister resides for the present here. What is the name of the Peale, who has the Museum? Rembrandt?[73] I want to open a correspondence with him touching that picture of Cecilius Calvert,[74] which he got away from Maryland, and which I am anxious to have restored. I will write to the Governor of Maryland upon the subject, but not until I have obtained from Mr. Peale the terms upon which he will part with it.—It is for this purpose I want his address. My situation here has given me ample opportunities to test the value of the Champagne system under which I am happy to testify to the wonderful efficacy of the treatment. I believe I am almost entirely well, and that without the help of the simple syrup, which, from mere neglect, I have

---

[71] McAran's Garden Theatre, Filbert Street between Seventeenth and Eighteenth, was opened June 30, 1840 (J. Thomas Scharf and Thompson Westcott, History of Philadelphia 2: 979, 1844).

[72] The Athenaeum of Philadelphia, founded 1813, opposite Washington Square on Sixth Street, has always maintained a good library.

[73] Peale's Museum, established in Philadelphia by Charles Willson Peale (1741–1827), was carried on by his son Rembrandt Peale (1778–1860), like his father, a famous artist with many talents.

[74] Cecil, Lord Baltimore, son of Sir George Calvert, founder of Maryland.

discontinued. This *Dunglisonian practice* is much admired and extensively adopted in the treatment of this City, and, I am glad to say, is likely to exalt its author even above the famous *Thompson*.[75] I hope for your sake (as the discovery is worthy to be rewarded by a speedy fortune) that you do not design henceforth to practise so cheaply as you have done in my case. If you would come on to see us here, I should take upon me to commend the chalice to your own lips in frequent doses of that same physic. Jackson won't write to me, so I must beg you to enclose him one half of the within *large* remittance, and to assure him, that I have a much more abundant store of goodwill for his use, and a lively hope of being furnished with an opportunity to shew it on some future visit to your city. . . . I have directed Lea to send Mrs. Dunglison a presentation copy of Rob ("Rob of the Bowl"—one of his novels). Can't you dissect such a subject for the benefit of your class? I think your Rowland de Sarcere must have been an ancestor of Tom Mollineux and Dick, whose characters are left us by Swift.—

> "Tom and Dick had equal fame
> and both had equal knowledge.
> Dick could write and spell his name
> But Tom had seen the College.
>
> Dick was a coxcomb. Tom was mad,
> and both alike diverting,
> Tom was held the merrier lad
> But Dick was best at farting."

Mr. Kennedy's amiable feelings towards his father-in-law—who fortunately still lives and is in the enjoyment of health—stand forth in strong relief in the following letter.

Baltimore Dec. 3, 1839.

Dear Dunglison,

I was carried to Philadelphia week before last by business, which so occupied me for the two days I remained as to leave me no time at my disposal to make you a call. Although a member of the Convention on Education I was but once, and that for a very short space, present at its deliberations. But for these impediments I should not have failed to have performed a duty of friendship, which would have been so agreeable to myself. I should sooner have answered your note of the 28th ult. but for the most unhappy illness of our friend Mr. Gray. He is now, I think *in extremis*, with scarcely a chance of recovery short of a miracle. His disease seems to be some strange exhibition of chronic gout—as we suspect. His bowels appear to be perfectly torpid;—his stomach does not even pass the liquids he drinks for nourishment. The abdomen is painfully distended, and so susceptible to the touch as to give him the acutest distress. His pulse is good; his skin moist and free of fever, and his mind quite unclouded—but, as yet, no stimulants of blisters or mustard poultices have had any effect to raise the energy of the stomach, or reduce the distension. Upon the whole, the Doctors (Mackenzie [76] & Buckler,[77] and Denny [78] today) look upon the symptoms

[75] Samuel Thomson (1769–1843), originator of the Thomsonian or Botanic System of Medical Practice, who acquired a tremendously popular and profitable vogue.

[76] John Pinkerton MacKenzie (1800–1864), M.D. 1821, University of Maryland, a physician of high honor and professional skill (Cordell, *ibid.,* p. 485).

[77] John Buckler (1795–1866), M.D. 1817, University of Maryland, a popular family physician of Baltimore (*ibid.,* p. 338).

[78] Possibly William Denny of Ellicot City, Md., or Theodore Denny of the Eastern Shore of Maryland (*ibid.,* p. 375).

with great anxiety, and scarcely allow us to indulge a hope of his recovery. The distress of the family, as you may imagine, for you know the relations of affection amongst them, is most intense. It is a consolation to us, that he contemplates the issue himself with extraordinary calmness and resignation, and seems scarcely to give utterance to a murmur, or even a wish to avert the end. Our Society will have never lost a man of purer heart or more generous impulses. I know not how long he may linger, but, today, he has a hiccough which seems to be considered as a very bad symptom. My poor wife and her sister are in deep dismay and it is not the least of my personal griefs in this most severe trial, that I am obliged to control my own emotions by way of cheering and comforting them.

very truly and faithfully your friend
J. P. Kennedy.

Dr. Dunglison

Mr. Kennedy has not carried into effect the intention conveyed in one of his letters, of spending any time in Philadelphia; and, as I have not since visited Baltimore; and only once or twice the watering places, especially Saratoga, at which he is in the habit of spending a portion of his summers, I have rarely met with him. A week or two ago, I had the pleasure of seeing, that he had been appointed a member of President Filmore's cabinet; and I could not resist the impulse of offering him my congratulations on the occasion in the following letter.

Philadelphia July 25, 1852.

My dear Mr. Kennedy,

I cannot resist the impulse, to state to you with what gratification I have heard of your appointment to the highly honorable post, which you now occupy. I have long anticipated that your country must still farther have the advantage of those various and commanding talents, which I have seen illustrated on so many occasions, and was not surprised therefore, when I found, that the distinguished individual who now presides so ably over the destinies of this great Republic should have invoked your valuable assistance. You know, that I am not a politician, and, consequently, no opinion of mine, connected with public persons and matters may be of much worth. Still, I may say to you, that I have been deeply impressed with the dignified and correct course of the President. On no occasion has he appeared to me, for personal or other considerations, to lose sight of what he considered to be required by the best interests of the country and he will, undoubtedly, retire from his elevated position with the warmest feelings of every disinterested patriot. Excuse me, my dear Sir, for this *explosion,* which I could not, and would not, resist, and believe me,

with the greatest respect and regard,
faithfully yours,
Robley Dunglison

Hon. J. P. Kennedy,
Secretary of the Navy . . .

To this letter of congratulation I received the following reply.

Washington Aug. 1, 1852, Sunday.

My dear Dunglison,

Your note of congratulation came very pleasantly with a gay company of good fellows on the same errand, and the only entertainment I could give to the party was to bestow upon them a special drawer in my office, which I have set apart for friends alone, and there hold them in quarantine till I had accommodated the host of *other fellows*, who are not friends but mere business men. So today, whilst everybody is at church, I turn in upon this select coterie, and make my first reverence to you—to you with whom I have had so many pleasant days in the past, with the hope of many others to come. I thank you, my dear Dunglison, for this remembrance, and I assure you I experience in it the fact, that some of the happiest incidents that belong to preferment are those which bring us evidence of the sympathy and approval of friends. You will make this expression of your good wishes still more acceptable to me, if you can find occasion to visit the seat of Government when I am fully established here, and in condition to bring you into the ring with some lads of your own complexion whom the world loves, and you would like to know. Shan't I see you in my short reign? Thanks again, my dear Dunglison, and believe me ever and truly yours.

John P. Kennedy.

Prof. Robley Dunglison.

(For farther notice of Mr. Kennedy, see Supplementary Ana p. 183.)

THE PHARMACOPOEIA OF THE U. S.

In the year 1840, I was appointed by the Jefferson Medical College a delegate to the National Medical Convention for the Revision of the Pharmacopoeia of the United States, which assembled in Washington on the first of January of that year. I was then Professor of the Institutes of Medicine, Materia Medica and Therapeutics, and, as the Faculty only proposed to send one member, the choice was properly directed to the Professor of Materia Medica. At the meeting of the Convention, a Committee of revision and publication was chosen, consisting of Drs. Wood, Bache, Dunglison, Cohen, Dunn, Stevens and Sewall; but as all the members with the exception of Drs. Wood and Bache, who had taken an important part in the revision of the pharmocopoeia ten years before, and who were, consequently, familiar with the duty, lived at a distance from each other and from the delegates from Philadelphia, Drs. Wood and Bache and myself were constituted a Subcommittee, of which Dr. Wood acted as chairman. The subcommittee were certainly very active, meeting at the house of Dr. Wood, and devoting a large amount of time to insure its accuracy, and adaptation to the wants of the profession, and the pharmaceutist, and the pharmacopoeia was, at length, published in 1842.

The Committee acted together with great harmony, and unity of purpose; but the duty was so absorbing, that I had no desire to form a part of the Committee of Revision for the year 1850—the next decennial period; and when it became necessary to appoint a delegate or delegates, I proposed, that, for one, the Professor of Materia Medica and Therapeutics should be chosen. As Dr. Bache, however, was desirous of still being a representative at the Convention, it was determined, that the Professor of Chemistry should be another delegate; and that the College should thus send

two as was the case with other Institutions. Dr. Huston, however, declined to serve; and a wish was expressed, that I should take his place, which I declined, as I considered the duty could be as well accomplished by one representative as by two; and I had no desire whatever to participate again in its actions; whilst there were others with whom it was a labour of love.

MR. JOHN VAUGHAN

It was on my return from this convention, that I found a communication from the Secretaries of the American Philosophical Society informing me, that I had been elected one of the Secretaries of that body, an office, which I have since continued to fill; but have been, over and over again, anxious to relinquish it on account of the feeling of duty I had to attend meetings of the Society, which subjected me, at times, to great inconvenience. About two years ago, I wrote to my friend Judge Kane a short time before the election, mentioning to him my desire to withdraw at the coming election, but he begged of me to hold on, and I have continued to do so. As a member, and Secretary of the Society I was thrown into very close relations with two of its venerable officers, the President, Mr. Duponceau, and the Treasurer and Librarian, Mr. John Vaughan. At the time of Mr. Vaughan's death, which occurred on the 30th of December 1841, he had been for more than fifty years an officer of the Society. He was 86 years old less a fortnight; [79] and had been for a long time almost identified with it. It was impossible for more honour to be rendered to anyone than was done to his memory, at a special meeting of the Officers and Council; and at one of the Society, convened on the occasion of his death. He was described "as the patriarch representative of the Society, its oldest member." "They can never forget"—says the minute of the officers and council, written by Judge Kane, who, although much younger, had long been his friend and counsel—

his zeal for science in all its departments, his sympathy with scientific men, and his unlimited devotion to the interests and honour of this Institution. They had proved the warmth of his social affections, and the constancy of his friendship. They have seen his active, unwearied yet discriminating benevolence as it extended itself through every circle; rejoicing with the happy, cheering the distressed, counselling the friendless, and succouring the needy. Like the rest of this community, they have venerated the moral beauty of his daily life, and they feel, that even in his peaceful death, he has not ceased to be a benefactor to the city in which he lived, bequeathing to it, as he has done, the rich legacy of his admirable example, and a memory without reproach.

Similar encomiums were passed upon him by Dr. Chapman; and it was determined by the Society, amongst other resolutions,—that a member should be appointed to prepare the biography of Mr. Vaughan

---

[79] John Vaughan (1756–1841) came to Philadelphia from England about 1790.

for publication under the auspices of the Society; and the duty was assigned to Alexander Dallas Bache Esq, one of the Secretaries—and farther, that the Society would "cordially cooperate with other Societies of which he was a member or individuals approving the design in erecting a durable monument over his grave." His remains were followed to the grave by the members of various societies, and by the numerous friends of the good old man. A sermon was delivered over his body by his friend and pastor, Mr. Furness,[80] who described him as "although not a man of science or literature, yet a dear lover of both." By residence in France and Spain, he spoke fluently the languages of both, and by travel and constant association with eminent scientific and literary men, his information was extensive, and he shewed himself a cultivated man. He loved to make men of science and learning acquainted with one another, and to assist all literary investigations, all sound scientific projects. To him the American Philosophical Society is greatly indebted, and in their resolutions upon the occasion of his death, well do they speak of their venerable and beloved associate, as at once the oldest member of their body, and one of the most diligent, faithful and efficient.

He filled a large space. Hundreds there are, scattered through the world, who owe to him their pleasant recollections of this city. He it is, who has given it a hospitable and honorable name with multitudes. Seldom a private individual makes himself so extensively known; and as we think of the intelligence of his death extending to our cities and villages, passing on to foreign countries, we can hear everywhere some one who knew him exclaiming: Ah! the kind old man is gone at last. I can never forget him. I rejoice in the remembrance of his friendship.

I had not seen much of him before I removed to Philadelphia in 1836, but betwixt that time and the period of his death, I saw him often. He was the Dean of the Wistar Association, of which I became a member in 1839, and have continued to be one ever since.[81] To this every stranger of distinction, with whom he was acquainted, was sure to be taken by him. A club was formed of several of his friends, which was called the "Vaughan Club," and of which I was a member. One of the party furnished the table and dessert, no meats forming part, and each member took with him his bottle of some wine prized for its age and excellence; and it can rarely happen, that so choice a variety will be met with at any table. But alas! soon after the death of the venerable gentleman, a searching and

---

[80] William Henry Furness, eminent author and clergyman, father of the famous Shakesperian scholar, Horace Howard Furness.

[81] In the palmiest days of the Wistar Party, Dunglison was one of the best talkers and most genial associates, for he had an easy flow of richly flavored conversation (Obituary, *Lippincott's Magazine* 3: 676, 1869). The Wistar Association was formed after the death of Caspar Wistar in 1818 by members of the American Philosophical Society in order to continue his Sunday soirees, the members taking turns entertaining in their own homes.

severe successor, who, I hope is as correct in all his dealings as he would seem to wish others to be, found, that the complicated accounts of the octogenarian did not square in every particular; and that sums were received by Mr. Vaughan for the society, which had not always been accounted for. Strange, if one who had been connected with an Institution, of which he considered himself, and was considered by others part and parcel, and who, at the time of his decease, had weathered eighty six summers, should not have left some errors in accounts spread over so long a space of time. No account was taken of the large sums, which he doubtless often spent for the Society, and the deficiency was sternly presented to view. The Vaughan Club no longer held its meetings; the biographer was arrested in his progress, and has never resumed the duty; and Mr. Vaughan's warmest friends were paralyzed, under the apprehension, that if they were to move actively in his favour, they might induce statements to be made which would cast a shade over his memory. At the meeting of the American Philosophical Society, held on the first of April, 1842 it is stated in the "Proceedings" of the Society, that

Dr. Dunglison drew the attention of the Society to the subject of a monument to Mr. Vaughan on which resolutions had been passed on the occasion of Mr. Vaughan's death; whereupon, on motion of Dr. Chapman, a committee was appointed to carry the resolutions into effect. Committee, Dr. Chapman, Dr. Dunglison and Mr. Kane.

Yet, from the causes I have mentioned, the monument was suspended, and has never been resumed. Well do I recollect the effect produced on my mind by these coldhearted rumours; and how well could I comprehend the last request of Mr. Jefferson, that Mr. Madison would take care of him after his decease. At a meeting of the "Five," as we called an association of *five* gentlemen, one of whom I was, and of which I shall have to say something hereafter, it was determined, that we too should look after each other under similar circumstances, but, fortunately, thus far (August 1852) no one has passed from our midst by death, although we have been sadly dispersed.

MR. DUPONCEAU

Mr. Duponceau survived his venerable friend about a year and a half. For the four or five years immediately preceding his decease, I was much in his society. He begged of me to be his attending physician; and the first time I was called upon to see him was after he had been knocked down by an omnibus in crossing the street opposite his house. He was scarcely hurt, however; and I had but little occasion to see him professionally until the time of his last illness.

He begged of me to dine with him every Sunday; which, although a great pleasure to me, involved likewise a great privation, with the strong domestic feelings I have always possessed; but when he told me, that, on the Sunday mornings, he clasped his hands together, and exclaimed, "I shall see my friend Dr. Dunglison today" my wife and myself agreed, that, as he could not expect to live long, and my visits were evidently a comfort to him, I should accept special, but not general, invitations for that day. Every Saturday, consequently, his messenger came to invite me for the day following. On one occasion, he failed; and I failed also. This led to an explanation. I told him circumstances might arise, which might render it inconvenient for him to receive me; and that he had better always send to me the day before. Until the day of his decease, no omission occurred afterwards.

In the latter part of March 1844, he was attacked with bronchitis, which was severe from the first, and, at his great age, made me very apprehensive as to the result. He himself participated in these apprehensions, but appeared entirely resigned to his fate. A few days before his death, he begged of me to ask Judge Kane, along with Dr. Patterson, to call upon him, as he was desirous of adding a codicil to his will. At the appointed hour they were there, when he playfully stated, that he was desirous of making me his *Executioner!* He was exceedingly accurate and precise in the language, which he employed on that occasion, to which Judge Kane has borne testimony in a letter to me, written after Mr. Duponceau's death, which was appended to the Memoir I wrote of him; and, until almost the last, was in full possession of all his faculties. On the first of April he ceased to exist, at the age of 84. A special meeting of the American Philosophical Society was called on the occasion of his death, when, amongst other resolutions, it was determined, that "a public discourse, in commemoration of him, should be delivered by a member to be appointed for that purpose."

On motion of Dr. Bache, it was resolved to proceed forthwith to the nomination of an orator under the third resolution; whereupon, Dr. Dunglison was nominated by Dr. Bache, and, on motion of Mr. C. C. Biddle, it was resolved, that the nomination be now closed, and that Dr. Dunglison be appointed to deliver the commemorative discourse.

This discourse was delivered by me on Friday the 25th of October, under the circumstances described in the following notice, which I ascribed at the time to Judge Kane.

DR. DUNGLISON'S DISCOURSE. We inadvertently omitted calling our readers' attention last week to the announcement of Dr. Dunglison's discourse in commemoration of Mr. Duponceau. It was delivered on Friday at the Musical Fund Hall before one of the most intelligent audiences ever assembled in Philadelphia, including the Philosophical and Historical Societies, the Professors of the University and Colleges, the Judges of the several courts. etc. etc. As might be expected from a writer so well practised and, at the same time, so critical as Dr. Dunglison, the discourse was a beautiful tribute to the memory of our venerable fellow-citizen. It traced him through his various walks of military and civic distinction, and vindicated abundantly his

claims to the titles of a great civilian, an energetic patriot, and a true philosopher, yet it was, in fact, a commemorative record, in which the undiscriminating praise and hackneyed phraseology of the epitaph found no place. It was a notice of the man, as the country and the scientific communities of Europe have long known him, and as posterity will remember him. The discourse did great honour to the learned gentleman by whom it was pronounced. It was apparent always, without any direct allusion to the fact, that he was himself the intimate friend of the individual he was describing, familiar with his habits of thought, and his personal characteristics, not unpractised in his favourite labours, and sympathizing with him in that zeal for diffusive usefulness, which formed so beautiful an element in his portrait of Mr. Duponceau.

It fortunately happened, that Mr. Duponceau had left behind him some imperfect autobiographical sketches, from which I could select what bore more immediately on his character; and in all cases I preferred permitting him to speak—my great object being not to illustrate myself but the subject of my memoir; and it was pleasing to me to find, both from the published notices and the private communications of my friends, that I did not fail to make this purpose manifest to the reader. The following communication from Professor Henry, then of Princeton, now of the Smithsonian Institution at Washington, is an evidence of this.

Princeton Nov. 25, 1844.
My dear Sir,
I have just finished reading your very interesting discourse in commemoration of Mr. Duponceau, and I cannot refrain while my mind is filled with the subject from thanking you for the pleasure it has given me. I fully agree with you in the opinion you have advanced, that the business of the Orator on such an occasion should be to illustrate another rather than to exhibit himself and his own opinions. The effect of your discourse on me has been just such as I presume you wish it to be on all who read it. You have exhibited Mr. Duponceau as he was, and, during the perusal of the account, I have seen nothing but the man, and it is only at the close that I am awakened to the reflection, that the pleasure and profit I have been deriving are to be accredited to your ready and felicitous pen.

With much respect and esteem,
truly yours,
Joseph Henry.
Dr. Dunglison
Prof. etc. etc.

My intercourse with Mr. Duponceau was of the most agreeable and familiar character, and he sufficiently shewed in various ways his excellent feeling to me. His young ————[82] had always been the source of great distress and anxiety to him, his habits, in all respects, being of the most disreputable kind,—idle, drunken, fond of cohabiting with coloured women, and with coloured persons of both sexes in preference to the white; so that it had become necessary to apply to the Court of Common Pleas, who had disposed of him as a vagrant; and, through my influence, he was kept, for a long time, in the insane department of the Philadelphia Almshouse.

[82] Identity deleted by editor.

By will, Mr. Duponceau bequeathed to him the interest of $12000 as long as he lived; but, at the same time, appointed his father and myself Trustees, with power to give him the whole or part of it as we might deem expedient. He went to New York, and conducted himself for some years after Mr. Duponceau's death better than he had done previously, so that the whole income was allowed him by his father, whom I permitted to act much as he chose in the matter; at length, however, he was attacked with signs of pulmonary mischief; went to New Orleans, and returned to New York where he died of tuberculosis consumption. His defective *moral* was, I conceived, connected with faulty cerebral conformation. His mother— ———— —had—as I was informed—given scarcely equivocal manifestations of mental aberration; and his sister was for a long time under my care for insanity from which she has fortunately remained free for a great many years, but by no means free from danger of a recurrence at some future period.

Under such circumstances, Mr. ———— —the son-in-law of Mr. Duponceau—was left in an almost isolated state. His son went soon afterwards to New York to live with an aunt, and his daughter speedily followed. Although, whilst Mr. Duponceau lived, I knew but little of him, except from meeting him weekly at his father-in-law's table, I was desirous of ministering to his comfort, and told him, that a knife and fork would be laid for him every Sunday at my table; and I have had the pleasure of seeing him there for the last eight years (1852) on all occasions, when he has not been prevented by other engagements.

Mr. ———— continued to take his place at my table until within a year of his death, which occurred in   . His last illness was a painful one, and probably dependent on organic disease of the stomach. He left me one of his Executors.

MR. WILLIAM B. WOOD

So, also has it been with my old friend, Mr. William B. Wood,[83] for a shorter period. Every Sunday he is expected, when not otherwise engaged, to form a part of our dinner table arrangement. Mr. Garesché is now about 76 years of age. Mr. Wood 74, yet both are vigorous, and as mentally active as they have ever been. I have known the latter about eighteen years. The first time I met him was in Baltimore, when I was in company with Dr. Macaulay, who introduced us to each other. Afterwards, I met him at table at Mr. Jonathan Meredith's and elsewhere, for there, as in this city, his associations have always been with the best. In Philadelphia, at an early period of my residence, I attended him professionally, in the absence of

Dr. Harlan; and, when Dr. Harlan left the city, on my own account, but never made him any charge, or received anything from him. As manager of the theatre, and as actor, he had filled a large space and, at one time, was well to do in the world; but he subsequently met with reverses, and had so much difficulty in obtaining means of subsistence, that several of his friends subscribed annually the sum of *ten dollars* for his support; and, last Christmas (1851–2) they insured his life for $500 for the benefit of his daughter.

Soon after this was written I became a subscriber, at my own request of Mr. J. W. Wallace, the Secretary, and have so continued.

In the year 1854, I urged him to publish Reminiscences of the Stage, which he had announced many years before, but had not been able to carry his intentions into effect, in consequence of the impracticability of finding any publisher who would give him a remunerating price for it. With Mr. Henry Carey Baird, who felt much interest in Mr. Wood, I succeeded in effecting a satisfactory arrangement, and the volume —a very interesting one—appeared in the autumn of 1854, under the title of "Personal Recollections of the Stage, embracing notices of Actors, Authors and Auditors, during a period of forty years."

The sale was, at first, rapid; but the times were difficult; and it fell off sooner perhaps than was expected; but not without yielding a return, which must have been pleasing to the Author. My family took five copies.

(See Mr. W. B. Wood in Supplementary Ana p. 180.)

He has always maintained an excellent character; and has been a decided ornament to his profession, and it has been a source of great pleasure to me, that I have been, and am, able to render the downhill of his life agreeable to him.

THE MEDICAL STUDENT

Besides the *American Medical Library & Intelligencer,* the first work I published after I resided in Philadelphia was *The Medical Student.*[84] The work originated—as I remarked in the preface—in the applications made to me for my opinions as to the best method of study for one about to enter upon professional life, as well as for one engaged in its prosecution. Parts of the work also formed portions of an introductory, and of a valedictory, lecture, delivered to the class in November 1836, and February 1837, copies of which were formally solicited for publication; solicitations which were, however, respectfully declined, as I was desirous, that they might appear in a more useful shape. The whole of the observations applied to Medicine as taught in this country. No speculations were entered into, as to what medical education ought to be. The work was intended simply as some guide

---

[83] William B. Wood (1779–1861), with his wife, Juliana Westray, noted on the English and American stage, managed theatre groups at the Chestnut Street and the Arch Street theatres in Philadelphia.

[84] Delightfully reviewed recently by Dr. William B. Bean (*Archives of Internal Medicine* 107: 952, 1961).

to the American medical student, "who too frequently is totally uninformed as to the course he ought to pursue—not only when he commences to read upon his profession, but when he enters a medical college for the prosecution of his studies there." A bibliography was appended, in which a short notice was taken of several of the books in the English language indigenous and imported, which are placed before the student in the ordinary bookstores or which he sees, from time to time, advertized in the catalogues. This department contained a notice of *one hundred and ninety five* different works; and the volume terminated with an account of the Medical Colleges of the United States; their history and existing conditions, which was reprinted with modifications, from an article furnished by me, at the request of Dr. Forbes, to the *British & Foreign Medical Review,* and of which I have before spoken (p. 77).

In consequence of the restricted sale, that such a work was likely to have, whatever might be the amount of useful matter contained in it, the copy money given to me by my booksellers was small. I scarcely, indeed, expected it to go into a second edition. It was published under the title of *The Medical Student; or aids to the study of medicine; including a Glossary of the Terms of the Science, and of the mode of prescribing; bibliographical notices of medical works; the Regulations of different Medical Colleges of the Union, etc. etc.* and the chapters—four in number—embraced the subjects of 1. Preliminary Education 2. Medical Education prior to attendance on Lectures 3. Medical Education during the period of attendance on Lectures, and 4. Medical Education after Graduation. The appendix contained the Bibliography; and an account of the Colleges.

In the year 1844, the first edition of 1000 copies was exhausted; but the work was so small, and the sale had been so slow, that my publishers did not feel disposed to go into another edition. I was desirous, however, that it should be kept before the profession—convinced, that it was calculated to be useful and they consented, under the conditions, as to copy money elsewhere stated. The work was subjected to an entire revision; much new matter was added in the different chapters; and it was issued in the autumn of 1844, under the less elaborate title of *The Medical Student; or aids to the Study of Medicine. a revised and modified edition.* It is now eight years since this second edition was printed, and I am more and more confirmed in the accuracy of the views expressed in it, especially as regards the plan of study to be pursued during attendance upon lectures. Nothing has appeared to me so prejudicial as the recommendations of certain teachers, who have not thought much or well on the matter, that the student should read over, in the evening, the subject matter of what he has heard in the lectures of the day, as contained in the pages of approved textbooks. The time for reading is assuredly not during

the session of lectures. He will be sufficiently occupied, if he attends, and listens well, to the various lectures; reflects upon them all in the evening, and, when at a loss, turns to the pages of an approved textbook.

The following is the number of copies and the amount of copy money paid me for the two editions of the *Medical Student,* reckoning the sum paid for the second edition in the same manner as that paid for the second edition of the Elements of Hygiene was estimated (p. 74).

A.D.	Edition	Number of Copies	Copy Money
1837	First	1000	$375
1844	Second	1000	178
		2000	$553

In the year 1838, Messrs. Marsh, Capen and Lyon of Boston, the publishers of a School District Library of Massachusetts, corresponded with me on the subject of writing for the series a work on Physiology, in two small volumes, which might be, if I chose, an abridgment of my large work on *Human Physiology.* I, of course, laid the proposition before my publishers —Messrs. Lea & Blanchard—who did not object to my undertaking the work, provided it was a real abridgment, and so understood to be—but they had doubts as to whether it might not interfere with the sale of the one in which they were interested. Before I came to any specific conclusion I took the liberty of asking for information of Mr. Sparks, who was concerned in the direction, from whom I received the following letter.

Cambridge Mass. Dec. 17, 1838.

Dear Sir,

I have received your letter of the 2d inst. making inquiry about the publication of the School District Library of Massachusetts. The Massachusetts Board of Education of which I am a member, supposed a great public benefit would be conferred, if a series of books, well selected, and adapted to the mass of readers in our country towns, should be published and circulated at a reasonable price. As an encouragement to a publisher to undertake such an enterprize, the Board agreed to examine *individually* every book that should be published, and to permit the series to be published under their approbation. The only conditions that the Board prescribed were, that the mechanical part of the work should be substantially and faithfully executed, and that the price, particularly to the school districts in Massachusetts, should be as low as the nature of the undertaking would admit. Hence the business is wholly an affair of the publishers; the Board having no other concern in it than what is here mentioned. As to the "security for payment," I can say nothing, for I have no knowledge whatever of the concerns of Marsh, Capen and Lyon. They are a respectable publishing house of some years standing, and I have never heard their credit doubted. But on this subject you may probably obtain as correct information from some of the Booksellers in Philadelphia as you could in Boston. Nor do I know what remuneration they expect to allow, but they will probably adopt some rule, which will apply to all the writers. I have been told, that Mr. Washington Irving has engaged to write a book for them; that he is to hold the copyright, and to be paid by a commission, or fixed sum, for each copy, which shall be printed. But what the amount is to be, I have not been informed. I hope, that on inquiry, you will be led to think favourably of the matter, and will furnish the treatise on physiology, as it would make a highly interesting and important part of the series.

With great respect and regard,
I am, dear Sir, your most
obedient servant,
Jared Sparks [85]

Dr. Dunglison

On consideration of the subject I ultimately decided on declining to prepare the work in question, as well as every other offer to make an abridgment of the original work, and I have had no cause to regret this decision.

NEW REMEDIES

I had long been impressed with the utility of a work which should incorporate the new discoveries in Materia Medica, upon a more extended plan than the Formulary of Magendie, in editing which I had been concerned whilst in England (p. 13). An extensive work had appeared in Germany, by Dierbach, under the title *Die neuesten Entdeckungen in der Materia Medica* [86]— "The newest Discoveries in Materia Medica,"—but it was far too voluminous, if translated, to meet with any success in this country. I had almost abandoned the idea of preparing such a work, when Riecke's little volume—*Die neuern Arzneimittel; ihre physischen und chemischen Eigenschaften, Bereitungsweise, Wirkung auf den gesunden und kranken Organismus und therapeutische Benützung,*[87] fell under my notice, and determined me to undertake a work of the kind, of which Riecke's might be the basis. An edition of this I offered to Mr. Waldie, for publication in the *American Medical Library*, provided he would allow me five hundred copies to dispose of. To this he assented, and, in the number of the *American Medical Intelligencer* for June 1, 1839, I made the following announcement.

In the Library department of the next number, we shall commence in alphabetical order an account of *New Remedies* in which will be comprised those of modern introduction, with certain older agents that have in recent times received novel applications. The experience of practitioners at home and abroad will be referred to, and illustrative formulae given at the end of each article. The country practitioner, especially, has some difficulty in obtaining precise information on the various uses of such agents as creosote, iodine in its various forms, etc, a deficiency, which we shall attempt to supply. At the termination of the volume, an index of diseases etc will be given. It is on the basis of Riecke (Stuttgart, 1837).

---

[85] Jared Sparks (1789–1866), Professor of History and President of Harvard University.

[86] Johann Heinrich Dierbach (1788–1845), 2d ed., 3 vols., Heidelberg, 1839–1843.

[87] Victor Adolf von Riecke (1805–1857), 1st ed., 1837; 3rd ed., 1842.

It was completed in the number of the *Library* for October first and in the number of the *Intelligencer*, of the same date, it was announced, that Messrs. Lea & Blanchard would speedily publish an edition of the work. This edition consisted of the *five hundred copies,* struck off for me by Mr. Waldie. These I offered to them, and received the following letter accepting the same.

Dear Sir,

We will take the *five hundred copies* of *New Remedies* deliverable in sheets, and pay *seven hundred dollars* therefor, at six months from the date of our publishing. The work to make about five hundred pages like the specimen sheet. But it must be understood, that the copies printed for the Library are only sold to the Subscribers to that work, and as a part of that work. You can readily see, that if any copies were offered separately, it would interfere with the value of the copies you offer.

Respectfully
Lea & Blanchard
Philad. July 15, 1839

Professor Dunglison.

It appeared under the title *New Remedies: the method of preparing and administering them: their effects on the healthy and diseased Economy* etc., etc. with the motto "prodesse quam conspici." [88] The following extract from the Preface will exhibit the objects I had in view in publishing the work.

To enable the profession to form an accurate estimate of the value of remedies of more recent introduction, or of the older remedies whose use has been revived under novel applications, the present volume was undertaken by the author. In Germany, several works exist on this subject, and that of Riecke, to which the author has repeatedly referred—served as a basis for many of the articles: his observations, however, do not come down farther than the year 1836. Some of the statements—especially in relation to the observations of certain of the German physicians are given on Riecke's authority, for he has rarely appended references, by which the correctness of his assertions could be tested. It has been a great object with the author to furnish exact references to works in which farther information may be obtained and the number of these will shew, that he has devoted no small amount of time and attention to the subject. He has likewise added the results of his own experience in public and in private. The motto, which he has selected—*prodesse quam conspici* conveys, in epitome, his feelings. His sole object has been, "to be useful"— and if he has succeeded, the reward is ample.

The five hundred copies, constituting the *second edition* of the *New Remedies,* were soon disposed of, and I received the following letter from my publishers.

Philad. Sept. 15, 1840.

Dear Sir,

It is understood between us, that we are to print and publish the *Third edition* of your work called *New Remedies* etc. to consist of *one thousand copies,*—the volume to form about five hundred pages, in octavo. For this edition we are to allow you *fifty cents* per copy. You are to revise the work, and carry it through the press, as soon as prac-

---

[88] To be useful rather than to be conspicuous.

ticable, and give us the refusal to publish all future editions. No other edition to be sent to press or published until we have sold off all of this Third edition. Will you acknowledge the receipt of this in a note to our address.

Very respectfully,
Lea & Blanchard

To Professor Dunglison.

This third edition experienced numerous modifications and additions; but, I confess, it appeared to me, as it did to my respectable publisher, Mr. Blanchard, that the work was not destined to retain its hold on the public long, as the very title *New Remedies* might lead to its early decadency. We were mistaken, however, as the following letter sufficiently shews.

Dr. Dunglison
    Dear Sir,
    We find it is now time to go to press with the fourth edition of your *New Remedies*. With your permission this fourth edition will consist of *1500 copies*—the price per copy, to be placed to your credit on publication, will be *50 cents* or $750 for the Edition, at six months credit. Of course, you will have the work fully revised and brought up to the day, and endeavour to make it up to the same size as the last edition. It is understood, that no other edition will be published until we have disposed of this edition, which we will use all due diligence in selling.

We are very respectfully,
Lea & Blanchard.
Philad. Nov. 7th. 1842.

This edition, also, was greatly enlarged and modified, as the following extract from the Preface exhibits.

Since the publication of the third edition of this work in 1841, the Pharmacopoeia of the United States has appeared under the revision of Professors Wood and Bache and the Author. This has rendered it necessary to modify somewhat the nomenclature, and, to a certain extent, the arrangment of the *New Remedies*. The Author has likewise endeavoured to embody all the new information, of a therapeutical or pharmaceutical character, contained in the different scientific journals, as well as in the *exprofesso* works on Materia Medica and Pharmacy, that have been published since the appearance of the last edition. Farther and varied opportunities have necessarily occurred for testing the value of many of the agents, and of the methods for preparing them. The results of these observations have been introduced. The labour, required to accomplish this, has not been trifling, the large amount of matter added—seventy or eighty pages—and the numerous alterations, that have been made, can only be accurately appreciated, however, by a close examination.

, This edition appeared under the modified title, *New Remedies: pharmaceutically and therapeutically considered*. It was exhausted in 1846, and a letter from my publishers, dated the 29th of May of that year, stated.—

We are about to go to press with a new edition of your work on *New Remedies*, and should be glad [if] you would prepare it for the printer. The quantity will be the same as for the last edition—1500 copies, and on the same terms, of fifty cents per copy, and at six months from the time of publishing.

This edition appeared under the title *New Remedies, 5th edit. with extensive additions*. It was exhausted in

1850; and in a letter to me from Messrs. Lea & Blanchard, dated November 14, 1850, they remark—"Of the *sixth edition* of the *New Remedies*, now in press, we are printing *seventeen hundred and fifty copies*, and will place to your credit the Sum of *$875*, at 6 months from the day of publication." The title was again modified as follows: *New Remedies with Formulae for their administration;* and, in the Preface, it was stated.

The last few years have been rich in valuable gifts to Therapeutics, and, amongst these, ether, chloroform, and other so called anaesthetics, are worthy of special attention. They have been introduced since the appearance of the last edition of the "*New Remedies.*" Other articles have been proposed for the first time, and the experience of observers has added numerous interesting facts to our knowledge of the virtues of remedial agents previously employed. To include all these, it has been necessary to add very greatly to the dimensions of the present edition.

To show, indeed, how much the work had been expanded, it will be sufficient to mention, that, whilst the first edition was in *four hundred and twenty nine pages*, the sixth comprised *seven hundred and fifty five*.

The following is the number of copies, and the amount of copy money paid me for the editions of the *New Remedies*, up to this date (1852).

A.D.	Edition	Number of Copies	Copy Money
1839	First	published in the Library	
1839	Second	700	$ 700
1841	Third	1000	500
1843	Fourth	1500	750
1846	Fifth	1500	750
1851	Sixth	1750	875
		6450	$3575

ROGET'S PHYSIOLOGY

In the year 1839, Messrs. Lea & Blanchard determined on publishing an American edition of two essays, which were communicated by Dr. Roget [89] to the *Encyclopaedia Britannica*, and they applied to me to edit the same, with such notes as I might deem proper. To this I consented, provided my name should not be connected with it on the title page. My preface exhibited the nature of the undertaking.

The contents of the present volume—as will be seen by the Author's Preface—form the articles "Physiology" and "Phrenology" in the *seventh edition of the Encyclopaedia Britannica*. These articles have been recently published, separately in Edinburgh, in two handsome volumes, and, from these, the present edition has been printed. Of the author's qualifications as a physiological writer, it is scarcely requisite to speak. The fact, of his having been

---

[89] Peter Mark Roget (1779–1869), M.D. 1798, Edinburgh, whose *Thesaurus of English Words and Phrases* (1852) still endures, wrote the articles on physiology and phrenology in the 6th and 7th editions of the *Encyclopedia Britannica*, separately in 1838. His Bridgewater Treatise, *Animal and Vegetable Physiology*, was republished in Philadelphia, 1836.

selected to compose the Bridgewater Treatise on animal and vegetable Physiology is sufficient evidence of the reputation, which he then enjoyed, and the mode, in which he executed the task, amply evinces that his reputation rested on a solid basis. The present volume contains a concise, well-written epitome of the present state of Physiology—human and comparative—not, as a matter to be expected, the copious details and developments to be met with in the larger treatises on the subject; but enough to serve as an accompaniment and guide to the physiological student. The attention of the American Editor has been directed to the revision and correction of the text; to the supplying, in the form of notes, of omissions; to the rectification of some of the points that appeared to him erroneous or doubtful; and to the furnishing of references to works in which the physiological inquirer might meet with more ample information. In Phrenology, the Author is a well known unbeliever, and his published objections to the doctrine have been regarded as too cogent to be permitted to pass unheeded. It will be seen, that farther examination in the interval of many years, which has elapsed since the publication of the sixth edition of the Encyclopaedia, has not induced him to modify his sentiments on this head. On the contrary, he appears to be as satisfied, at this time, of the fallacy of the positions of the Phrenologist, as he was at any former period.

No bargain was made with the publishers for the revision and notes, but they placed to my account the sum of *one hundred dollars*. The work was by no means successful as a commercial speculation, whatever might be its merits as a literary and scientific composition.

TRAILL'S MEDICAL JURISPRUDENCE

Another article from the *Encyclopaedia Britannica* had been printed, in separate form, in Edinburgh: *Outlines of a course of Lectures on Medical Jurisprudence* by Dr. Traill,[90] of Edinburgh; and I was requested to revise, add notes, and superintend an American edition, which I consented to, under the same conditions as the work of Dr. Roget, and for which they placed to my account the insignificant sum of *fifty dollars*. I undertook it, however, as a pleasing task; as the work contained an epitome of the leading general principles of Medical Jurisprudence, and also of Medical Police. "It therefore"—I remarked in the Preface—"cannot fail to be a useful manual of reference both to the physician and the lawyer, and a valuable accompaniment to the Medical Student." "The American Editor"—I added—"has restricted himself to the correction of errors, and to the adding of such notes as appeared to him to throw additional light on the topics discussed in the text."

The work, although useful as a manual, and as an accompaniment to Lectures on the subject, did not meet with an extensive sale.

---

[90] Thomas Stewart Traill (1781–1862), M.D. 1802, Edinburgh, President of the Royal College of Physicians of Edinburgh, first published his *Medical Jurisprudence* separately in 1836.

PRACTICE OF MEDICINE

In the spring of 1839, I received the following letter from my publishers, Messrs. Lea & Blanchard.

Dear Sir,

We have long contemplated the publishing of a *Practice of Physic*, to form two volumes 8vo, and, long since, made arrangements with Professor Geddings to prepare it. He has delayed its preparation, and is willing we should engage others to carry out our plan. It is essential to its success, that it should now be under way. The object of this is to present you the opportunity of preparing such a work for speedy publication. The terms we can arrange when we learn your willingness to undertake it.

Very respectfully,
Lea & Blanchard

To Prof. Dunglison

On expressing my willingness to undertake the preparation of such a work, I received the following letter from them.

Dr. R. Dunglison,
Dear Sir,

Our view, as regards the *Practice of Physic,* which in accordance with our note of the 24th, you are willing to undertake, is this. The work to form two volumes octavo of not over six hundred pages each, solid or leaded matter as you may suggest. The edition to consist of, at least, *two thousand copies*. The future editions to be not less than that quantity. To have the work ready for publication in the spring of 1840, to constitute a complete *Practice of Physic* up to the day, and that shall have reference to a permanent work for extended circulation. We to pay you for this *first edition* of *two thousand copies fourteen hundred dollars,* by note at six months from the day of publication—the right to publish all future editions to be with us, and, for such future editions, you are to receive *seventy five cents* per copy, at 6 months from publication. If this meets your views address us a note assenting thereto.

Yours respectfully,
Lea & Blanchard
Philad. May 9th 1839

PS. You will observe, that the difference between the price per copy for the first and future editions will not more than cover the cost for alterations, while we must bear the difference between the cost of manuscript for that edition and printed matter for future editions.

I did not assent to this proposition as is shown by the following note.

Philadelphia,
9 Girard St.
May 11, 1839.

Gentlemen,

In answer to your favour, offering me *fourteen hundred dollars* for the first edition of a work on the *Practice of Physic,* adapted to the present state of Medical Science; and *75 cents* per copy for future editions,—no edition to consist of a fewer number of copies than 2000, I have to express my willingness to accept your propositions, provided you allow me the very moderate compensation for the labour of *75 cents* per copy from the first, the work to be finished for the press at as early a period as is practicable.

I am, Gentlemen,
Very truly yours,
Robley Dunglison.

Messrs. Lea & Blanchard.

This proposition was accepted by them.

Dear Sir,

We have yours of the 11th, accepting our proposition of the 9th relative to the *Practice,* and accede to the modification there stated, which requires us to pay, for the first edition, at the rate of *seventy five cents* per copy. We will embrace some favourable opportunity to make the announcement.

May 13, 1839.

Yours respectfully,
Lea & Blanchard

I now went to work vigorously in preparing my *Practice of Medicine* for the press; and not until the autumn of 1842, was I able to issue the first volume; the second volume, completing the work, was published in the following spring, or rather at the end of January 1842. It was dedicated, "affectionately,"

To the Gentlemen
who have honoured
the Author by their attendance on his Lectures
in the course
of the
last sixteen years.

The objects, had in view by the publication of this work, were frankly stated in the following preface.

The improvements and modifications incessantly taking place in the departments of Pathology and Therapeutics, render it advisable, from time to time, to incorporate them, so as to furnish those, to whom the different general treatises, monographs and periodicals are not accessible, with the means of appreciating their existing condition. Perhaps at no time has it been more necessary than at present to bring together those various elements: certainly, within the last ten or twenty years greater activity has been exhibited amongst observers than at any former period, and the researches of recent pathologists have greatly altered the face of the science, in regard to certain lesions more especially. Different views are still entertained on some of these; but they ought all to be familiar to the observer, in order that his own investigations may receive the proper direction, and—what is all important— that he may know when to remain in doubt. The departments of Special Pathology and Therapeutics have necessarily occupied a large amount of the Author's attention, engaged, as he has been, for upwards of sixteen years as a Medical Professor, and, for a much longer period, as a practitioner. His opportunities, too, for witnessing the phenomena presented by disease in both hemispheres, have been varied. During a long service, as a medical student, in the North of England, in Edinburgh, London and Paris; during a practice of six years in London; of eight years whilst he was Professor in the University of Virginia; three years as Professor in the University of Maryland, and upwards of five years as Professor in the Jefferson Medical College of Philadelphia, he has carefully noted the modifications that appeared to be produced by climate and

locality. Moreover, his service for three years, as Physician to the Baltimore Infirmary; and, for a longer period, as Physician to the Philadelphia Hospital, one of the largest charities in the country, has equally enabled him, to appreciate the differences presented by the same malady, according as it may fall under the care of the private practitioner or of the medical officer of an eleemosynary institution; and to pronounce, as the result of such observation, that the great principles of Pathology and Therapeutics are the same everywhere, and that one, who has been well grounded in those principles, can exercise his profession with as much satisfaction to himself, and advantage to the sick, in the scorching presidencies of British India, as in the more temperate regions of our own country. As in the case of epidemics, differences are observable; but those differences are readily seized and appreciated by the well educated physician, and the appropriate treatment suggested accordingly. Hence, the medical officers of our army and navy and especially the latter, whose duties carry them to every part of the globe, are found to be as successful in the management of the cases, that fall under their care in distant regions as they would be in the treatment of those, that prevail in the spot where they received their medical education. In regard to the execution of the work, the author would merely remark, that he had endeavoured to give a faithful exposition of what he considered to be the existing views in relation to the subjects of which it treats. He is not conscious of possessing any exclusive opinions, and has endeavoured to be essentially eclectic. Neither is he aware of having any undue prejudices. It has been his good fortune to pass, thus far, through life, without imbibing unpleasant feelings towards any honourable member of the profession; he has, accordingly, throughout the work, felt a pleasure in referring to the labours of observers, everywhere, and it has been no little satisfaction to him, that he has been called upon so often to make mention of the investigations of those on this side the Atlantic. In the preparation of such a work, a large amount of labour and of reflection has been necessary; and the Author humbly hopes, that it may not be found to have been bestowed in vain.

This Preface was dated on the 4th day of January, 1842.

From almost every Journal, at home, and from all those abroad, which noticed the *Practice of Medicine* it received high encomiums; more than enough to satisfy the most fastidious author. In the *British & Foreign Medical Review,* it was lauded in the different editions in various articles; and in the *American Journal of the Medical Sciences,* a highly liberal and commendatory notice appeared from the pen of Dr. Stewardson, of whom I have spoken elsewhere (p. 75) with which I had every reason to be satisfied. It certainly shewed, that he had no *counter* prepossessions against the Author; and that if any prepossessions whatever existed, they must have been of the opposite character. In that well written article, Dr. Stewardson depicts the duties devolving upon the author of a *Practice of Medicine,* strictly in accordance with the views I then possessed, and still sanction. "Dr. Dunglison" he says,

opens his subject by observing, that the modifications incessantly taking place in the departments of pathology and therapeutics, render it advisable, from time to time, to incorporate them for the use of those, to whom other means of information are not fully accessible, and although students of medicine are not expressly mentioned, we presume that one principal object of the publication was to furnish them with a complete text-book. To present to the reader a correct view of the state of the Science, both in reference to facts and opinions; to spread before him a panoramic representation of its several parts, and of the architects and workmen employed in their construction, that he may see how far the structure has progressed, what parts are at present advancing with the greatest rapidity, and who are employed in accomplishing the task, is the legitimate object of a work like that before us. Its author should appear chiefly as the scenic painter, unless when really engaged in assisting in erecting some part of the edifice, in carving a block or rearing a column, and not with bustling vanity endeavour to display himself at every point, claiming to be as thoroughly versed, and as competent to decide upon every question, as those who have made them severally especial objects of laborious research. He should be satisfied with the merit, and that no inconsiderable one, of arranging and adjusting the products of other men's labours, so as to form a connected whole, all the parts of which should be displayed in their proper relative position and importance. We do not make these remarks in view of any supposed deficiency of the work before us in these respects, but the rather because it strikes us, that it contains a more complete, and just account of the state of our knowledge, in reference to the various diseases of which it treats, derived from a multitude of sources, than any similar production, which we have met with.

And he concludes his review as follows—

Enough perhaps has been said, however, to convey to the reader a pretty just conception of the general plan and execution of the work before us. Our examination of it has necessarily been hasty, and we are confident, that we are very far from having done it all the justice which it deserves. Its author has not only given evidence of the most extensive research, but of much judgment in the choice and arrangement of his materials, and whilst he carefully recounts the various opinions of medical men upon numerous points, he equally avoids the admission of anything like confusion into his descriptions. In the midst of conflicting opinions, the opinion of his own views is marked by much clearness of conception and justness of conclusion. The whole character of the work is eminently practical. It has no tendency to lead the reader to the adoption of exclusive views; but, on the contrary, presents him with a remarkably just estimate of the state of opinion on most points. With these and other features previously noticed, of the work before us, we have been highly gratified, and we have little doubt that our readers, when they shall have an opportunity of forming their own opinions by a perusal of it, will agree with us, that it contains a vastly more complete digest of the state of our knowledge, in reference to pathology and therapeutics, than any previous treatise of a similar character.

If the Author had endeavoured to lay down a plan for a work on the Practice of Medicine, he could have had no better model than that given by Dr. Stewardson, in the early part of this extract. To endeavour to make such a work entirely original must end in an utter failure. The condition of a progressive Science is formed from the observations of *savans* of all periods—past & present—and any work, which professed to give a picture of such condition, must necessarily embody the collected wisdom of every age. Yet the objection that has often been made to this, as well as to others of my works, has been, that they were not "original." If by this was meant, that the facts and reflections, contained in them, were not wholly the creations of my own brain, the charge is true. I have, in all cases, endeavoured to give the existing state of science, and have not felt myself at liberty to discard the phenomena and deductions recorded by any trustworthy observer. I have considered the objection, indeed, as a compliment. But, most commonly, I have known it urged by persons, who have never done me the honour to peruse the works on which they animadverted. The liberal and too encomiastic observations of Dr. Stewardson are in striking contrast with those of the Editor of the same Journal, Dr. Hays, of whom I have spoken already (p. 86) in his notice of the second edition of the same work (*Amer. Journal of the Medical Sciences,* July 1844 p. 195) who, after giving a bibliographical account of the *Cyclopedia of Practical Medicine,* without the slightest notice of myself (the American Editor) merely remarks on the *Practice of Medicine* "The early call for a second edition of this work is sufficient testimony, that it has been an acceptable one to medical students," and in equal contrast are the liberal and laudatory remarks of Dr. Condie [91] in the Number of the same Journal for January 1848, p. 221.—

The work of Dr. Dunglison is too well known to require at our hands, at the present time, an analysis of its contents, or any exposition of the manner in which the Author has treated the several subjects embraced in it. The call for a third edition within five years from the appearance of the first, is, of itself, a sufficient evidence of the opinion formed of it by the medical profession of our country. That it is well adapted as a textbook for the use of the student, and, at the same time, as a book of reference for the practitioner, is very generally admitted; in both points of view for accuracy and completeness, it will bear a very advantageous comparison with any of the numerous contemporary publications on the practice of medicine, that have appeared in this country or in Europe. The edition before us bears the evidence of the author's untiring industry, his familiarity with the various additions, which are constantly being made to our pathological and therapeutical knowledge, and his impartiality in crediting the general sources from which his materials have been derived. Several pathological affections, omitted in the former editions, are inserted in the present, while every portion of the work has undergone a very thorough revision. It may, with truth, be said, that nothing of importance, that has been recorded since the publication of the last edition, has escaped the attention of the author; the present edition may, therefore, be regarded as an adequate exponent of the existing condition of knowledge on the important departments of medicine of which it treats.

I do not purpose to adduce the various opinions given in the Journals, of this or any other of my books. They will be found, in epitome, appended to the adver-

[91] David Francis Condie (1796–1875), M.D. 1818, University of Pennsylvania, fellow of the College of Physicians of Philadelphia, member of the American Philosophical Society, highly respected Philadelphia physician.

tizing lists of the time, issued by my zealous and enterprising publishers. I shall merely remark, that in the *British and Foreign Medical Review* so ably edited by Dr. Forbes, a short notice was given of the first edition, in the number for July 1842, p. 220; and two extended notices in the numbers for October 1844, and October 1845; in which last two, it was subjected to a comparison, analytically and critically with the *Guide du Médecin Praticien* of M. Valleix.[92] A portion of the first of these notices, I extract, in consequence of its bearing upon a charge to be noticed hereafter—of injustice to Authors, whose works I had employed in its preparation, in not assigning them due credit.

In the volumes before us, Dr. Dunglison has proved, that his acquaintance with the present facts and doctrines, wheresoever originating, is most extensive and intimate; and the judgment, skill and impartiality with which the materials of the work have been collated, weighed, arranged, and exposed, are strikingly manifested in every chapter. The author may truly say, that "he is not conscious of possessing any exclusive opinions; and has endeavoured to be essentially eclectic; neither is he aware of having any undue prejudices." The consequence of this is that the different parts of the work bear a more equable proportion to one another than is apt to be the case with treatises compiled by writers of great originality, real or imagined. Great care is everywhere taken to indicate the sources of information, and, under the head of treatment, formulae of the most appropriate remedies are invariably introduced. In conclusion, we congratulate the students and junior practitioners of America, on possessing, in the present volumes, a work of standard merit to which they may confidently recur in their doubts and difficulties. *Brit. & For. Med. Rev* July 1842, p. 220.

CHARLES CALDWELL AGAIN

A striking exception to the favorable notices occurred in the *Western Journal for Medicine and Surgery*, for June 1842.[93] I have before referred briefly to this matter, and stated, that I might have occasion to recur to that article as well as to one that had been previously written by Prof. Gross (p. 28), of a very different tenor. This last article was contained in the number of the same Journal for April of that year; and is more a kind notice of myself than of the work, the title of which was placed at the head of the article. It was as follows:—

Professor Dunglison is certainly a man of the most prodigious industry, to say nothing of his talents, which are, doubtless, of a high order. The present treatise constitutes the seventh that has been furnished by his prolific pen within the last twelve or thirteen years. His first production, entitled *Commentaries on Diseases of the Stomach and Bowels of Children*, appeared in England in 1824, a short time before he emigrated from that country of which he is a native, to the United States, to assume the duties of a teacher of medicine in the University of Virginia, to which he had been invited by President Jefferson, the distinguished founder of that excellent Institution. Like the early efforts of Cooper and Bulwer, the work in question fell stillborn from the press; it was severely criticized by the reviewers, and no reprint of it has ever appeared in this country. All his other works, however, certainly the majority of them, have been well received by the profession, and one of them *Human Physiology* has already reached the fourth edition. Scarcely forty five years of age, he has published a greater number of works than any other physician on this side of the Atlantic, whether dead or living. To him writing seems to be no labour; books come to him almost at a call; he has only to say, "let there be a book, and a book there is." In addition to his various medical productions, Professor Dunglison has been editing for some time past, a medical periodical, one of the most valuable, by the way, in the country, and has contributed some very learned articles to several of our literary reviews. From the immense amount of labour, both intellectual and physical, which he has accomplished, one would suppose, that the Professor was a thin, emaciated, Smike-like [94] personage, or, still worse, prematurely old, and grey-headed, with a body bent like that of poor Scarron in the form of a Z.[94a] Not so, however, He is really a fine looking man, whose whole exterior is indicative of the most vigorous health, whose embonpoint would do honour to an alderman, or even "my lord mayor"; who has none of that irritability which so much worries and frets the poet, especially those in high places—the garret and the attic, and whose social qualities. are as rare as they are delightful. Notwithstanding his immense labour as an author, he possesses a sort of ubiquity truly remarkable. Go where you will in Philadelphia, you find the Professor. Does a stranger of distinction visit the city, he is the first to call upon him. (I fear I do not deserve this credit); in the morning he is at the Blockley Hospital or in attendance on his private patients; in the afternoon, he saunters forth in pursuit of health or recreation; in the evening, he is at a party,[95] or perhaps at the rooms of the American Philosophical Society of which he is the corresponding Secretary, and one of the most useful and active members. But when, asks the reader, does the Professor find time to compose his works? Not certainly in the morning, nor in the afternoon, nor yet early in the evening; no, these hours are all occupied in other pursuits; it is late at night, at midnight if you please, when other men are wrapt in sleep, and dreaming, *not of what they have done, but of what they intend to do*, that he produces those works, which have made his name so honorably known in America and Europe, and so closely identified it with the history of medical science and literature in the nineteenth century. Having said so much about the *man* we need not make any particular comments on his recent *production*, as our object merely is to attract to it the attention of the profession. Like its predecessors, it bears the impress of sound judgment, profound research, and extensive acquaintance with the labours of others in the same department. Next to the *Physiology* it is decidedly the best and most finished of his works. The arrangement is admirable, and the whole is executed in a manner highly honorable to the talents and learning of the distinguished author. If we mistake not, it will be

---

[92] François Louis Isidore Valleix (1807–1855) published his *Guide* in 10 vols., 1842–1848.

. [93] See Emmet Field Horine, *Charles Caldwell*, item 194, pp. 76–77, 1960.

[94] Smike was a broken-spirited character in Dickens's *Nicholas Nickleby*.

[94a] Paul Scarron (1610–1660), French humorist, at the age of thirty, in consequence of a rheumatic attack in which he was treated by a quack, became an invalid for life—deformed and contorted (Walter Miller, *Standard American Encyclopedia*, 1916).

[95] For an intimate description of one of Dunglison's student parties, see Squibb's *Journal* (pp. 247–248).

extensively sought after by the profession, and adopted by many of our schools as a text-book. That it will supersede most of the treatises on medicine now in use among us, there can be no doubt. The mechanical execution is worthy of the matter, the paper and typography are excellent; and the *tout ensemble* is such as to induce the conviction, that, in bookmaking at least, we are but a step in arrear of Old England herself.

Whatever objection may be made to this notice, on the score of undue encomium, there can be none as to the excellent feeling towards the author and his works displayed in it. It did not, however, satisfy my *quondam* friend Dr. Caldwell, for reasons I have stated elsewhere, (p. 101) and, still more, for the opinion I expressed in the Preface, that it signified not, where a man was educated. If properly instructed in the principles of his profession, he would be able to practice his profession understandingly any, and everywhere. This view was opposed to the expressed and published sentiments of the venerable gentleman on, what was called, "sectional medicine," who maintained, that, in order to practice medicine properly in the West, it was necessary that the practitioner should have been educated in the West, and, as he was a Professor in a Western Medical School, the argument, so industriously promulged by him, led to the suspicion, that interest might have lain at its foundation. Be this as it may, the excessive self complacency, self sufficiency, and want of tolerance in his essays on the subject cannot be doubted. Under these feelings, combined, he prepared a most bitter notice of my *Practice of Medicine,* which the Editors—Drs. Drake, Yandell and Colescott [96]—accepted and inserted in the *Western Journal of Medicine and Surgery,* for June 1842; and that Journal, accordingly, exhibited the singular phenomenon of a highly laudatory article in one number; and one diametrically opposite in another number, published two months afterwards, and under the same editors. Of the matter and management of this article by Dr. Caldwell, a sufficient judgment may be formed from a few extracts. At the time it first reached me, in 1842, I was confined to my house by a severe attack of sickness; and its effect upon me was very different from that which its energetic Author would have anticipated. The sustained, overstrained, effort to censure was irresistibly ludicrous to me; and when on the same evening, I was visited by my friends Dr. Robt. M. Patterson and Judge Kane, the amusement it afforded us was not a little. Judge Kane was, indeed, so tickled with it, that he took it home, and returned it to me with marginal observations, not at all favourable to the taste or the English of Dr. Caldwell, and which would have mortified the vanity of that gentleman considerably, had they been subjected to his inspection.

On his favorite topic of sectional medicine, Dr. Caldwell has the following remarks, suggested by the ob-

servations in the preface to the *Practice of Medicine,* given before.

Few things in Medicine have surprised us more than the inveterate pertinacity with which Professor Dunglison, and perhaps some other writers and teachers in the Atlantic States, adhere to their notion, that they are themselves actually competent to the exposition and treatment of the herculean diseases of the West & South, *which they have never witnessed*—and (still more erroneous and assuming) that they can communicate to their pupils a thorough practical knowledge of that, which they have never themselves had an opportunity deliberately to study, nor even momentarliy to inspect! The Professor, as we are told, is a sturdy disbeliever in *mesmeric clairvoyance,* and even imputes to those, who are holders and advocates of the opposite belief, a condition of mind scarcely short of *imbecility* and *foolishness.* Yet does he labour, with the zeal of a fanatic and the perseverance of a martyr, to impress on the world an implied admission of his own *omnivisient clair-voyance* (may we coin an extraordinary name for so extraordinary a power)—a faculty so clear in its function, and so boundless in its sphere of action in him, that he can, by means of it, while seated in his study in Philadelphia, explore, successfully, the specialties, as well as the general characters, of the complaints of the Mississippi valley, and learn how to cure them! *"Risum teneatis amici!"* [97] So easy and natural is it for gentlemen, of a certain organization, temperament and caliber, to strain at gnats as respects the performances of others, and to swallow camels and moose deer themselves. p. 445.

Yet all the "zeal of a fanatic and the perseverance of a martyr" ascribed to me by Dr. Caldwell, is wholly imaginary; but in regard to the being a disbeliever—"a *sturdy* disbeliever," if he prefers the phrase—in mesmeric clairvoyance he does me but justice. I certainly have not—so far as I can recollect "imputed" to those who are holders and advocates of the opposite belief a condition of mind scarcely short of "imbecility and foolishness"; and yet if epithets were necessary in these and similar cases, it would not be difficult to find some, that might be appropriate for those who entertain the wild beliefs, expressed in the following quotation from a work entitled *Facts on Mesmerism, and Thoughts on its Causes and Uses* from the pen of this same Dr. Caldwell. Alluding to *"the contest"* then in progress, respecting the truth and usefulness of mesmerism—the Author—remarks:

I declare that contest to be as susceptible of an immediate, easy, and certain decision as would be a dispute about the product of the union of sulphuric acid with soda, zinc or any other substance. Of either question, the solution must be drawn from the result of experiments, alike simple and easily performed. And in each case, *ten* experiments *correctly* performed, and *identical* in their issue, are as conclusive as *ten thousand.* I have myself done, in a single hour, what ought to convince, and did he witness it, *would* convince any unprejudiced, candid, and intelligent man, of the *entire* truth of mesmerism, etc.

Never has there been before a discovery, so easily and clearly demonstrable as mesmerism is, so unreasonably and stubbornly doubted, and so contumaciously discredited and opposed,—opposed, I mean *in words;* for the opposition is

---

[96] Thomas W. Colescott became junior editor in 1842.

[97] You should restrain your laughter, my friends (Horace).

but a mass of verbiage; while the defence is a body of substantial facts. Yet never before has there been made in anthropology, a discovery at once so interesting and sublime; so calculated to exhibit the power and dominion of the human will; its boundless sway over space and spirit. . . . For one person completely to identify another with himself—sense with sense—sentiment with senti-ment—thought with thought—movement with movement—will with will—and, I was near saying, existence with existence—and to gain over him so entire a control as to be able to transport him, in his whole mind & being, over mountains, seas and oceans, into distant lands, and disclose to him there the objects and scenes, which actually exist, of which he was utterly ignorant before, and becomes alike ignorant again, when restored to his usual condition of existence; and, higher and grander still, to waft him, at pleasure, through space to any or all of the heavenly bodies, of which we have any knowledge, and converse with him about them, such deeds as these may well be called amazing, yet are they as easy, certain and speedy of performance as many of the most common transactions of life!

But, to return to Dr. Caldwell's Review of my *Practice*. In reference to my works in general, he remarks, that all, "hitherto published, have been the product of the inferior order of the intellectual faculties—of those which, by observation and perception, re-ceive knowledge, and communicate it by language. Of the fruit and finish of the higher intellectual faculties, they are altogether destitute." p. 434. I have else-where said (p. 102) that Dr. Caldwell informed me that from an examination of my head the *organs of perceptivity* predominated, whilst *he* enjoyed a pre-ponderance of the organs of *intellectuality*. My works necessarily confirmed his phrenological inferences.

On the "literary standing" of the volumes of the *Practice*, he passes judgment as follows.

It is *ordinary* in all parts, except where it is *low*. The style of the work is everywhere loose, crude and ungraceful, with frequent outbreaks of pomp and bombast. Without pronouncing it positively obscure and muddy, we are fully justified in saying, that it is wanting in that cloudless perspicuity, which, in every kind of composition is the most important quality, and, in elementary treatises in science, is altogether indispensable. Criticism might, *in charity*, deal gently with it, were it the production of a *freshman or a sophomore*, but coming, as it does, from a grave Pro-fessor, in the very meridian of mental life, and who prides himself on being a *British scholar* (educated, we mean in England) (I certainly never gave expression to any such pride) strict justice should be made to visit it; and, all circumstances considered, that is *condemnation beyond reprieve*. To the simplicity and purity, conciseness and vigor of well prepared composition in the Anglo-Saxon tongue, the style and manner of the work are as utter strangers, as if they were not the product of a native pen. As respects many of the writer's customary forms of ex-pression, though we do not pronounce them actual *foreign-isms*, yet have they more of a foreign than of a *native* cast. They are, at best, but a hybridous product, between the language of Great Britain and that of sometimes one and sometimes another, of the nations of continental Europe. In most instances, however, the adulteration comes from France. Yet does the writer bear the character of a ripe and high bred English scholar. p. 435

He afterwards adds,—

That these free and unceremonious strictures on Professor Dunglison's *Practice of Medicine* will be attributed, by some persons, to unfriendliness of feeling toward him on our part, is hardly to be doubted. The charge, however, if preferred, will be groundless. We are not unfriendly toward him (?). But, in the words of Brutus to Cassius. "We do not like his faults." And his style is not only faulty in itself; it is calculated to do mischief by the pro-duction of similar faults in others. With the junior portion of the medical profession in our country, more especially with pupils, the Professor is a popular—we might perhaps say, a *favourite* writer. And favourites are likely to be followed and copied, particularly in the attributes which render them favourites—and in none of these more so, provided they be writers, than in that of style and manner. It is more than probable, therefore, perhaps even certain, that the style of Professor Dunglison will become a model for imitation by young medical writers, whose taste and judgment are yet immature and easily vitiated and led astray. And if so, the consequence is plain. Such imita-tion will operate to the deterioration of the style of the medical writings of our country. Hence, to contribute our *part* toward the prevention of the evil, we have deemed it our duty to pass on the Professor's composition the fore-going strictures.

One more extract I may make, which is more con-solatory to others than complimentary to myself.—

It presents to the physicians of the United States a very valuable and praiseworthy example of literary research, ambition, and perseverance, and were that example judi-ciously analyzed, and deciphered, improved and followed out, it might be rendered the most profitable attribute of the book. It furnishes a lesson, replete with instruction, as to what may be accomplished by labour alone, without the aid of *genius* or *talents*. For, in the preparation of his system of *Practice*, the Professor has not made the slightest manifestation of either. He has shewn himself to be, in the strict and entire meaning of the phrase, a "working man," possessed of the excellent and essential qualities of assiduity, faithfulness, and perseverance, but, in our view, of little or nothing higher. And the example he has thus put forth ought to prove exceedingly useful and encourag-ing to all moderately gifted men, who constitute an immense majority of the Caucasian race itself, and the entire mass of all the other races:

—and he adds in the same strain.

By the compilation of his *Practice of Medicine*, then, we repeat, Professor Dunglison has done much to gratify, and encourage, the great body of moderately gifted men. He has practically shewn them, what it is important they should know, that by *industry, zeal* and *perseverance alone*, without the aid of mental powers above *mediocrity*, much may be done to acquire reputation and reward ambition. For that, in the estimation of a great majority of that portion of the Faculty of our country, who may read his *Practice*, his reputation will be enhanced by it, is hardly to be doubted.

Yet, amidst all this severe censure, occasional ad-missions were made, which satisfied me. I had not, even in Dr. Caldwell's estimation, failed altogether in the object I had in view in writing the work. These admissions were extracted by my publishers, and, greatly to the annoyance of their Author, were employed in their advertisements as offering favorable testimony to the work. Nor ought they to be blamed for this course. The object of Dr. Caldwell, whether instigated by private pique or public duty was doubtless to injure the sale of the book, and I was informed afterwards by Dr. Flint, of Louisville, that he boasted of having "Killed" it; and it certainly was justifiable in them to employ any of his expressions, which might serve their purpose. Such was my impression at the time, and such is it now; and perhaps a little wickedness entered into my feeling, determined, as I was, to take no notice of the review myself, and knowing the ire it would produce in the mind of Dr. Caldwell, when his own expressions were brought forward to subserve a purpose diametrically opposite to that which he had in view—

The commendatory sentences extracted by the publishers were the following. "It"—the work—

makes a part of the great scheme of *labour-saving* and *power-giving* means and contrivances of the day. It brings together, and presents to its readers, under a succinct and condensed form, a vast amount of medical matter, which no physician could collect for himself, without a most tedious and toilsome examination of detached works, and which from a want of leisure and books, the great body of physicians in the United States could never collect. By its copious references to authorities, it serves to those, who are solicitous of farther information on given topics, as an index to libraries to which they may have access. It presents to the physicians of the United States a very valuable and praiseworthy example of literary research, ambition and perseverance.

I could scarcely desire testimony more favourable than this. The employment of it, however, by my publishers, gave occasion to an immediate effusion of rancour, not in the pages of the *Western Journal*, the Editors of which were unwilling, perhaps, to insert anything more of the same tenor from my venerable antagonist, but in the *Louisville Journal*, a political newspaper. The character of this effusion is foreshadowed in the title *Literary peddling and piracy, or a new way to gain reputation as a Bookmaker, without deserving it*: the "reflections" suggested "by a recent manoeuver of Professor Dunglison, of Philadelphia, executed by himself in person, in his prurient eagerness to acquire reputation as a *maker of books*, or by somebody else in his behalf or under his influence." This review in the Western Journal, he says, was

considered the very *reverse* of flattering or favourable to the work and its compiler. . . . Yet in the face of all this, has the Professor himself, the learned editor of the American Journal, or both united—or some other person, by permission or commission—extracted from said review the highest compliment to said *Practice* it has ever received, and very probably the highest it is destined to receive. And the deed was done in a spirit of *artifice* and *trickery*, which some people *may* laugh at as an *ingenious device*, but which others *must condemn*, if not *reprobate*, as an exceedingly *dishonourable*, not to call it an *unprincipled stratagem*.

He adds,—

The design of the article is obvious. On those readers of the *American Journal*, who have no other source of information respecting our review of *Dunglison's Practice* it is intended to act as a low but impudent deception, by palming on them the belief, that the extract before them is in harmony with the tenor and spirit of the whole critique. That this, however, is not the case, but that the expedient, petty in its character, is as beggarly an act of literary knavery as stands on record, or as can well be imagined, the following extracts from our review will abundantly demonstrate; and the number of them might be so multiplied as to include a large portion of the article, for, in its condemnation of the Professor's works, that article is quite unparsimonious.

And he concludes.

On one point, however, *and but one*, our feeling, more perhaps than our judgment, induced us to dole out as much of commendation as might silence the craving of charity, and as much certainly as conscience could tolerate. And behold the consequence! which should operate on both reviewers and readers as a memorable warning. Professor Dunglison himself, or some friend or agent in his behalf, with the eagerness of a drowning desperado grasping a straw, seized on the gratuity, and pompously emblazoned it on one of the pages of the American Journal, to proclaim in triumph, throughout the wide sphere of the circulation of that periodical, the momentous fact, that Dunglison's *Practice of Medicine* is acknowledged to constitute a portion (*good, bad*, or *indifferent*—nobody has said which) of the "labour-saving" and power-bestowing machinery of the day! Such are the character and amount of the encomium on his *compilation, virtually purloined* or *swindled* from the *Western Journal* by Professor Dunglison or someone in his behalf. And may he gain as much in useful experience, from the rebuke and lesson he has here received, as he will necessarily lose as relates to candour and manliness of reputation, by the miserable stratagem. In preparing our review of the new system of *Practice*, our object was the *promotion of truth*, and its *necessary concomitant*, the *advancement of public good*. And it shall not be our fault, if by an effort of *grovelling cunning* and *trickery*, that object be made to pander to falsehood and mischief. While we shall never, of choice, put the shadow of an obstacle in the path of the Professor's praiseworthy designs (*and we believe he has many*) neither shall we be willingly made, by petty finesse, to minister to the success of any he may conceive of an opposite character.

This article, like the Review—received no comment from me; and I had no communication whatever from Dr. Caldwell, until a son [98] of his colleague, Dr. Miller, [99] came to attend a course of lectures at Jefferson Medical College, a few years ago, when Dr. Caldwell gave him a letter of introduction to me, which I duly honoured, as much so as if he had never penned the articles, which I have mentioned.

In the year 1844, the first edition of the *Practice* was exhausted; and another was issued in the month

[98] Probably Edward Miller, who became a prominent surgeon (D. P. Hall, Early Surgeons of the University of Louisville, *Kentucky Medical Journal* 35: 151, 1937).

[99] Henry Miller (1800–1874), Professor of Obstetrics and Gynecology, Medical Institute of Louisville, and President of the American Medical Association, 1860.

of May of that year. The only alteration made in it, was to omit the references to Authors in parentheses, which interfered with the reading, and to introduce the names of the Authorities, as often as was practicable, in the body of the sentences, so as to preserve their continuity.

In the middle of December, 1847, the *third edition* was published, of the same number of copies as the first, and second—2000. Some months after the appearance of this edition, the work was severely handled in the *New Orleans Medical Journal* of which Dr. John Harrison [100] was editor, but, at the time, lying on a bed of sickness—and as the event proved—a bed of death. The initials appended to the review, were W.M.B. known to be those of Dr. Wm. M. Boling of Montgomery, Alabama; a physician of observation, and of much information, who, I afterwards found, was a student of the Jefferson Medical College, the first year of my service there; and graduated at that Institution.[101] At that time, unpleasant feeling existed between certain members of the Faculty; and I have deemed it probable, that some unpleasantness that occurred to him, the nature of which neither myself nor the Janitor, Mr. Watson—recollects—although we remember that there was some—may, without cause, have been referred to me, and rendered him not as friendly to me as he might otherwise have been, and may have acted in some measure as an incentive to him to notice my work unkindly; for it is rare to find an alumnus of a school press forward to review unfavourably, even if justly, the production of one to whose instructions he has listened, and whose name is attached to his diploma; and still more rare to have the work of that professor placed in contrast with one on the same subject by a professor in a rival school. In the Article in question, the work of Dr. Wood, of the University of Pennsylvania is noticed with encomiums, whilst my own is severely animadverted on; and a charge laid against me of not properly referring to the works, which I had consulted, and employed, in the preparation of my own—a charge, which, I had presumed, applied as little to me as it could possibly do to anyone, as I have always been anxious to render justice, in my works, to every honorable contributor to the science, and had repeatedly referred to the contributions of Dr. Boling himself; and, as I have already shewn, by extracts from the *American Journal of the Medical Sciences* (p. 139), and the *British and Foreign Medical Review* (p. 140) had received credit for my assigning to others what properly belonged to them. On the

appearance of the offensive article—offensive because unjust—I had decided, as on all other occasions, to pass it by without notice; but as my friend and publisher Mr. Blanchard, was of opinion, that the article might prove *commercially* injurious, I determined with his advice, to write a *private* letter to Dr. Harrison, without any expectation, that he would publish it in the Journal, of which he was Editor; but still being of opinion, that if he ventured upon the insertion of a *private* letter, he would give it *entire*. How much surprised, then, was I to see, in the next number of the Journal, this letter introduced; but with the commencement and termination, both of which were essential for its being fully understood, omitted.

I now determined to move no farther; and it confirmed me in my previous impression, never to reply to any strictures, that might be passed upon me, but to trust to my strength of character to rebut them; and, if this was insufficient, to submit. The following is a copy of my letter to Dr. Harrison—the passages infralined being those omitted when it was published in the Journal. The Editor stated, that he had received it as a *private* communication from Professor Dunglison, but thought it due to me to publish it, leaving it, of course, to be inferred that he had inserted the whole of it.

Philadelphia June 2, 1848
Dear Sir,

In the last number of the "New Orleans Medical Journal," of which you are one of the Editors, a notice is taken of the second edition of my "Practice of Medicine," published four years ago, on which I am desirous of making a few observations, not with the view of replying to any of the positions or opinions expressed by your correspondent, but of stating to you the principles upon which I have acted, in the preparation of that and other systematic works, and which you are so well able to appreciate.

In the Preface to that work, you will find I remarked, that "the improvements and modifications incessantly taking place in the departments of Pathology and Therapeutics render it advisable from time to time, to incorporate them, so as to furnish those to whom the different general treatises, monographs and periodicals are not accessible, with the means of appreciating their existing condition." In the fulfilment of this object, I laid every essay of value under contribution, and so far as my ability permitted, endeavoured to render the work a consistent whole, to which the reader might refer with satisfaction, to learn the existing condition of the branch of science on which it treated. Where facts or histories of disease were concerned, I regarded them as common property, unless where the descriptions, which was rarely the case, were entirely original, and in all cases, I have striven to do justice to the labours of observers, who have, in any way, enlarged, or seemed to me to have enlarged, the boundaries of science. Especially have I considered this important, where opinions were stated, which belonged, apparently or really to an individual. Accordingly, in the first edition of the *Practice*, I placed in a parenthesis after a description that could in any manner be esteemed to belong to anyone, his name, and, in the part of the work on Diseases of the Eye, which has been selected by the Reviewer for especial comment, this was done repeatedly, and in cases where the name of Dr.

[100] John Hoffman Harrison (1808–1849), M.D. 1831, University of Maryland, established, with Carpenter, the *New Orleans Medical and Surgical Journal* in 1845; Vice-president of the American Medical Association, 1847; died of pulmonary consumption.
[101] William M. Boling, M.D. 1838, Jefferson Medical College, Professor of Obstetrics, Transylvania University (1849), was favorably known in the South.

Taylor ought scarcely, perhaps, to have been added alone. When the second edition of the work was called for, the plan of parenthetic insertions was objected to by my publishers, and it was regarded by the reader as destroying the continuity of the sentences. It was, therefore, abandoned; but still, as you will find, the *Library of Medicine* was specially referred to; and the name of Dr. Taylor repeatedly introduced, wherever I deemed it advisable, and so often, that the idea of concealment of authority on my part must be out of all question. In the number of the Journal to which I have referred, passages are taken from my *Practice,* and from Dr. Taylor's articles in the *Library of Medicine,* and sentences, detached from the rest, are placed in such juxta-position, and so animadverted on, as to lead of the inference, that other and extensive passages have been adopted without acknowledgment; and a special reference is made to the diagnosis of Iritis as having been taken without acknowledgment from Dr. Taylor.

Now, the main object I have in view, in this communication, is to shew you, that the passages in question are not original with Dr. Taylor, and have been regarded by him so completely as common property, that although they, and numerous others, have been adopted with little or no verbal alteration by him from his predecessors and contemporaries, *he* has not considered it necessary to refer to them. If you will take the trouble to examine the Chapter on Iritis in Mackenzie, and certain paragraphs on the same subject in Lawrence, you will find them almost identical with those in the *Library of Medicine*—many of the paragraphs on the Purulent Ophthalmia of infants are scarcely changed from those in Lawrence; whilst the article on Corneitis differs hardly at all from that of Mackenzie. All these works—with the best of France, Germany etc—were before me when I wrote the articles in question and it was difficult, if not impracticable to say, to whom to give the priority for the analogous and identical passages, but certainly the *Library of Medicine* was not entitled to it. I would draw your attention, also, to the divisions of purulent ophthalmia by Dr. Taylor, cited by the Reviewer, which are taken almost *verbatim* from Lawrence, to whom, indeed, he refers, and you will find, that a similar division is adopted, on like principles, by Dr. Watson, and most modern writers. I do not indeed, know of a single systematic work that has appeared of late, which does not contain passages against which the same objections might be made as against mine. Descriptions of natural history, and, as a part of the natural history of disease, have never been regarded as belonging to individuals; nor, so far as I know has it ever been esteemed improper to use them. It is otherwise, as regards thoughts and opinions.

I have been anxious to make this statement to you, not as a reply to your correspondent, nor by way of RECLAMATION *in your Journal, for an Author is public property, but* PRIVATELY, *in order, that you and your colleagues may be satisfied, that I only followed a custom adopted by the writers of systematic works on all subjects, and without which, in a science of progression, they could not well be regarded as adequate exponents of the existing state of the science. It would always be easy to alter the language, so as to conceal the source, but this could scarcely be commendable.*

> I am, dear Sir,
> yours respectfully
> Robley Dunglison.

Prof. Harrison,
New Orleans.

By examining the underlined portions, that were wholly omitted in the published matter, purporting, of course, to be my whole letter, it will be seen how much injustice was done me by the Editors of the New Orleans Journal; and how little satisfactory the portion that was published must have been. Perhaps, however, the conduct of the editor of the *Southern Medical and Surgical Journal*—Dr. Paul F. Eve, then of Augusta, was more displeasing to me than even that of Dr. Harrison. Articles in commendation of an author are sometimes copied into other periodicals; but, unless envy, jealousy, or some other unpleasant feeling is entertained by an Editor, it is not usual for notices of an opposite character to be transferred from one Journal to another. This was done, however, by Dr. Eve, who, subsequently, took credit to himself for having afterwards copied into his Journal my private letter, mutilated as I have shewn it to have been, as if that materially mitigated the offensive action. The appearance of the copied Review in his Journal gave encouragement to the unjust attack on my *Human Physiology,* to which I have already referred (p. 49); but of neither of those did I take the slightest notice. I had already seen enough of the inutility of any sort of reclamation, public or private, by the manner in which my letter to Dr. Harrison had been treated.

It was somewhat singular that at the very time when these attacks were made upon me for not having sufficiently acknowledged the sources of information to which I had had recourse, similar attacks were made on my distinguished friend and brother physiologist, Dr. Carpenter; who had adopted the same rule as I, and I believe every other systematic writer, had done, of considering all descriptions of Natural History as common property; and on the two sides of the Atlantic, Dr. Carpenter and myself must have been writing out replies almost synchronously.

In the *London Athenaeum* for May 20, 1848, there is the following editorial.

We have received a letter from a correspondent, in reference to our notice of the *Cyclopedia of Natural Science* by Dr. Carpenter, in which the writer complains that the latter gentleman has used extensively the labours of others without acknowledgment. Our correspondent's remarks apply to the volumes on Botany and Zoology. With regard to the first he has supplied us with a series of parallel paragraphs from Dr. Lindley's *Ladies Botany,* and Dr. Carpenter's book, which would undoubtedly lead to the inference, that the latter had borrowed from the former. But, we would remark, that in works which profess to be mere compilations, the same amount of freedom is very generally taken with the labours of others. Such quotations should, however, undoubtedly be acknowledged. The second charge is that the volume on Zoology is little more than a translation of M. Milne Edwards's *Élémens de Zoologie.* We have taken the trouble to compare the two works, and although it is very evident in many parts, that Dr. Carpenter has made free use of Milne Edwards's book, we can most decidely state that his own is not a mere translation of that writer's work. Dr. Carpenter's volume embraces a wider field than M. Milne Edwards's—and contains much matter not to be found in the latter work.

At the same time, we do not observe, that Dr. Carpenter has acknowledged to whom he is indebted for his matter. We repeat, that in compilations like his, it is proper, that the sources, from whence information is obtained, should be distinctly avowed. p. 518.

This course, adopted by the Editor of the *Athenaeum* is in contrast with that of the Editors of the *New Orleans Journal,* and even more with that adopted by the *Southern Medical & Surgical Journal,* both of whom published the article as transmitted· to them, without the slightest inquiry into the correctness of the allegations contained in it. In the following number of the *Athenaeum* is contained a letter from Dr. Carpenter, dated May 22, in which after stating, that he does not know *how* he could more explicitly, or *where* more appropriately, have acknowledged his obligations to the *Zoologie* of M. Milne Edwards, as well as to other systematic works than he had done in the Preface to his "Treatise," which seemed to have escaped the notice of the Editor of the *Athenaeum* as well as that of his correspondent,—he adds—

Even if I had committed the fault imputed to me, I might have sheltered myself under the example of some of the most distinguished writers of Systematic Treatises, such as M. Milne Edwards himself, who, in his larger *Élémens de Zoologie,* has incorporated, with little modification, whole paragraphs from the *Regne Animale,* of Cuvier, without thinking it necessary to make the slightest reference to that vast storehouse of information, now the common property of all naturalists. In regard to the use made of Dr. Lindley's *Ladies Botany,* in the Treatise on Botany in the same series, I freely admit, that I derived assistance from it, as from other works of the same distinguished botanist; and here, too, a reference to the Preface will shew, that I have endeavoured to discharge my obligation to it, by strongly recommending it to my readers, in a manner, which will, I think, prove, that I had no intention of concealing the use, which I had made of it. I would farther remark, that coincidences in particular passages of two *descriptive* works often result merely from the necessary fact, that the same objects are being described and the same sources of information employed, in both cases. The following paragraph with which the preface to my *Zoology* concludes, is, of course, equally applicable to the *Botany.* A little reflection will shew, that any general Zoological treatise must necessarily be, in great part, a compilation from the works of other Naturalists; and the merit of an Elementary work like the present must consist, rather in the judgment shown in the selection and arrangement of the materials than in the originality of its contents.

The striking similarity in the line of defence adopted by Dr. Carpenter and myself, and almost at the same moment, sufficiently shews, that the same principles guided us, as they must all writers of systematic works of every kind. To attempt to make such a work original would, as I have remarked before, be an absurdity.

The following is the number of copies, and the amount of copymoney paid me for the editions of "Practice of Medicine" up to this date (Aug. 25, 1852).

A.D.	Edition	Number of Copies	Copy Money
1842	First	2000	$1500
1844	Second	2000	1500
1848	Third	2000	1500
		6000	$4500

CYCLOPEDIA OF PRACTICAL MEDICINE

In the latter part of the year 1843, Messrs Lea & Blanchard were anxious, that I should undertake the editorship of an American reprint of the *Cyclopaedia of Practical Medicine,* edited in England by Drs. Forbes, Tweedie and Conolly; and that my name should appear as the American editor. I had always objected to be ostensibly connected with the reprint of any book in this manner; but inasmuch as it would be necessary to bring up the work by numerous original articles as well as by interpolations, I consented, at length, to assume the undertaking, and the following letter exhibits under what circumstances.—

Dear Sir,

In relation to Forbes and Tweedie's *Cyclopaedia of Practical Medicine* it is understood, that the work is to be brought up to the day of publication by such additions or modifications of the original articles as you may deem advisable, or by the substitution of better articles from other sources for such as you may reject, particular reference being had to such diseases as are peculiar, or prevail in this country, so that the work may be creditable to you as editor. In this revision, all articles relating to Materia Medica are to be excluded. It is understood, that the first volume is to be issued about the first of March next, and that the others will follow at intervals of about three months each—the whole to be included in four volumes. The copyright, and all control over the work, to be exclusively with the undersigned. For the editorship and preparation of this work we are to pay you *one thousand dollars,* at six months credit from the completion of the whole work. It is understood, that you are not to be required to read the proofs of the printed matter, provided, that it is so prepared, that a good printer can with correctness read it. The new matter will, of course, have your revision.

In addition to the one thousand dollars, stated above, we are to furnish you with Books for the use of this work to the amount of from fifty to one hundred dollars.

We are, very respectfully,
your obedt. Servant,
Lea & Blanchard,
Philad. Oct. 31, 1843.

To Prof. Robley Dunglison,
S. 10th St.

On the second of November following, I accepted the proposition contained in the above letter, with the modification, that *two hundred and fifty dollar*s should be due six months after the publication of each volume, thus making in all, *one thousand dollars.*

The first Vol. appeared in July; the second in September; the third in December 1844; and the fourth and last, in March 1845. It was stated in the Title page to have been "thoroughly revised, with numerous additions" by me. In the Preface, I allude to the idea, entertained by Dr. Copland, Dr. Gordon Smith[102] and myself, of preparing an encyclopedic Dictionary of Practical Medicine, to which I have already referred (p. 17), which probably gave the impulse to the appearance of the *Cyclopaedia of Practical Medicine*, and I added:—At the request of my publishers, I "consented to superintend the republication, to revise the various articles, and to make such additions and modifications as I might deem desirable." "To this determination he" (the undersigned)

was impelled in part by his knowledge of the accomplished editors, and his consequent anxiety, that the different articles should be preserved, as far as practicable, in their pristine integrity. Generally speaking, they were so full and comprehensive, that but little modification seemed to be necessary. In all cases, however, the undersigned has endeavoured to add the facts and opinions, which form part of the medicine of the day, and to omit nothing calculated in his view, to throw additional light on the subjects canvassed. . . . Besides these, he has inserted different topics, which had been fully omitted in the English edition, some of which are essentially American in their character. To these additions his signature is appended. For one article on the History of American Medicine before the Revolution he is indebted to Dr. J. B. Beck, of New York. In making these additions, the undersigned has endeavoured to be as brief as perspicuity would permit; and where his opinions have been fully expressed elsewhere, to avoid repetition, he has referred to them for farther details. On every occasion, he has been anxious to cite the views and observations of contemporary, and especially, of American writers, where such views and observations appeared to him to merit especial notice. Appearing, as the work originally did, in parts, articles were necessarily omitted in their proper places, and many were thrown into a supplement. This defective arrangement has been rectified as far as was practicable, so that the American edition will, in this respect, be found to possess advantages, as a book of reference, over its English prototype. . . . One subject only has been omitted—Medical Bibliography—which formed an appendix to the last volume of the English edition, and was subsequently published in a distinct form. This the publishers determined to discard, in consequence of the additional cost to the purchaser, which its insertion would occasion, and of their conviction that notwithstanding its intrinsic value to a few, and the high character of its author—Dr. Forbes—it would be considered wanting in interest and importance to the many, on this side of the Atlantic, who will seek the work as a guide to Practical Medicine.

Besides the interstitial additions, a great number of articles were added by me all of which have been enumerated in Vol. 1. p. 20 of these *Ana* (p. 197).

Summary of the Number of Volumes published, & of the Copymoney paid me for works published in this country up to this date—Aug. 26, 1852.

---

[102] John Gordon Smith (1792–1833), M.D. 1810, Edinburgh, editor of the *London Medical Repository*, taught at the Gower Street School, the Royal Institution and the Royal Westminster Establishment in London.

Work	Number of Copies	Copy Money
p. 49 Human Physiology	11000	$ 7333.34
47 Medical Dictionary	17250	10700
74 Elements of Hygiene	3000	1372
76 General Therapeutics	6250	4600
134 Medical Student	2000	553
136 New Remedies	6450	3575
137 Rogets Physiology	—	100
137 Traill's Med. Jurisprud.	—	50
146 Practice of Medicine	6000	4500
146 Cyclopaedia of Medicine	—	1000
	51950	$33783.34

As the first edition of the *Dictionary* was in two volumes, and the editions of the *Therapeutics*, except the first, as well as the *Human Physiology & Practice of Medicine;* the number of volumes would be as follows.

Brought over	51950
Additional vols. of Physiology	11000
Therapeutics	4750
Practice of Medicine	6000
First edit. of Dictionary	1000
Making	74700 volumes

The amount, thus far paid me by the firm of Lea & Blanchard, is $32583.34.

EDITIONS PUBLISHED SINCE AUG. 26, 1852

Work	Edition	Number of Copies	Copy Money
Oct. 1852 Med. Dictionary	Ninth	1500	$ 900
June 1853 Do	Tenth	1500	900
October—Genl. Therapeutics	Fifth	1750	1400
Nov. 1853 Med. Dictionary	Eleventh	1500	900
Sept. 1854 Do	Do	1500	900
June 1855 Do	Twelfth	1500	900
May 16, 1856 New Remedies	Seventh	2000	1000
24 Human Physiol.	Eighth	2000	1333.33
Feb. 11 Med. Dictionary	Thirteenth	1500	900
Oct. 20 Do	Fourteenth	1500	900
1857			
Oct. 1 Genl. Therapeutics	Sixth	2000	1600
Sept. 5 Med. Dictionary	Fifteenth	1500	900
copies to England		100	25
1858 Do before		200	50
May 5 Med. Dictionary	Sixteenth	1500	900
Dec. 24 Do	Seventeenth	1500	900
1859 copies to England		100	25
Sept. 8 Med. Dictionary	Eighteenth	1500	900
1860 copies to England		100	25
Mar. 3 Med. Dictionary		1500	900
Oct. 16 Do Do		1500	900
1861 Copies to England		100	25
Dec. 31 Med. Dictionary		1500	900
1862			
Nov. 12 Do Do		1500	900

EDITIONS PUBLISHED SINCE AUG. 26, 1852—*Continued*

Work	Edition	Number of Copies	Copy Money
1863 additional copies		70	$ 42
June 12 Med. Dictionary		1500	900
1864 March 6 Do – Do		1500	900
Oct. 14 Do Do		500	300
1865 Feb. 11 Do Do		1500	900
— Oct 8. Do Do		1500	900
1866 April 11. Do Do		1500	900
Dec. 18 Do Do		1500	900
1867 Oct. 17 Do Do		1500	900
1868 Sept 25–28 Do Do		1500	900

NUMBER OF COPIES OF THE FOLLOWING WORKS SOLD AND AMOUNT OF COPYRIGHT RECEIVED TO SEPTEMBER 1865

	Copies sold	Copyright
Human Physiology	13000	8666.67
Medical Dictionary	44750	27392.00
Human Health	3000	1372.00
General Therapeutics Materia Medica	10000	7600.00
Medical Student	2000	553.00
New Remedies	8450	4575.00
Practice of Medicine	6000	4500

[table on previous page, sale, as follows to Dec. 1, 1875

Dictionary sold to
    April 1869 (death of R. D.)
        $31892.00 from 52,920 copies
    add. April 1st to Dec. 1.
        1869   1875   12,500 copies

No. of Volumes sold (of all his works) to Dec. 1, 1875.
                                    —— 134,875.][108]

DUNGLISON'S INDUSTRY

In the notice of me, given by Dr. Gross, he alludes to the time employed by me daily, in the production of the works, which I have issued, and makes the mistake of supposing, that they were written at midnight; "when other men are wrapt in sleep, and dreaming, *not of what they have done, but of what they intend to do*" (p. 140). It has been a question, often put to me, whether I did not consume, nightly, the "midnight oil." Yet, through life, I have never deprived myself of rest for purposes of study. At an early period, I accustomed myself to make use of the minutes; and to leave off even in the middle of a sentence, so that I might resume the train of thought when the cause of the interruption ceased. By custom, also, I have been able to compose with my family placed around me; so that, in their presence, I never experienced difficulty in employing those short periods, which are generally spent away

[108] Inserted in another person's handwriting, probably that of his son, Richard.

from the desk. I look upon these habits as having enabled me to execute what I have done; and the practice, which I have never neglected, of never postponing, has enabled me to find time for all things. On a visit, which I paid to Buffalo in 1850, Dr. Flint [104] was kind enough to present me with a copy of the 3rd vol. of the *Buffalo Medical Journal* (1848) of which he was Editor, in which he had inserted with preliminary remarks, a friendly notice of me by Dr. McPheeters [105] of the *Saint Louis Medical and Surgical Journal,* but not entirely accurate, in regard to those business habits of which I have spoken.

Professor Dunglison. The following deserved tribute to Prof. Dunglison by one of the Editors of the St. Louis Medical and Surgical Journal, may be perused with advantage by those, who would avail themselves of the secret of attaining a like position of eminence and usefulness. It is imputed to Hogarth, that on some reference to his *genius,* he replied, "Genius, I know not what it is if it be not industry." That differences do exist as respects general and special mental endowments, we do not deny, but talent without application can accomplish but little in any department of effort, and unremitting industry will often compensate for the want of extraordinary natural parts. As the writer observes, Dr. D. is not to be regarded as a man of ordinary ability, yet, as he himself declares, it is by means of systematic and constant industry, that he has been able to perform, and with apparent ease, such a vast amount of intellectual labour.

We have long regarded Dr. Dunglison as a remarkable example of one who accomplishes more in the way of science, and of professional and general literature; and has more time to devote to the cultivation of polite letters, and to the demands of society and his family, than anyone within the whole circle of our acquaintance. We speak from knowledge derived from a personal acquaintance with his habits, now some six or seven years ago—at the time Dr. Dunglison was professor (as he still is) in one of the first Medical Colleges in the country, and filled a most important chair with distinguished ability—was visiting physician to the Philadelphia Hospital, in which institution he delivered clinical lectures to a small class of private pupils during the summer, and to a large class from the College in the winter—was constantly issuing from the press new and standard medical works, as well as fresh editions of former publications—conducted a medical periodical—attended to his private practice—was also a prominent member of the American Philosophical Society, and of the Academy of Natural Sciences (Dr. McPheeters is in error here. I had not then been a member of the Academy) as well as similar institutions for which Philadelphia

[104] Austin Flint (1812–1886), fourth, in succession, of medical ancestry, founder of the Buffalo Medical College, reached the apogee of his career after he was rejected by Jefferson Medical College in 1859 and became Professor of Medicine at Bellevue in New York. He was an associate fellow of the College of Physicians of Philadelphia, President of the New York Academy of Medicine and of the American Medical Association (Norman Shaftel, Austin Flint, Sr., Educator of Physicians, *Journal of Medical Education* **35**: 1122, 1960).

[105] William Marcellus McPheeters (1815–1905), M.D. 1840, University of Pennsylvania, Professor of Materia Medica and Therapeutics, St. Louis Medical College (1842–1862) and Missouri Medical College (1866–1874), while a resident at the Philadelphia Hospital served under Dunglison.

is so celebrated, and many of whose able reports he himself prepared. In addition to all this, he kept pace with the progress of our ever advancing science, and with the current literature of the day—attended dinner parties, and evening parties of literary men, and always found leisure for any new or extraordinary demand on his time, which might arise. All this he did, and did it well. We had the curiosity to *enquire*, on *one occasion*, how he effected so much, and yet never seemed hurried about anything. His reply was, "by a systematic employment of those fragments of time, which are usually thrown away." We mention these facts to shew what system and well-directed talent will accomplish. It must not be inferred, however, that we regard Dr. Dunglison as a man of ordinary ability. Far from it. But we do believe, that he owes very much of his success in life to a systematic industry and a proper application of his talents, which all must admit are of high order. McP.

In my *Medical Student*, (edition of 1844, p. 177) I recommend the same course which I had myself found so advantageous.

In the way to study,—as in the way to wealth,—fractions must not be disregarded. It is a trite, but a wise maxim, that if we take care of the pence, the pounds will take care of themselves, and the parody is no less just,—that if we take care of the minutes, the hours will take care of themselves also. It is surprising what may be accomplished by seizing upon every interval for study, and by disciplining the mind to the effective exercise of its powers. If the student succeeds in this, but little nocturnal application will be necessary to treasure up the materials of science.

My valued friend—the Revd. Dr. Geo. W. Bethune [106] —pays me the distinguished compliment in his oration *on the Duties of Educated Men*, delivered before the Literary Societies of Dickinson College, Carlisle, Pa. July, 1843, of selecting me as an example of one who, although much occupied, found a time for all things. "You may have read"—he remarks—

of a zealous scholar, who finding himself daily called to his food some little time before it was ready, improved the moments, which must otherwise have been lost, in preparing three huge quartos for the press. A learned friend of my own (Dr. Dunglison of Philadelphia), whose elaborate works have won for him genuine and exalted fame in both hemispheres, and who, while skilled as a general scholar, is a most voluminous writer, as well as an indefatigable and eminently successful teacher of his peculiar science, has often assured me, that he was never so busy, but he could find time for something more; to the truth of which assertion those who know the delight of his instructive and ever cheerful society can bear witness (*Orations and Occasional Discourses*, by Geo. W. Bethune, D.D. p. 290, New York, 1850).

CLUB OF FIVE

In was in consequence of the reunions at the Athenian Institute and the American Philosophical Society, that a few of us determined to form a small Club, without

any regular organization excepting, that it should consist of *five* members, who should assemble, when it might be convenient, without any ceremony, at each others houses. It was suggested by the occasional coming together of five congenial individuals :—Dr. Robert M. Patterson, Judge Kane, Professor A. D. Bache, Dr. Bethune and myself; and I well recollect the propriety of such an association being first talked of by Prof. Bache and myself about the year 1843. At one of our accidental reunions at my house in Spruce St. about the year 1842, not an article was to be found, that could be brought on the table, excepting fresh eggs, and we were too remote—at the corner of Schuylkill 8th Street [107] to send out readily for oysters or any other eatables. Boiled eggs formed consequently, the staple, which were washed down by champagne and hock; [108] and yet that simple repast has probably been more strongly impressed on the memory, than others that were more varied; and this mainly in consequence of the equanimity with which my excellent wife received us, and furnished us with the little that the house afforded.

At these different meetings, what "a feast of reason and a flow of soul"; what a satisfaction for us, engaged in different pursuits, to meet together occasionally, after the affairs of the day had been transacted—What excellent feeling was engendered; and what a fund of information informally imparted to each other!

But alas! this reunion of "the five" has become an affair of memory only. At this date (Sept. 1, 1852) Professor Bache is living, and has lived for many years in Washington, ably fulfilling the duties of a most responsible office—that of Superintendent of the Coast Survey—: the Revd. Dr. Bethune has removed to New York, to assume the pastorship of a church at Brooklyn; Judge Kane has left the city to take up his residence in the country, although about to return to town again; and Dr. Patterson's mind and body have become so enfeebled, that he could not take part in any reunion should this be attempted, which is not now probable. So that I, alone, of the "Five" remain in the city and happily as vigorous, I hope, mentally and corporeally as I was when we first began to assemble together. It is not more than two months since my friend Dr. Bethune and myself, in New York, feelingly deplored the dispersion of the "Five," which he looks back to with the greater regret as can find no substitute for it in New York; and I despair of any such being found in Philadelphia; sed *"olim meminisse juvabit."* [109]

A reference to "The Five" is made by Judge Kane in an Obituary Notice of Dr. Patterson delivered before the American Philosophical Society and also in an obituary Notice by me of the Reverend Doctor Bethune

---

[106] George Washington Bethune (1805–1862), American reformed Dutch clergyman, man of great learning, edited Walton's *Compleat Angler,* was a frequent visitor at the home of Dr. Dunglison.

[107] Now Fifteenth and Spruce.

[108] Hock is white Rhine wine.

[109] But it will be pleasant to remember these things in time to come (Virgil's *Aeneid,* Book II).

delivered before the same Society and published in its *Proceedings.*

For a farther notice of my excellent friend—now no more, see Supplementary *Ana* p. 170.

BARON VON RAUMER

In the year 1844, I had the pleasure of becoming acquainted with Baron von Raumer, the distinguished Prussian historian, who informed me, that he had come to this country to witness, for himself, the success of the "experiment of Jefferson," for such he regarded the existing condition of the country as illustrating. He informed me with what pleasure he had pored over the Memoir and Correspondence, published by Mr. Jefferson's grandson, and what a veneration he had for the character of that illustrious statesman. Baron von Raumer had visited several portions of the United States before he came to Philadelphia, and had heard many unfounded statements in regard to Mr. Jefferson, and especially as to his private character, which I had great pleasure in contradicting. It happened, that on the very day, on which I called upon him, Mr. William Norris [110]—the Engineer—was to have an entertainment at his house in the country; and I ventured to take Baron von Raumer with me, who was much pleased with the Company he met there, which consisted of some of the scientific gentlemen of Philadelphia. His presence was especially agreeable to Mr. Norris, who was about to set off for Vienna to fulfill a contract with the Austrian government to establish railroads, and build locomotives in Austria. In his work on *America and the American People,* which he published on his return home, Baron von Raumer refers to his visit to Philadelphia in the following letters:—

Philadelphia 24th August

This has been equally a day of enjoyment and instruction. Mr. R. (Mr. Benjamin W. Richards, whom I introduced to the Baron) came for us in a carriage, and we visited with him, first, the engine manufactory of Mr. Norris. He employs about 300 persons, who are paid from five to eight dollars per week. Yet he is able to furnish steam engines to Austria, and wants no high duties. Of the great and much talked of Prison and House of Refuge I have given an account elsewhere. The water works here deserve the most honorable mention beside those of New York. A mighty dam restrains the waters of the Schuylkill, which are raised by means of immense wheels to the reservoirs above, whence it is then distributed throughout the city in a highly appropriate manner. A cemetery near the Schuylkill, formed by the exertions of Mr. R. extending over hills and slopes, and abounding in beautiful trees, monuments, and views is, next to Père la Chaise and Greenwood, the most beautiful I have seen.

Philadelphia 28th August.

Yesterday, through the uncommon kindness and attention of several gentlemen, and of Dr. D. (Dr. Dunglison) in particular, proved a highly interesting and instructive day.

First, Dr. D. took us to the Athenaeum, a scientific institution [111] possessing a good library and a reading room. The Philosophical Society has existed already one hundred years, and has performed meritorious services of various kinds. We saw there a number of curiosities; immense mammoth bones; rude works of art from Central America; the original of the Declaration of Independence; and a picture of Jefferson, which represents him older, but much handsomer and more intellectual looking than other portraits. In the State House, we saw the hall as it was when the Declaration of Independence was signed in it; and obtained from the cupola an extensive and delightful prospect over the great city and its environs. Dr. D. then took us in his carriage, successively, to the Insane Asylum, the Poor House, the Institution for the Blind, and the Gas Works. These establishments are not only large and well adapted to their objects, but the two first are so magnificently appointed, that they look like palaces. (Translation by William W. Turner, p. 473, New York, 1846.)

I had much pleasure in proposing Baron von Raumer as a member of the American Philosophical Society. To my communication, announcing his election, he replied to me, in a letter of which the following is a translation.

Much honoured Sir,

Your announcement to me, that the American Philosophical Society had elected me a member has given me great pleasure, and I return my heartfelt thanks for the same in the German language, as it is so generally known in Philadelphia. I venture further to consider my election as a favorable omen of the reception of my book on the United States. Howsoever much there may be to find fault with, with truth, and to correct, it certainly is far from being one-sided; and is the result of sincere adoption and admiration. I certainly am by head and heart more a friend and adherent of the Americans than many of the English travellers especially. You will very much oblige me, if you and your friends will in future, favour me with your honest and rigorous judgment for my instruction and improvement, as so many have already been kind enough to instruct me in Philadelphia, and to assist me in the right way. I have become acquainted here with the Coolidge family. You may imagine what pleasure it gave me to hear the Granddaughter of Jefferson speak of her excellent and admired Grandfather. She perused the chapter of my book, which treats of Jefferson in Manuscript; and her approval of it leads me to hope, that the difficult undertaking of depicting such a man has not been a failure with me. By anticipation let me thank you for your promised communication on Mr. Duponceau. Johannes Müller recollects you with great cordiality and begs me to warmly greet you. I know with great certainty, that my book on America will give much dissatisfaction in Europe, and occasion me bitter reproaches in my native country. Were I younger I would return to America and serve (like Jacob surrounded by his loved ones) until I obtained the greatest of all Orders and Titles—the American citizenship.

Yours most devotedly,
v. Raumer.
Berlin, 2 May, 1845.

Dr. R. Dunglison, etc. etc.

---

[110] William Norris (1802–1867), designer of steam carriages, founded the Norris Locomotive works.

[111] The Athenaeum of Philadelphia is a literary, rather than a scientific, society; by 1856 more than 30,000 non-member visitors had signed its register.

JUDGE STORY

I have this day (Aug. 31, 1852) accidentally fallen upon the *Life and Letters of Joseph Story*, edited by his son William W. Story in which Judge Story, in a letter to his wife dated Washington, Jan. 27, 1833 thus alludes to our meeting at the President's table, which I have mentioned elsewhere. (See p. 68.)

On Thursday we dined with the President at an equally fashionable hour. There were several ladies at the table, and of course, I was called on as one of the Court, to hand a lady to table. Accordingly it fell to my lot to be in attendance upon Miss McLemore a niece of the President, who was pleasant, well bred and companionable. But on my other side was Dr. Dunglison, a scientific English gentleman, now a Professor at the University of Virginia, with whom I entered into a very free and agreeable conversation. He was full of what Dr. Johnson would call good talk, and I was quite gratified to meet with one with whose mind I could so well sympathize. How he could content himself with a banishment from the elegancies of English life, to which his education entitles him, for the occupation of a college life in the interior of Virginia, is to me most marvellous. He is already the Author of some Medical books of high character. And, by the by, I may say, that his visit here was to see a Canadian soldier,[112] who was shot during the late war, through the rib, diaphragm and stomach, and is yet living and in good health. By one of the most extraordinary efforts of nature, though a considerable aperture was made in the stomach, it adhered to the sides of the abdomen, and there is now a cavity from the outside of about an inch in diameter, through which the operations of the stomach in digestion may be seen. He is sometimes fed by and through this orifice as well as by the mouth. I am told, that it is one of the very few cases, which have been known to exist, and that it was brought to light some extraordinary medical facts. *Life etc. of Judge Story.* vol. 2 p. 118, Boston 1851.

SCIENTIFIC MEDICINE

Although devotedly attached to the science of my profession, I have always been dissatisfied with the mode in which it is generally practised, as I have elsewhere remarked in these *Ana* (see pp. 74 and 80) and I have myself endeavoured to avoid the errors to which I have objected in others. Perhaps, in many of these cases, the public are as much to blame as the profession. They are rarely, indeed, satisfied unless the practitioner prescribes; and usually they demand a degree of activity from him, that fosters the tendency to *heroic* agencies, which has done so much harm, and against which I have never failed, in my teachings, to protest. It is gratifying to me to be able to state, after many years spent in the practice of my profession that I have endeavored on all occasions to practice conscientiously; and, in no case, to prescribe drugs unless I saw clearly their adaptation. The well informed physician, as years pass over, becomes less and less satisfied of the curative powers of his "remedial agents"; and more and more satisfied of the benefit to be produced by placing the system, or any part of it, in the most favourable circumstances for

[112] Alexis St. Martin, Beaumont's celebrated patient.

curing itself. Having made the subject of the *modus operandi* of remedial agents one of close study, I have not been disappointed in them; and whilst many even of my professional brethren would, perhaps, charge me with having too little faith in drugs, and in their adaptation to special morbid conditions, I have had and still have, a much greater confidence than many of them in the great principles of therapeutics, as I have been in the habit of expounding them; and whilst others—active and energetic "druggers"—may speak of the great uncertainty of medicine, such a remark would rarely or never escape me, in consequence of my not anticipating results from articles of the Materia Medica, which ought never to have been expected to flow from them. They, who are best acquainted with the science of medicine must necessarily be the safest and most satisfactory practitioners; although, unhappily, this is not the prevalent belief; and never will be, as long as the notion exists, that a man may be "born a physician," or may have some gift, that will enable him to practice medicine successfully in the absence of all proper education. Yet there is not much probability of this notion passing away; for if we look at the condition of medicine around us—its regular and irregular practitioners.— homeopathists, hydropathists, kinesipathists etc. with the legion of quacks, it presents a general picture indicating as little progress as any that could be formed of any period, perhaps. Moreover, in this country, the so called "practical" predominates, and it is constantly contrasted with the "scientific." I well recollect, when attending a convention at Charlottesville, in Virginia, held for the purpose of the internal improvement of the State, Genl. Fenton Mercer, maintaining the doctrine, that he preferred greatly the *practical* engineer to the *scientific* thus making them antagonistic to each other, instead of urging that the two qualities ought to be combined. The same feeling prevails extensively in regard to the medical practitioner and is ruinous in its results, as it leads to the employment of the most ignorant members of our own profession, and the practisers of every form of empiricism. In this respect, the feeling, amongst the well informed, differs greatly in America and Great Britain. In the latter country, the physician, who exhibits his knowledge—his science—by the publication of a successful book on medicine, is sure to be run after, and the evidence, thus afforded by him of his acquaintance with the great principles of his calling, is, in many cases, the cause of an extensive practice. Armstrong, Watson and other English writers have frankly communicated to the world the great increase in their daily employment, produced by their works on the Practice of Medicine.

In this country, on the other hand, to be considered highly scientific and to have published largely on his own profession and more especially to be eminent in general literature is positively detrimental, by encouraging the idea, that the physician may be very learned and very scientific and literary but, for that very reason, probably not practical, and I am sorry to say, that this popular feeling has been occasionally taken advantage of by the members of his own profession, when an eminent member has come amongst them and it has been insidiously remarked, when his qualifications have been inquired into by the laity, that "he is undoubtedly a man of *science* and *learning* in his profession, but they do not know, whether he is a *practical* man." This, I know, has been repeatedly said of me, both in Baltimore and Philadelphia and even by one or more of my personal friends! and it has been farther remarked, that "they did not think I cared for practice; that I was more fond of the teaching of my profession and of my literary labours in it." Such reports, with the feeling to which I have referred as prevailing amongst the laity, cannot fail to be most injurious to a practitioner; and by an honorable, high minded man they cannot be corrected; as he cannot stoop to have his qualifications advertised either by himself or by others. The same circumstances apply to the lawyer, but not perhaps to the like extent. "It is a most singular circumstance," says Judge Story

—that eminence in general literature should, in the public mind, detract from a man's reputation as a lawyer. It is an unworthy prejudice, for certainly the science of jurisprudence may borrow aid, as well as receive ornament, from the cultivation of all the other branches of human knowledge. But the prejudice exists; and yet one would think, that the public had witnessed so many examples of men, who were great scholars and great lawyers likewise, that the prejudice might be at this day disarmed of so much of its quality as is apt to do injustice to the reputation of living men. Lord Mansfield was a most eminent scholar in general letters; but he was also unsurpassed in jurisprudence. Sir William Blackstone was so elegant a scholar, that his commentaries are models of pure English prose; but they are none the less the invaluable mine of the Laws of England. Lord Stowell, the friend and executor of Dr. Johnson, was, in various attainments, exceeded by few; but his knowledge of general jurisprudence was greater than that of any man of his day. Some of the proudest names, now on the English benches, are some of England's best scholars. But there, as well as here—though certainly it is far greater here—the public prejudice almost denies to a great scholar the right to be eminent as a jurist. *Life and Letters of Joseph Story*, ii, 454

This prejudice in England, doubtless, applies to the possessor of extraneous accomplishments, but his success is insured in his profession, if he has given evidence, that he is deeply learned in it; whilst, in this country, as I have remarked above, the general impression is often the reverse.

SIR JAMES CLARK

From Sir James Clark, physician to Queen Victoria, I received one or two kind letters. I had sent him my *Human Health,* and he had transmitted to me, by the hands of Dr. Gibson, a copy of his work on *Climate,* which, by the way, was not put into my hands for some time after Dr. Gibson's return from England, for the

reason—as he himself stated to me, "that he had been so much occupied he had not had time to ask where I lived! !" In one of his letters, Sir James urged upon me to visit England, as he was afraid my application to my profession and to my literary labours, would injure my health, and that I might prematurely die a martyr to the cause. I had much pleasure in procuring his election as a member of the American Philosophical Society; some time after the announcement of which, I received the following letter.

London, April 2, 1846.

My dear Sir,

I scarcely know how to ask you to present my letter of acknowledgment to the American Philosophical Society, after delaying so long to perform that duty. The delay occurred from one of those accidents, which, as a busy man, *you* well understand. It has vexed me much, lest the Society should consider it as a mark of disrespect, my having so long delayed to acknowledge the honor they conferred on me. I beg you to assure the society (if you think they require such an assurance, which, I trust, may not be the case) that nothing could be farther from my intention, and that on the contrary, I am quite aware of the high standing of the Society, and deeply sensible of the honour of being elected one of its members. I will feel greatly obliged to you to put me right with the Society on this point. I have to thank you, my dear Sir, for this honour, as well as for many other acts of kindness in sending me your valuable works. I am sorry I can send you so little in return. A new edition of my little work on Climate is now in the press, and I hope to send you a copy soon. There is nothing relating to our profession of sufficient interest, I think, to notice. The loud outcry for medical reform which has excited the whole profession, and confounded the minister, has ceased for a time at least, and is not likely to be renewed again, I think, this year.[113] I hope you may pay us a visit. You must want a little relaxation, and nothing would do you so much good as to *change the Climate* of America for that of England, for a few months. I shall be delighted to welcome you here, and, in the meantime, I remain, my dear Sir,

very truly yours,
Jas. Clark

In 1854 I had the pleasure of seeing Sir James in London & receiving his hospitalities. From his home, as mentioned in my letters from England I went to a conversazione at Lord Rosse's.[114] Through, I doubt not, the kindness of Sir John Forbes I was indebted to him for many acts of civility.

I had likewise much pleasure in having elected into the Society my valued friend Dr. Copland, of London, and my correspondents—Dr. Johannes Müller of Berlin, and Dr. Henry Holland of London. I had the satisfaction to procure for Professor Müller, at his re-

---

[113] A bill brought up in Parliament for medical licensure and regulation in 1844 was defeated in 1845, but in spite of the prediction here, was brought up annually until the much needed Medical Act of 1858 was passed (Charles Newman, *Evolution of Medical Education in the Nineteenth Century,* pp. 134–193, 1957).

[114] William Parsons, 3rd Earl of Rosse (1800–1867), astronomer, President of the Royal Society, 1849–1854.

quest, a specimen of the *Amia calva* [115] of the Northern Lakes; through the kindness of Dr. Kirtland, of Cleveland,[116] to whom I had written on the subject, as well as to Dr. Engelman [117] of Saint Louis; the receipt of which he acknowledged in a letter to me dated Berlin, April 20, 1847.

SIR HENRY HOLLAND

With Dr. Holland, I became personally acquainted in the Autumn of 1845, when he visited this country, *semi-incog.* having entered his name, amongst the passengers of the Steamer, as *Mr. Holland.* He had become known to Dr. Meigs, when he visited England a short time before. He was only a day or two in Philadelphia, when I had an opportunity of dining with him at the house of Dr. Meigs, along with some of the prominent gentlemen of the profession. The same evening, or the evening before, he went to a whig political meeting, where he heard some of the prominent politicians address the assemblage and was much struck to hear the English language spoken as at home— 3000 miles distant. He was pleased with Mr. Josiah Randall—not one of the most correct of our speakers— and animadverted on one word only and that not English—*nucleus,* which he fancied Mr. Randall pronounced *nooclus.*

On leaving us at Dr. Meigs's, he wrote me the following note expressive of his feelings on leaving Philadelphia.

Jones's Hotel, Wednesday
Sept. 24, night

My dear Sir,

x x x x x x "Let me again, my dear Sir, at this moment of my reluctant departure from Philadelphia (and I know not that I ever left a place with more regret, or more earnest desire to be there again) thank you again for your great kindness and attention, and for the very valuable works you have put into my hands. I feel the same gratitude to the other physicians of Philadelphia, whom we met at Dr. Meigs's this evening, and who have so much honoured me in the expressions of their esteem. I find, on returning to my hotel very kind notes from Dr. Meigs and Dr. Mitchell, accompanied by presents, which are most welcome to me as tokens of their regard. I have already imposed much trouble upon you but I would fain add to this the request, that you will thank them on my behalf, either in words or by putting this note into their hands. I would have written them thanks myself but it is now nearly midnight, and I have had a day of untiring activity, with the need before me of rising at 6 tomorrow morning.

Farewell, my dear Sir, and believe me ever yours, with great regard,

H. Holland.

---

[115] The bowfin or mudfish, remarkable as the only living representative of the suborder Cyclogonoidei.

[116] Jared Potter Kirtland (1793–1877), physician, teacher, and eminent naturalist, widely known among scientists of the United States and Europe, at the age of eighty-two was elected a member of the American Philosophical Society.

[117] George Engelmann (1809–1884), M.D. Würzburg, Germany, president of the Academy of Sciences of St. Louis, author of *Flora of the United States.*

Soon after his return to London, I received a letter from Dr. Holland of which the following is an extract.

London, 25 Brook St.
Dec. 3, 1845.

My dear Sir,

I willingly find a cause for writing a few lines to you that I may, at the same time, again express to yourself, and, through you, to others, the continued feeling I have of the great kindness I received at Philadelphia, and the earnest desire I entertain to cultivate farther, at some future time, the friendships thus happily begun. Whether I may ever be able to visit the United States again, circumstances must hereafter determine; but I much wish, that it may be possible to me. The more immediate cause for writing now is merely to mention, that I am sending by my friend, Mr. O'Sullivan, of New York, (who sails for America by this packet) a parcel directed to you, which he will leave at Philadelphia on his way to Washington. This parcel contains some copies of the 2nd edition of my *Medical Notes,* etc. One of them is addressed to you on the title page, as some slight token of recollection and regard. With respect to the others, will you let me ask it of your kindness, that you would present them severally in my name to my medical friends in Philadelphia, Dr. Meigs, Dr. Chapman, Dr. Mitchell etc. I see the bookseller has only put up six copies, which are insufficient as an expression of my feeling of the many courtesies I received; but I may find a future opportunity of sending others; and, meanwhile, I ask the favour of your judging for me in the presentation of these, —putting my name into each volume. If I recollect rightly I begged you to present a copy, which I happened to have with me, to the Library of the American Philosophical Society. If I am wrong in this, will you be good enough to give one copy that destination? I am sorry to trouble you this far, but I trust to your kindness to excuse me. . . .

Farewell and believe me ever yours with great regard

H. Holland.

I gave to the *Medical Notes* etc, the destination which he indicated, and added to the names he had mentioned those of Drs. Jackson & Wood.

Dr. Holland is a liberal and enlightened gentleman; holding the highest rank in his profession in London; and associating with persons the most elevated in rank and position. His sentiments in this country were evidently favorable; but he did not remain long enough to judge, with any degree of accuracy, of us, either in our political or social relations. On a remark being made by me, that there was a satisfaction in knowing, that here, no one arrogated to himself the possession of a loftier position than another, especially if that other belonged to a learned profession, and was himself possessed of talents and acquirements—he stated, that the highest persons in the kingdom—the Duke of Wellington included—met at his table—and I knew, that his associations were of the most lofty character. Still, I could have replied, that by this, he was not placed nor considered on an equality with his guests; but was only tolerated, as it were. I did say, however, that there ought only to be one aristocracy—of talent and virtue; and I could have added with sincerity, that to a place in such an aristocracy no one would deny that he himself was eminently entitled.

Of late, I have not heard from Dr. Holland.

In reference to Sir Henry since the above was written see Supplementary Memoranda (p. 176).

At the commencement of the year 1848, I received the following communication from Dr. R. M. Patterson, the Director of the Mint of the United States informing me, that I had been appointed a Commissioner to inspect the assay of the Mint.

Mint of the United States
Jan. 31, 1848.
Sir,

I have the honor to transmit to you the enclosed communication from the acting Secretary of the Treasury, stating, that you have been designated by the President of the United States, as one of the Commissioners to attend the next annual assay at the Mint. I pray you to inform me whether it will suit your convenience to accept this appointment; and if (as I sincerely hope it will), I have respectfully to request your attendance at the Mint on Monday the 14th prox. at half past nine in the morning.

Very respectfully
your faithful servant,
R. M. Patterson
Director.

To Prof. Robley Dunglison,
Jefferson Med. College

The following is a copy of the communication referred to.

Treasury Department
Jan. 27, 1848.
Sir,

I am authorized by the President to appoint you a Commissioner to act conjointly with Professors James B. Rogers and E. N. Horsford, to carry into effect the 22d Sect. of the act supplementary to an act establishing a Mint, and regulating the Coin of the United States. The Commissioners are to meet on the 7th (14th) February next.

Very respectfully,
McC. Young.
Acting Secretary of the Treasury.
Prof. Robley Dunglison

This appointment I accepted, and attended to the duties that appertained to it.

LEAD POISONING

In the Autumn of the same year, I received a communication from Professor Horsford, with whom I had served most agreeably on the Commission, of which the following is a copy.

Cambridge, Aug. 10, 1848.
Prof. Dunglison—My dear Sir,

There is now great excitement in this vicinity and especially in Boston upon the subject of lead service pipes. The waters of a very considerable lake are about to be introduced into that city through iron mains, and whether or not leaden pipes for distribution may be employed is a question of great interest. May I ask from your valuable time a moment to write me, whether there have occurred from the use of Fairmount water, served through Iron and Lead as I know, any instance of lead disease in Philadel-

phia? By so doing you will confer a favour not upon me only, but upon the Authorities of the city of Boston.

I am very respectfully
and truly yours
Eben N. Horsford
Prof. Dunglison

My reply to this letter was published in the *Report of the Water Commissioners on the Material best adapted for distribution Water pipes; and on the most economical mode of introducing Water into private houses, submitted to the City Council, Aug. 14, 1848* Boston, 1848: and is particularly referred to in a pamphlet by Professor Horsford, entitled *Service pipes for water, an Investigation made at the suggestion of the Board of Consulting Physicians of Boston*: published in the *Proceedings of the American Academy of Arts and Sciences*. (See my collection of Pamphlets (bound) for 1848 & 1849. The reply was as follows.

Philadelphia Aug. 12, 1848.
My dear Sir,

I have never witnessed the slightest effect from the use of the water of the Schuylkill, conveyed in leaden service pipes, which could lead me to suppose that there was any injurious impregnation. I have lived, too, in other cities in which water was conveyed in tubes of that metal, and with the like impunity. Whilst it cannot be denied, after the experiments of Christison, Taylor and others, that minute portions of lead are taken up by pure water, sufficient protection would appear to be afforded against any deleterious effect by the minute saline impregnations, which generally exist. I recollect Professor Hare stating, that he had used the Schuylkill water conveyed in leaden pipes to his laboratory in the University of Pennsylvania, for more than twenty five years, and had never perceived the slightest indication of the presence of the metal in it. Had there been any, the reagents which he had been accustomed to employ must, he conceived, have rendered the impurity evident. The results of all my observations in Philadelphia, and elsewhere, would lead me to express very confidently the belief, that leaden service pipes, constantly filled, as they necessarily are, are entirely innocuous.[118] You may recollect, that some years ago it was suggested by hygienic purists, that wooden pavements may exhale malaria in sufficient quantity to render the neighbourhood unhealthy; and although no "sufficient reason" was presented for the opinion, and certainly nothing in the shape of resulting malarious disease, no little excitement was, for the time, occasioned.

I am, my dear Sir,
very truly, your obed. Servt.
Robley Dunglison.
Prof. Horsford

DR. PEREIRA

In the year 1849, Dr. Pereira [119] was so good as to send me the first volume of the third edition of his

[118] The startling statistics on the poisonous properties of water from lead service pipes in Lowell, Mass., by Dr. Samuel Luther Dana, in an appendix to his translation of Tanquerel on lead diseases in 1848, alerted the public to the dangers of lead poisoning, known from remote antiquity.

[119] Jonathon Pereira (1804–1853), M.R.C.S., published the first edition of his *Elements of Materia Medica and Therapeutics* in 1839.

excellent *Materia Medica and Therapeutics*, the receipt of which I duly acknowledged, and, at the same time, sent him a copy of my *New Remedies*, and of my *Medical Dictionary*, on the receipt of which I received from him the following letter.

Finsbury Square, London,
May 5, 1849.

My dear Sir,

Accept my best thanks for a copy of your *New Remedies* and also of your *Medical Dictionary*. Both are most useful and valuable to me. I have a copy of a former edition of your Dictionary in two volumes, and have found it a very excellent aid on many occasions. You have, I see, omitted the Bibliographical notices. I assure you I have frequently found them useful. I should like to see from you a biographical medical Dictionary in one volume, containing short notices of ancient and modern medical men. Your early editions of your Medical Dictionary contains notices, which would do. Except Callisen's voluminous and expensive work, there is no book on the subject that I am acquainted with. Pray think of it. I believe such a book would sell here, and, in America, still better. You might publish it here and in America simultaneously.

Believe me, my dear Sir,
most faithfully yours,
Jon. Pereira

Dr. Dunglison.

To this I replied, that I did not agree with him, that the sale of such a work would be such as to encourage my engaging in it; and that this was not only my own opinion but that of my publisher, Mr. Blanchard; but if he would apply to Messrs. Longman & Co. or some other responsible firm, and they should be of a contrary opinion, and would make me a remunerating offer, I would be glad to enter upon the preparation of a work, which I really considered to be wanting in medical literature.

The following letter from Dr. Pereira shews, that I was correct in my anticipations.

Finsbury Square, London,
July 8, 1849.

My dear Sir,

Your estimate of the probability of the sale of a bibliographical and biographical Medical Dictionary is, I suspect, more accurate than mine.

I have communicated with two London publishers, namely Messrs. Longman & Co. and Mr. Highley. I took them the two volume edition of your Dictionary, & the last edition. I read to them some extracts from your letter to shew the nature of the work, and endeavoured to make them fully acquainted with its object. Both declined to undertake the risk. They much doubted whether such a work would have a remunerating sale. I had sundry misgivings when I read your history of the Medical Dictionary. If a book of the kind does not succeed in America, it is scarcely probable that it will in England, where the readers of medical works, or rather the buyers, are much less numerous than in the New World.

Poor old Dr. Ant. Todd Thomson [120] is just dead. He

suffered with diseased lungs. I hear, that he also had empyema.

Faithfully yours,
Jon. Pereira.

Dr. Dunglison.

On transmitting Dr. Pereira a copy of the new edition of my *General Therapeutics and Materia Medica* I received from him the following letter.

Finsbury Square, London.
Aug. 15, 1850.

My dear Sir,

I have received, through Messrs. Lea and Blanchard, a copy of your very useful and valuable *General Therapeutics and Materia Medica*, which, they inform me, you were kind enough to request them to send for my acceptance. Their letter arrived in London while I was in Switzerland, where I went for a week's recreation. I am much obliged to you for your kind remembrance of me. Your work is really a very interesting one; and, by the number of editions already published, is obviously well appreciated in America. I hope to make frequent use of it in finishing the second volume of my *Elements*, as I have done of your *New Remedies*. I don't know how it is, but it is obvious, that Medical books sell vastly better in America than in England. Americans have more of the "go-ahead" spirit in physic than the English have. No doubt, the lower price at which you get out books has something to do with it, but only to a very limited extent. I am working hard at the third edition of my book, and have a good deal of it finished; but the calls on my time, for hospital and private practice, prevent me proceeding so fast as I should like.

Believe me, my dear Dr. Dunglison,
very faithfully yours
Jon. Pereira.

Dr. Dunglison.

On sending him the last—*sixth* edition of my *New Remedies*, I received from him the following acknowledgement.

Finsbury Square, London.
July 3, 1851.

My dear Sir,

Many thanks for your kind attention in sending me a copy of the new edition of your very useful work, *"New Remedies."* I shall gladly avail myself of its aid in the completion of my *"Elements of Materia Medica."* I suppose the people of the United States take more medicine than we do here. At any rate, your works on all matters appertaining to drugs sell much faster than our works on similar subjects here. You must have a much larger number of readers than we have. Shall you visit Old England this summer? I think the Exhibition is really worth the cost, the labour and the danger of a voyage from America here.

(Caetera desunt).[121]

SYDENHAM SOCIETY

In the year 1843, the establishment of the "Sydenham Society of London" was determined on.[122] It was "in-

---

[120] Anthony Todd Thomson (1778–1849), one of the early editorial group of the London Medical Repository, Professor of Materia Medica, London University.

[121] The rest is lacking.

[122] The Sydenham Society, a sort of medical "Book-of-the-month" club, terminated about 1856, but a New Sydenham Society then published contemporary works from 1859 to 1907.

stituted for the purpose of meeting certain acknowledged deficiencies in existing means for diffusing medical literature, which are not likely to be supplied by the
efforts of individuals," and the following application
was made to me to give my assistance to the undertaking.

24 Finsbury Place,
London Dec. 16, 1843.

Dear Sir,

By the direction of the Council of the Sydenham Society
I beg to enclose you a Prospectus, and to request your attention to the objects of the Society. The Council are anxious,
that the Society should be known in the United States, and
should receive support from the profession in America. If
the constitution and objects of the Society meet with your
approbation, I am directed to ask if you will be good enough
to favour us with your services as *Local Secretary* for
Philadelphia. You would be the channel of our communication with your part of the States and would receive
and transmit subscriptions to the Treasurer, and our Books
would be sent to your care. Beyond this, your duties would
consist simply in making known the existence and objects
of the Society in any mode that might appear to you desirable. I am writing to make the same request of other
gentlemen in others of the principal cities of the United
States. The Council feel assured, that your name and influence will materially serve the interests of their association, and will be much gratified if you will oblige me with
a favourable reply to this communication, at your earliest
convenience. Our numbers now amount to nearly 1400 and
we have two works in the press, which are expected to
appear very shortly. Others are in the hands of Editors
appointed by the Council, and will appear in the course of
a few months.

I am, Sir,
yours truly,
Jas. R. Bennett [123]

To Professor Dunglison MD etc, etc, Philadelphia.

I accepted the office, and up to the present time have
exerted myself not a little to promote the interests of
the Society. Twice, in 1847 and in 1852, I addressed
the American Medical Association, and my letters appeared in their published minutes. I drew the attention of the profession, also, repeatedly to it in the pages
of The *Medical Examiner,* and yet, notwithstanding all
this trouble, I have annually paid my *five dollars* subscription to it. I have always regarded it as a most
useful Society and have continued to act as *Honorary
Local Secretary,* fearful, that its interests might not be
as carefully and zealously attended to, if transferred
to other hands. In the year 1849, the Society gave me
the purely honorary position of *Vice President,* and
continued it until the annual meeting of 1852, when
a law requiring "that of the Vice Presidents four shall
retire every second year." I was superseded. (end
of Vol. 7 except for the Index which follows.) [124]

---

[123] Sir James Risdon Bennett (1809–1891), F.R.C.P., consulting physician to St. Thomas' Hospital.
[124] The original index by Dr. Dunglison is incorporated into
a general and more elaborate index at the end of this work.

FIG. 14. The venerable Professor Dunglison. From a daguerreotype in the Historical Collections of the College of
Physicians of Philadelphia.

VII. LIFE'S ZENITH

Cholera. Human burial. Death of daughter. Death of
wife. Honorary degree LL.D. Voyage to Europe. Death
of Dr. Patterson. Appointed Dean of Jefferson Medical
College. Supplementary Ana. American Philosophical Society. Hugh Blair Grigsby LL.D. Hon. Edwin M. Stanton. Sir Henry Holland. Professor Tucker. Revd. Dr.
Bethune. Judge Kane. Dr. Elisha K. Kane.

CHOLERA

In the year 1849 much apprehension existed, that
there would be an epidemic of cholera in Philadelphia,
an apprehension, which was subsequently realized.[1] A

---

[1] In the cholera epidemic of 1849, there were 1,049 deaths
in Philadelphia, considerably less, proportionately, than in the
memorable epidemic of 1832.

subcommittee of the city councils was appointed to inquire into "The best means of securing the health and cleanliness of the City." This Committee addressed a series of questions to several of the prominent physicians, and to myself amongst the number. The answers were given and published in pamphlet form (See the bound volume of *Medical pamphlets* for 1849, in my library). Mine were perhaps more fully given, and, therefore, to many, more satisfactory than the others, and it so happened that they were printed in an early part of the pamphlet. This—I was informed by Mr. Benjamin Gerhard—the chairman of the Committee— gave occasion to the inquiry from some small mind, why so much prominence was given to my answers, to which he replied, that D was an early letter of the alphabet, and that the desire of the subcommittee was to arrange the answers as nearly as possible alphabetically. The alphabetical arrangement was, however, most imperfectly followed. Mr. Gerhard was forcibly impressed with this exhibition of *petitesse* on the part of the querist; and was kind enough to express himself highly satisfied with the character of my answers. The queries embraced the best method of cleansing or rendering innoxious privies and other wells under the existing circumstances,[2]—whether there were any factories within the city limits, which were prejudicial to health during the summer,—whether it was proper to water or sprinkle the streets before cleaning them, or at any other times, and what precautions, if any, it would be desirable to employ, during the summer, whilst cleaning the streets and gutters;—whether the docks ought to be cleaned during the summer, or how else should those, full of offensive matter, be rendered innoxious, and if cleaned, what precautions if any, should be used;— whether it is desirable to use lime, chloride of lime, or other, and what, cheap disinfecting agents in the streets, alleys, houses or wells of the city; which are the best of those agents, and how they should be used;—whether fires are desirable, and would be useful for the ventilation of houses during the summer, and how they should be recommended to be used; and lastly, whether I could state anything else of interest upon the subject of inquiry, to wit, "the health and cleansing of the city at the present time."

To all these questions I replied *seriatim;* concluding with the remark, that

all the municipal authorities can do is to attend, as far as possible, to the thorough ventilation and cleansing of the city, and to be careful not to change too suddenly circum-

[2] By 1850 there were a few sewers under the streets in Philadelphia, but the river backed into them at high tide and they often overflowed. Most of the filth was carried off to the river by surface drainage in the gutters. Although water closets that flushed into the sewers were already coming into use, privies situated in the yard were still almost universal. These had to be emptied by the euphemistical "odorless" excavators when they became filled to an inconvenient height (R. La Roche, *Yellow Fever* 1: 21, 1855).

stances long connected with a locality, even when such locality may seem to them to require material modification. The period of a threatened attack of cholera is not the one best suited for rapid and thorough revolutions, either as regards places or persons. In times of spreading sickness, a sudden and total change of inveterate habits adds, no doubt, greatly, to the extent of the calamity. The drunkard becomes alarmed, abandons his accustomed excitant, and, under the depression that follows, readily receives the morbific influence, and sinks a victim to incautious and untimely reformation.

The measures taken by the municipal authorities were, I think, judicious, and proved eminently salutary. The disease certainly prevailed less than in New York, and many other places; and a part of the good result was probably owing to the hygienic cares that were bestowed by those in power under judicious advice. A full account of the epidemic is contained in a pamphlet, bound up in the same volume to which I have referred, entitled *Statistics of cholera with the Sanitary Measures adopted by the Board of Health prior to, and during the prevalence of the Epidemic in Philadelphia in the summer of 1849, chronologically arranged prepared by the Sanitary Committee, approved by the Board, and ordered for publication, October 10th, 1849.*

HUMAN BURIAL

About the same time, the subject of intramural interments was engaging the attention of the City Councils, who were desirous of the opinions of different professional gentlemen thereon. The following complimentary letter was addressed to me by Mr. Waterman, Member of the Council.

My dear Sir,

I take the liberty of enclosing you the following circular, in the hope, that you may find leisure to give me an early reply. If Mr. Tyson, the Chairman has sent you a circular, your answer may be sent to him. If not, please direct it to me, and I will go *on my own hook* with it, with a perfect consciousness, that it will be more to the point, and give us more useful information, than any other. I am a member of the Special Committee, and very desirous for *your* views on the important questions propounded.

With very great regard,
Your friend
A. G. Waterman,
Monday June 4, 1849

R. Dunglison, MD

The following is a copy of the Circular, alluded to in the letter of Mr. Waterman.

Philadelphia, May 21, 1849.
Sir,

At the request of a *special committee,* appointed by the Select and Common Councils of the City of Philadelphia, to inquire, whether any legislation is necessary in regard to the depth of graves within the city limits, and also, whether it is expedient to prohibit interments within the said limits, I have the honour to send you the subjoined queries, which, it is hoped, your leisure may permit you to answer at an early day. Please to direct your communication to the Chairman, Job. R. Tyson, Esq., No. 2, Prune St.

I am your obedient servant,
Henry Helmuth
Secretary Special Committee

1. Please furnish us with any information you may possess, as to the effect upon health, of the interments of deceased persons in crowded communities or cities?

2. Are such interments compatible with the public health?

3. What, in your opinion, is the effect upon the general health, nervous system or otherwise, of living in the vicinity of graveyards, if any?

4. What is the effect of burying, in cities, in vaults, and under churches? Are they or not injurious to the public health of cities, and how do they compare, in their probable effect upon health, with the ordinary mode of interments?

5. Do you know of any graveyards in Philadelphia the interments in which are within six feet?

6. What depth of graves is necessary to prevent the escape of noxious gases from animal decomposition?

7. Will a depth of 8 or 9 feet prevent that escape in our soil?" [3]

To these Queries I furnished replies, but unfortunately kept no copy of my answers, the more especially as the originals could not be found, when I asked for them, in the Archives of the City Councils.

DEATH OF DAUGHTER

When I arrived in Philadelphia in 1836, my family consisted of my wife, my daughter Harriette Elizabeth, born Oct. 29, 1825; John Robley, born Dec. 16, 1829; William Leadam, born 18 July 1832, all Virginians, & Richard James, born in Baltimore, Nov. 13, 1834. On the 10th of March 1837, Thomas Randolph was born, and on the 28th of January 1840 Emma Mary. The happy chain was not destined, however, to remain long unbroken. My daughter Harriette, in the Winter of 1841, was attacked with Endopericarditis, the foundation of which, I fear, was laid in an organization derived from her progenitors, her mother having suffered from the same malady many years before we were married. She died on the 11th of March after an illness of three weeks; and unexpectedly to me then, for although I looked upon her as in danger, I had witnessed the recovery of her mother on several occasions from the same disease, when the symptoms were much more urgent, and, therefore, did not expect the fatal result when it took place. I had been out on the evening of the 10th, and, on my return, found her suffering under an increase of the difficulty of breathing, and some pain in

the region of the heart, for which I immediately applied the dry cups. She was perfectly sensible; and stated that she had never felt in that manner before. Her mother and myself were supporting her, and on the baby crying in the crib near her mother's bed, she expressed her apprehension that it might fall out, soon after which she appeared to faint, and ceased altogether to breathe. All the family had gone to bed, and my wife and myself were alone to receive her last breath.

Her loss was a sad affliction to us both. She was affectionate, intelligent, pure, and, although only in her sixteenth year, had become imbued with a strong religious feeling. Although an Episcopalian she was fond of the ministrations of the Revd. Dr. Bethume, whom she was accustomed to see at my house, and to whom she was much attached, not infrequently attending divine worship at his church. The anguish, which we experienced in recalling her many virtues, and in deploring her loss, made her mother and myself determine to suffer in silence, and not to add to the poignancy of each other's afflictions by recurring to the subject.

Although perfectly satisfied of the correctness of the diagnosis in her case, I was anxious, that a *postmortem* examination should be made, which was kindly done by my friend Dr. Pancoast, in the presence of my friends Drs. Patterson and Huston, both of whom saw her in the progress of her disease. The following report of the appearances presented was given me by Dr. Pancoast.

On raising up the Sternum and cartilages of the ribs, vascular injection of the cellular tissue in the anterior mediastinum and of the pleural coating of the pericardium observed. The pericardium on its left side attached by resisting, but reddish false membrane to the pleura, covering cartilages and ribs of the same side. Sac of the pericardium greatly enlarged, measuring six inches across at its base just above the diaphragm. Left margin of the sac extending two inches beyond the junction of the fourth rib and cartilage of the same side. Right margin one inch beyond the junction of the corresponding rib and cartilage of the right side. Lungs attached to pericardium on each side by false membrane which was of recent formation and easily torn. Costal and pulmonary pleurae united by false membrane, pretty generally over both sides of the chest, firmer than that over the pericardium but easily torn with the finger. A few old adhesions obvious on the upper and front part of each lung. Seven ounces of ser-sanguineous fluid found in the left thoracic cavity, five in the right.

Sac of pericardium opened. Pericardium everywhere adherent to the outer surface of the heart, so as to leave no vacant space. The adhesion was by the intermedium of false membrane, which, at its thinnest part over the right auricle, was about a line in thickness, and, at its thickest, over the left ventricle and base and appendage of the left auricle was fully a quarter of an inch. The right ventricle was covered by false mem-

---

[3] Graveyards were situated within Philadelphia city limits for a long while after its original settlement, usually around churches, sometimes in detached spots. In the course of years it became necessary in some to place stratum on stratum of earth to make room for additional bodies. In the 1830's cemeteries were established outside the city (*ibid.*). New York City prohibited burials within the city in 1823.

brane to a depth varying from one and a half to two lines. The false membrane seemed to have been deposited in successive layers; that nearest the pericardium was reddish and possessed considerable firmness,—that next the heart, especially over the left ventricle and auricle, was red and gelatinous—Both from the heart and pericardium false membrane could be peeled off, leaving the natural serous surfaces smooth and shining,—Drawing upon the pericardium, the false membrane broke, leaving the largest part adherent to the heart and presenting a honeycomb or villous appearance like the bonet (stomach) of the calf.

Cavities of the Heart—Right auricle—Lining membrane presented generally an opaque pearly appearance —was readily torn up in long strips, leaving still a smooth shining surface below. Doubt its being an effusion of false membrane; think the opacity owing to a deposit in the subserous cellular tissue. Cavity large —*Right ventricle*. Lining membrane over tricuspid valves exceedingly thickened and opaque. Around the free margin of the valves a rounded, bluish semi-cartilaginous deposit in oblong sections. Walls of right ventricle thicker than usual. Valves of pulmonary artery natural. *Left auricle*—Large like that of the right; capacity of each seemed much superior to that of the corresponding ventricle. Same opaque and pearly appearance of lining membrane as described in the right auricle, but more strongly marked. *Left ventricle*. Mitral valves three times the natural thickness. Cartilaginous deposits along the free margins; chordae tendineae thickened and semi-cartilaginous— *Aortic valves*—Cartilaginous deposits on the corpuscula Aurantii, rather larger in size than a grain of wheat and which they resemble somewhat in shape. The reflexed fold or festoon of the lining membrane forming these valves was thickened, rugous, fibro-cartilaginous, and stood up in relief. The aortic festoon seemed natural, thin and flaccid, as usual. The lining membrane of aorta seemed nearly natural. It tore up freely in strips an inch long, and had a cast of colour rather more yellow than natural. Coronary arteries healthy. The limitation of the morbid changes to the inner festoon of the aortic valves seems to indicate clearly the endocardial origin of the disease. The walls of the ventricles were pale in colour and rather soft.

Lungs. Much bloody mucus in the bronchiae of both sides. The upper lobes of both lungs and middle lobe of right distended with air and crepitating *loudly* on pressure or when cut. No appearances of interlobular emphysema. A serosanguinolent effusion throughout the upper lobes. Lower lobes of both sides in a state of splenification, dark blue colour, nearly solid, very easily torn. Blood mixed with the tissue so as not to be forced out by pressure. No tubercular deposits in any part of the lungs.[4]

---

[4] Dunglison believed there was a hereditary predisposition to acute rheumatic fever, the basic cause of death here,

Her remains were deposited in Laurel Hill, in a lot, which was kindly presented to me, some years before, by the Proprietors, with all of whom I was on intimate terms,—the gift being probably suggested by an article on "Rural Cemetaries," which I wrote for the *American Medical Intelligencer* for July 1, 1837, of which I was the Editor. In that article, I alluded to the Laurel Hill Cemetery in the following language:—

How often has it happened in the progress of our own city to its present population, that places of worship have been disposed of; their cemeteries desecrated, and ashes, which, at the period when they were deposited there, it was presumed, would ever remain free from violation, been exhumed and scattered to the winds. These and other considerations have given rise to the beautiful cemeteries of Pere la Chaise, near Paris, of Mount Auburn, near Boston, and of Laurel Hill near this city. The preceding remarks have, indeed been suggested by a recent visit to the last of these. Situated at a convenient distance from the city of Philadelphia, yet so far from it as to almost preclude the possibility of future molestation in the progressive improvement of the city or from other causes, on a sylvan eminence immediately skirting the Schuylkill, and commanding a beautiful view of that romantic river; embellished in a manner most creditable to the taste and liberality of spirit of the respectable individuals under whose management it has been projected, and carried into successful execution, it is indeed a hallowed place where affection may delight to deposit the remains of those on whom it has doated;

"a port of rest from troublous toyle. The world's sweet In, from paine and wearisome turmoyle."

The whole article is copied into the *Guide to Laurel Hill Cemetery* published by the Managers.[5]

DEATH OF WIFE

Twelve years after this I was doomed to suffer a more desolating affliction, in the death of my most affectionate wife, one of the most devoted of mothers; and the loveliest of women; who had left father and mother, brothers and sisters to accompany me to Virginia; and who had gladdened my home, and rendered it always most dear to me for nearly thirty years.

At all times her bodily condition had been to me a source of anxiety; and I fear, that my solicitude was often the occasion of privations to her, which may at times have been unnecessary. At the age of thirteen, her life had been despaired of. During an attack of acute rheumatism, at that early age, she was affected with, what was regarded as, rheumatic inflammation of the heart; which had not then received attention to any extent from medical men. A communication on the subject had been written in the *Medico-Chirurgical Transactions* of London, by Sir David Dundas [6]—a

---

although the immediate cause was probably the pneumonitis with adhesive pericarditis.

[5] See also: *Hints on the Subject of Interments within the City of Philadelphia; addressed to the Serious Consideration of the Members of Councils, Commissioners of the Districts, and Citizens Generally* by Atticus, Philadelphia, 1838.

[6] David Dundas, Serjeant-Surgeon to the King, gave the first careful description in English of mitral stenosis, peri-

distant relation of the family, who visited her during her sickness; and, as far as I can recollect, Dr. Babington[7] and Sir Astley Cooper—then Mr. Cooper[8]—were also in attendance with her father. After long suffering, she recovered, and when I first knew her, in her sixteenth or seventeenth year, she was apparently in good health, with the exception of irregularity of the circulation induced by slight causes of excitement. From that time until a few years after our arrival in Virginia, she had excellent health. In the year 1829 or 1830, she had a most violent attack of her old malady, —acute rheumatism—from which she suffered excessively and, not long after its onset, endopericarditis supervened, with the greatest irregularity and intermission of the pulse; and the most manifest physical signs of coagulation of the blood in the cavities of the heart, accompanied by anhelation and respiratory distress to such a degree, that, in company with my friends Dr. and Mrs. Patterson I, on different occasions, watched over her under the expectation, that every moment would be her last. In this apprehension she participated; and, on one occasion, took what, she conceived to be, her last leave of me and my family, at that time very small. One of the most distressing symptoms in her case was the sound produced during the contraction of the heart, which resembled that of the croaking of a frog, and was audible in every part of a large bed chamber. She gradually recovered, however, with some infiltration, for a time, of the lower extremities; and, after her restoration, the persistence of a bellows murmur over the seat of the semilunar valves accompanying the first sound of the heart, indicated the mischief produced by the endocarditis.

Another very severe attack of acute rheumatism, accompanied, again with endo-carditis supervened soon after our arrival in Philadelphia. The affection of the joints retained her in bed, with intense suffering, for several weeks; and the endocarditis for some weeks longer. On this occasion, her recovery was very tedious. When she was able to be moved, I took her to West Chester, where she was still affected with great anhelation, irregular and intermittant pulse, and copious infiltration of the lower extremities. In the same hotel, there was a lady—a Mrs. Renshaw—labouring under the same malady—acute rheumatism succeeded by cardiac dropsy—who died soon afterwards. She was a patient of Dr. Caspar Morris, and my wife informed me that her (my wife's) recovery was according to the expression of Dr. Morris, "a monument of

my skill." That skill consisted in the watching of every phenomenon, the satisfying of her every want— more than in the employment of what has not improperly been called "meddlesome" and too often certainly—"perturbating" medicine. My apprehensions, here again, were of the most melancholy character; yet she gradually recovered; and remained in her ordinary health, enjoying life as usual until the spring of 1841,—the year of my daughter's death, when she was attacked with acute bronchitis, and it was whilst labouring under this disease, that she had to support that sad bereavement. We were then living in Girard St.[9] but the residence became the source of constant distress to her and we determined to seek another; and, not long afterwards, we moved to the corner of Schuylkill 8th & Spruce Streets.[10] I took every precaution to prevent unnecessary fatigue on her part during the moving, but soon after our change of residence, she was again attacked with her old malady. This time, however, although the heart was greatly implicated the attack was of shorter duration, and after some weeks of anehelation, restless nights and infiltration of the extremities, she was again restored to her ordinary state of health. Nor did the narrowness of the orifices of the heart appear to have increased, but there was, ever afterwards, more irregularity and intermission in the pulsations, with a "catch" in the heart's action, always accompanied by a cessation of the pulse—a loss of beat—and unpleasantly frequent, whenever she was affected with any severe corporeal ailment.

All these phenomena made me, if possible, still more watchful over her; and I cannot but feel, that it was owing to the cares I bestowed upon her, that I was blessed for many more years with her happy influence. When, however, she had visitors with her, she would exert herself so much more than I deemed safe, that, excepting in the case of those whose presence did not materially add to her exertions, I always felt apprehensive of consequences. In the winter immediately preceding her death, these considerations induced me to throw obstacles in the way of the reception of resident visitors; and it was fortunate that I did so, as I should have been disposed, perhaps, to refer to their agency, what was produced by other causes. In the propriety of this action on my part, she wholly coincided, appearing herself to dread the effects of any unusual exertion.

In December 1852, soon after Miss Cornelia and

carditis and endocarditis in a paper read before the Royal Medical and Chirurgical Society of London, 1806 (*Medical Chirurgical Transactions* 1: 37–46, 1809). He delivered the Hunterian Oration in 1818.

[7] William Babington (1756–1833), L.R.C.P., of Guy's Hospital, teacher and life-long friend of Richard Bright.

[8] Astley Paston Cooper (1768–1841), one of the most popular British surgeons and medical teachers, created a baron in 1821 (Bransby Blake Cooper, *Life of Sir Astley Cooper, Bart.,* London, 1843).

[9] Mr. J. Alden Tifft, a descendant of Dr. Dunglison, in August, 1959, wrote: "My grandmother used to tell me about my great-grandfather . . . that Dr. Dunglison entertained many savants and friends at his home on Girard Street. This street I saw when I was a boy. As I remember there were a number of marble fronted dwellings. The street was between Chestnut and Market and ran for one square between 11th and 12th. . . ." Richard Dunglison was still listed at this address in the *Philadelphia Directory* for 1868.

[10] Schuylkill 8th Street is now Fifteenth Street. The population shift was carrying the more affluent to pretentious homes rising on Walnut and Spruce Streets west of Broad Street.

Miss Mary Randolph—who had been staying with us for a short time, and were never the source of any anxiety or trouble—had left us, my wife was violently attacked with acute bronchitis—which was, at the time, epidemic in Philadelphia. Towards the end of the month, the symptoms were so severe as to confine her to the house, but it was difficult for her to be convinced, that the attentions I wished to have bestowed upon her were necessary. On the first of January, she was so indisposed, that I could not permit her to be at the dinner table. Her absence on the first day of the year, she remarked, was of bad omen, but she submitted. On the following Sunday, the second of January, she was at the dinner table for the last time. On the sixth, she was so chilly as to be constantly at the hot air register or the fire in the small room, and when I returned from my lecture, at 6 o'clock, finding her feverish, I begged her to go to bed to which she reluctantly consented, so anxious was she to hope, that the step was scarcely necessary. By care and attention, she was much improved, but on the 12th, her bronchitic symptoms became greatly aggravated; with considerable irregularity and intermission of the pulse; superficial and clear beat of the heart, and other phenomena indicating the supervention of endocarditis. Gradually, however, the bronchitic and cardiac symptoms abated; and on Tuesday the 25th inst. I considered her decidedly better. The same night, however, the fever and dyspnea became greatly aggravated, and, on Wednesday morning she was so ill, that she scarcely noticed any of us. I was then suffering under a slight attack of gout,[11] and she permitted me to sit up all that night and the following, without expressing any solicitude, which, under ordinary circumstances, she would have been so anxious to do. In the morning, I had great doubts, whether she could live through the day, and my friends Drs. Huston and Meigs, who saw her, and were most kind in their attentions to her and to me, had the greatest apprehensions for her. In the afternoon, however, she became less distressed; and gradually improved so much, that for about ten days before her last exacerbation, which occurred on the 23d of February, she sat up in her chair, and moved gently about the room; for the sake of change, went, indeed, into the next chamber, and the day before she became worse, she was borne down stairs into the parlour, and, in the evening, was carried up again. During the whole time, however, the disturbed action of the heart continued; and although I for the time considered, that immediate danger had passed away, I was apprehensive that cardiac dropsy would supervene. She was restless, sleepless at night,

but comparatively comfortable during the day, excepting, that there was more shortness of breath at all times than there ought to be. Gradually this became more urgent; and, on the night of the 23d of February, she was attacked with harassing nausea, which nothing would relieve, the febrile phenomena returned, with the most distressing anhelation, and irregular attacks of neuralgia, occasionally over the region of the heart, and, at other times, over the sacrolumbar and sacro iliac junctions, which were almost always removed by counterirritants—as sinapisms, dry cupping—the local application of chloroform etc.

On the third and fourth of March, she expressed her conviction, that she could not recover, but said little to me for fear she should distress me. The wretched sleepless nights she had passed could not it seemed to her—be continued without fatal consequences, and on the afternoon of the fourth, she felt satisfied she could not survive the following night. About ten o'clock, although the dyspnoea was so distressing as to compel her to sit up, she appeared to take no farther notice of those who surrounded her. From that time until her death, I supported her, and when I addressed her, asking her to lean on me, she obeyed, but her consciousness seemed to be momentary. About half past three in the morning of the 5th, she expired in my arms, after a severe struggle—although, manifestly, the moans, to which she gave utterance a few minutes before dissolution, were not the evidences of conscious suffering—in the presence of her children, with the exception of Emma, whom I had sent to bed to spare her the painful scene, but who was awakened by the loud moans of her dying mother.

The attentions, which were bestowed on her during her illness by our numerous friends were the source of much pleasure to her, as they must always be of deep gratitude to me.

The kindness of Drs. Meigs and Huston was extreme, and she was occasionally visited by Dr. Mitchell. Dr. Meigs saw her the day before she died, and although she said to him—not in my presence—that she could not live through the night, she asked him in my hearing, to call and see her the next day. She had a great liking for him, and deservedly, for a more kindhearted, generous individual does not exist. The more I have known him the more I have learned to estimate his excellent qualities.

The visits of my medical brethren were a great consolation to me. The burthen of the treatment necessarily rested on me, but it was important for me to hear their suggestions from which I often profited. It was, moreover, gratifying to the invalid, who knew, that my own cares would be diminished thereby. Since her decease, I have looked back at the whole course of management, and it has been pleasing to me not to be able to discover anything done, or omitted to be done, that I could regret. My poor wife, a day or two

---

[11] Dunglison mentions his affliction with gout frequently in these *Ana*. Dr. Squibb, in his Journal (p. 188), noting Dunglison's lecture on October 28, 1851, states: "At the termination of these remarks he begged the class to excuse him as he was suffering from catarrh and gout, and closed the lecture a quarter of an hour before the usual time."

before her dissolution expressed to me how fortunate she was, compared with many poor creatures, to have everything she wanted, with persons surrounding her to anticipate her wishes, and to carry into effect all her desires.

It was fortunate for me, that, throughout the whole of her protracted illness, I was enabled to attend upon her; and, except on two occasions, to fulfill my college duties. For some years, I had been compelled, in consequence of my repeated attacks of gout, and the inconvenience I felt, when compelled to get up at night, to abandon all but consultation practice; and my disturbed nights, during my wife's illness, made me naturally, anticipate a recurrence of my old malady. I had, indeed, more than one demonstration, but had recourse to large doses of colchicum—thirty drops of the wine of the root, which I imported from Apothecaries' Hall, three times a day—and I was enabled, although under much suffering, to attend to her and to my accustomed duties. The session of the medical school terminated on the last day of February, when I took leave of my class, under sad forebodings, and, every morning of the month of February, for an hour and a half, I was occupied in examining candidates for graduation. My constant watching, however, induced a severe attack of acute bronchitis, attended with much fever and copious expectoration, owing to which I was prevented, on two occasions, from meeting my class; and I could not help remarking to them, in my last lecture, that, although liable to attacks of a painful and confining malady, it was a subject of congratulation, that, for the 16 years I had taught in the Jefferson Medical College I had only been compelled, by sickness, on five occasions to omit a lecture, and never by any other cause; and that since the year 1847, when gout prevented me on one or two occasions, I had appeared before them without fail, a regularity of attendance perhaps without example.

My bronchitic attack became milder before the death of my wife, but did not finally cease for some weeks afterwards. Under it, the disturbed rest, and the intense mental anxiety I had experienced, it is not surprising, that I should have lost twenty or twenty five pounds in weight.

The sympathy, every where exhibited on the announcement of the death of my wife was as striking as it was consolatory to me. On learning, on the night of the 4th of March, that she was dying, Judge & Mrs. Kane, and Dr. Huston came to the house, and remained until past one o'clock in the morning, when I insisted upon their retiring, and promised to call upon them in the morning, should the fatal event transpire in the course of the night, of which there was little or no doubt in my mind. My kind friends reluctantly consented, and subsequently expressed their regret, that they had not staid with me, as they might—they fancied—have been able to render me some assistance.—

Such was especially the considerate feeling of Mrs. Kane. There is, however, on such occasions, with me, a consolation in mourning unseen; and the feeling is strongly impressed upon me, that we should avoid, as much as possible, distressing others with the intensity of our own private griefs. To my valued friend—Judge Kane—who, throughout the whole of my wife's illness, had been most devoted in his solicitude for her recovering—I immediately and unhesitatingly applied for aid in the melancholy condition in which I was placed. All the arrangements were left to him; and all were executed in the most satisfactory manner to me. The only request I made of him was, that the funeral should be as private as possible and that, in particular, no idle ostentation should be permitted. On the 7th of March, the remains of her—the dearest to me of mortals—were deposited alongside those of her daughter whom she loved so much, in the cemetery of Laurel Hill,—the pallbearers being Judge Kane, Professor Tucker, Dr. Franklin Bache, Dr. Meigs and Dr. Mitchell.

On the evening of Saturday—the day on which she died, Dr. Meigs was to have held his Wistar party, and of course, all his preparations were made; but he at once issued a printed circular to his invited guests, stating, that the party was postponed in consequence of her death. All my colleagues immediately closed their houses until after the funeral, and many of the houses of our private friends were equally shut up. The graduation party, which was to take place on the following Wednesday, two days after the funeral, was omitted; and the students of the college sent me a letter of condolence. My considerate colleagues went even farther in their tokens of respect; and declined, in April, having the annual supper, on the settlement of the Dean's accounts. I could not possibly have construed the observance of regular and stated affairs of the College as any want of sympathy on their part towards me, but it was not the less grateful to me, that they should have been so delicately pretermitted. The feeling appeared to be unanimous, and certainly they could not have done more to shew their respect for me. My excellent and generous friend Dr. Meigs called upon me before the day of the funeral to press upon me to accept a large inclosed lot belonging to him in the Woodland's cemetery; but I was compelled to decline his liberal offer, knowing that it would have been more satisfactory to my wife to believe that she would be placed alongside her daughter.

From several of my valued acquaintances I received the kindest letters of condolence, all testifying to the rare excellence of my lost angel. Unknown to me, my old friend Professor Tucker, who had known us intimately ever since our arrival in the country, in the year 1824, inserted the following feeling tribute to her memory in the Pennsylvania Inquirer of the 7th of March [1853]—along with the announcement of her decease.

For the Inquirer.

OBITUARY

DIED, in her 51st year, of a heart disease, Mrs. HARRIET DUNGLISON, wife of Dr. Robley Dunglison, of this city. The worth of this excellent woman entitles her to more than a passing notice. Mrs. D. was born in London, and was the daughter of an apothecary * of estimable character and extensive practice. Soon after her marriage in 1821, she came to Virginia, on the opening of its University, to which Dr. D. had been appointed a Professor. They remained there nine years, and then removed to Baltimore on the Doctor accepting a Medical Professorship in that city. A few years afterwards, he received the appointment in the Jefferson School, which he now holds. In all these places of residence, Mrs. Dunglison won the esteem and love of all who knew her, and was often characterized as a model in the exercise of the domestic charities. Her self-devotion, as a wife and a mother, were never surpassed. Warm, generous and affectionate, the great purpose and happiness of her life was to make those happy about her. The writer of this fleeting tribute has known her for near thirty years, the greater part of which he has been her neighbor, and in all that time, he never witnessed in her aught of word, act or look, that did not indicate kind feelings and a good heart. In her attachments she was unchangeable, and a friendship once formed, lasted for life. She has been known to make daily visits to a friend ill with the varialoid [sic], at the risk of taking it herself, or what would have been far worse, in her eyes, of communicating it to her children. With all these gentle virtues, she possessed great energy of character, and was as indefatigable in managing her family as in serving a friend.

Engrossed, as she was, by domestic duties, home was the chief theatre of her virtues. It was there that those gems diffused most of their mild radiance, and it was only those friends that had access to the casket, who could fully appreciate their purity and lustre. But alas! those eyes that lately beamed with benignity, are now closed in darkness— that voice which was ever ready to say something soothing or kind, is now hushed for ever! But they will be long present, in fancy, to numerous friends, as well as to her afflicted husband and children. May the daughter and four sons she has left profit by her bright example, and imitate the virtues of a mother whom they can never forget.

The following letter is from my excellent friend the Revd. Geo. W. Bethune, D.D. who as a member of the "Five" (see p. 149) and a most cherished acquaintance, had numerous opportunities for appreciating her sterling worth.

New York, March 10/53

My very dear friend,

I have just heard of the overwhelming sorrow which has come upon you, and cannot restrain myself from expressing at once the grief and sympathy I feel. I had known that Mrs. Dunglison was ill and anxiously from time to time, without troubling you, procured information. It is but a few weeks since that Mrs. Elwyn wrote me most cheerful news congratulating me on the hope I might cherish of Mrs. Dunglison's recovery. I thanked God for you both and for your children's sakes. You know, dear Doctor, how esteemed—the word is too cold—how beloved your

* The Apothecary in England is the general practitioner, not, as in the United States, a seller of drugs. He compounds only his own prescriptions, or those of physicians with whom he is in consultation—as a general rule.

admirable wife was by all who had the happiness of seeing her in her home which she made so pleasant to her friends and where we saw her fulfilling every duty with such cheerful tact and consideration. You know too that of those friends no one could have been more attached to Mrs. Dunglison, as well for her own kindness as the blessing she was to the life of my dear friend the father of her children. I must rely on your knowledge of my heart for assurance of my deep sense of your desolation and of my suffering for you—words cannot express what you will believe that I feel. Never in all my observation of people have I known man and wife so fitted to make each other happy or more devoted to each other's happiness, and how you are to bear your bereavement God only knows. To God only can I go with my anxiety for you, and most devoutly have my poor prayers gone up, as they will often, that He who has smitten will sustain you. The world is valueless at such a moment, but He who made the heart and sees its inmost bitterness has commanded us through his son Jesus Christ, the man of sorrows and the God of comfort, to cast ourselves upon him, that we may find support on his bosom. Dear Doctor, we are passing away, our lives fail—we go gradually to the grave—let us look beyond the present scene, and assure ourselves through the Grace of God, of a better inheritance where death cannot reach us, and sorrow cannot come because there there will be no more sin. It was but a day or two ago I was thinking of your dear daughter who left you for heaven, and remembering thankfully that I had been of some use to her in preparation for a better life. Mrs. Lawrence (Benj. Richard's daughter) came in, and some not unpleasant tears were shed by us both while speaking of that dear child whose face was so often upturned to mine as I preached the Gospel to her willing ears. Now the mother and daughter are united. Let us try to follow them, my friend. I am pained that I did not know soon enough to be among those who were near you when this precious dust was laid in its resting place. Had I known, nothing short of absolute inability could have prevented me from going on. Mrs. Bethune joins me in assurances of sympathy.

Your greatly attached and affectionate friend,
Geo. W. Bethune

Robley Dunglison, MD.

The above letter I sent to Judge Kane for his perusal, who returned it to me with the following:—

My dear doctor,

I have read to my wife Dr. Bethune's letter: all that he says is true. We all of us loved Mrs. Dunglison. We knew her beautiful character by heart for there was no concealment or disguise about it. Her loveliness spoke out to every one and every where. All your friends felt that she was our friend too. We shall none of us meet so hospitable a spirit again as her's was, nor commune with one more catholic in its charity towards others. I never heard her speak a word that could give pain, or that anybody could have wished unsaid.

But you were the radiating centre of her thoughts and affections. I never saw a wife more devoted to her husband or more sedulous in sharing with him all the duties of a parent. It was not a sense of duty: it was, like her fearless kindness to Mrs. Tucker in her sick room, one continuous admirable impulse, that sympathized with duty, but went before it. She was the model of disinterestedness, as devoid of selfishness as an angel. This was indeed her most marked characteristic. It impressed itself on all her virtues, and make their beauty.

I return to you Dr. Bethune's letter with this poor tribute of my own to a memory that we cherish alike.

Most sincerely
and faithfully yours
J. K. Kane.
16 Mar. 1853.

The following letter is from Professor Lomax, one of my excellent colleagues in the University of Virginia to whom I have referred elsewhere (p. 42).

Fredericksburg, March 22, 1853.
My dear Doctor,

My wife and myself have heard with deepest affliction of your heart-rending bereavement. We dare not intrude upon the sacredness of your sorrows under this crushing dispensation with the vain attempts at consolation. We cannot, however, suppress the feelings of our sincerest condolence. It has long been our delight to cherish in our recollections the surpassing excellencies of the best of wives, the most affectionate of mothers, the sincerest of friends, the loveliest of her sex. To the last moment of our lives, her memory will be sacredly in-urned in both our bosoms. May God of his infinite goodness soothe the pangs of a dispensation which no human sympathy can alleviate, in the hearts of the surviving husband and children, and may He pour down upon you his richest blessings is the prayer of

Your sincere friend
Jno. Tayloe Lomax.

Between the family of Mr. Nicholas P. Trist (see p. 22) and my own there had always been the most intimate relations, from the time of our first becoming acquainted with them as part of the family of Mr. Jefferson, and during Mr. Trist's residence in Cuba, I had taken the guardianship of his son—who was deaf and dumb—and an inmate of the Institution in Philadelphia. The character of his feelings, as well as those of his most estimable parents and sister, and of the other grand-daughters of Mr. Jefferson, who had visited us a short time before my wife's last illness (see p. 161), will be shewn by the following letters. Mr. Trist, when he first heard of her sickness, and, subsequently, of her death, was, at the time, in England, on a visit to Mr. Pendarves, member of Parliament for Cornwall.

London March 8, /53
25 Queen Anne St. Cavendish
Square
My dear Dunglison,

Willie's letter to Browse [12] (which overtook us at Creaton in Northamptonshire, after having travelled all the way to Landsend almost), although no surprize to me, was, I need not say, a great shock; for these are things one can never be prepared for, however familiarized with the painful fact of their liability to occur at any moment. This has been the state of my mind for many years past in regard to your dear wife; but I was not any the less sensible to the imminency of the blow, on becoming aware that it had been impending over your heads in so fearful a way. It made me realize more fully than ever how truly she stands to me in the place of a sister, and how strong are the

[12] Browse, the middle name of Trist's father, Hore Browse Trist, refers here to one of Nicholas P. Trist's sons with the same name.

sympathies, my dear friend, which bind me to you and to those who would have been involved with you in the dreadful calamity of losing such a wife and such a mother. However reassuring were Willie's closing words—"Since then she has been improving daily, and today father says she could not do better," and notwithstanding the continuance of this improvement down to the 7th, the date of his P.S., I could not but feel how precarious must her condition remain for some time, and how great the liability to relapse.

Ever since, I have been hoping, from one steamer's arrival to another for fresher tidings, but, although many letters came, both from Virginia and from Louisiana, and all devoted in part to the same subject, yet none of them brought any intelligence but was much older than that conveyed by Willie. At length one arrived from Jeff, a week later than W's, closing his account of the matter with "I am very happy to say that she is entirely out of danger, and recovering slowly." This affords substantial ground for the hope that the next letters will be in harmony with the prayer put up by my wife in one which came to hand yesterday. "God grant her precious life may be spared many years to her family and friends."

With my best affection for Mrs. Dunglison
and everyone of you.
yrs most faithfully
N. P. Trist.

London April 7, 1853.
Dear Dunglison,

You need not, any of you, be told that this deepest of all possible afflictions for you is not unfelt by me. Nor need you be told, that it is not from sympathy alone that my tears have been made to mingle with yours. In my life also, does this blow make a void—a void which will remain ever: for she was to me as a beloved sister, doubly so, a beloved sister to my wife and through my wife, a beloved sister to myself.

Desolate as your hearth must ever hereafter be for me, there will, however, be a soothing virtue in the feeling which the presence of her shade must ever so vividly revive in my heart; the feeling how pure and generous and ingenuous was her nature, how warm and gushing her tenderness.

Farewell my dear friend. With the truest love for you all, with the feeling towards dear Emma that she has, in my wife a mother and in my child a sister.

Yours always,
N. P. Trist.

Bowdon [13] March 20/53
My dear Willy,

When the mail bag was opened yesterday morning the sight of a letter with a black edged envelope and black seal shot a pang through my heart. I knew well the sad tidings it was the bearer of. Your dear mother was to me like a sister, and as such I have ever loved her. I cannot tell you how I deplore her loss, how lovely I have always thought her. My heart aches for you all, my dear friends. I hope when this reaches you, the keenest anguish will have subsided in your hearts. May God grant you resignation to this blow. Time must soothe your grief, and in a family so affectionate and disinterested, so like your angel mother, I feel convinced that efforts will be made by each one to restore cheerfulness to the others.

My poor dear friend the Doctor, how is he? I hope that he will go to visit his friends in England this summer. I think the change of scene and change of climate will be of service to him. A bright angel in Heaven watches over

[13] Bowdon is in Cheshire, England.

him, and if she can feel sorrow where she is, it will be in seeing him suffer. Pattie and myself have lamented the great distance we are from you all, often this winter, since we heard of the illness of dear Mrs. Dunglison. The letters were so long in reaching us that even when bearers of more cheering accounts of her, our hope was shadowed by the knowledge that twelve or fifteen days had passed since they were written. We have grieved that we could not go to see her before our departure for this distant country. Life is so uncertain that we never part with a friend without feeling uncertain whether we shall ever meet again. When we came to Louisiana, it was with the expectation of returning in May to the north, but we are to stay here a year longer. . . . We were wishing this morning that dear little Emma was with us. Kiss her most affectionately for both Pattie and myself, and tell her how dear she is to us. Give our love to your dear father of whom I think often, also to Dick and Tom. I hope you will all write to us, we are so far away and think so much of you.

Browse sends his kind regards to the Doctor and all of you. Adieu my dear Willy, believe ever in the sincere affection of your friend,

V. J. Trist.

W. L. Dunglison

Bowdon March 22/53

My dear Dick,

It is impossible for me to tell you the pain Willie's letter to mother has given me; the moment I caught sight of it my heart sank, for although our hopes had been raised by the last accounts of your dear mother's health contained in Willie's letter of Feb. 13th, still I had a constant dread of what the next letter might bring. Since first hearing of her illness, my thoughts have incessantly dwelt on her, and on you all, and I have prayed that so precious a life might be spared. Since God has willed otherwise and taken this angel to himself, Oh! my poor dear Dick, may he give you all strength to bear the terrible loss, and in his own good time send you the consolation which he only can give. Would that I were near you at this time to tell you and show you how much I sympathize in your sorrow. To tell you how deeply I mourn the loss of your sweet lovely mother is not possible. I am most thankful to have been allowed the privilege of being with her last winter; and the recollections of her motherly love and kindness then, and on all other occasions since I can remember, will be treasured in my heart. The Dr. and dear Mrs. Dunglison have always been to me like the kindest uncle and aunt, and if I had not long ago loved you all, I should do so now for the sake of your sweet mother.

Give my best love to your father and to Willie and Tom, to dear little Emma many kisses, and tell her how very much I love her, and how anxiously I shall always watch for any occasion of showing her some return for the affection that has been given me. God bless you my dear Dick, believe in the most sisterly love and sympathy of your friend

Pattie.

R. J. Dunglison

New York March 9/53

My dear Dick,

I cannot express the feelings of sorrow with which I received your letter informing me of the death of your dear mother. Last Monday evening I went into the Steward's office to see if any letters for me had come from the Post Office, and found one from yourself. I was very much startled by the appearance of the black seal and edges of the letter, so that I had not the courage to break the seal at first, for it told me that some death must have occurred in your family, and that if so, it must be that of your much

beloved mother. I cannot yet realize her death, it was so sudden and unexpected. I was about to write to congratulate her upon her recovery, when I received the distressing intelligence of her death. How hard it is for me to reconcile myself to the loss of one who had, for years, taken a mother's place toward me, as you know, and whose untiring attentions and sweetness of temper endeared her very much to me. I shall never forget the motherly welcome with which I was always saluted, whenever I came to spend my holidays with your family during my school-days. For four years I have rarely seen her, but have consoled myself by the hope that I should, in future have the pleasure of meeting her oftener; but alas! I shall no more have the gratification in this world. Nevertheless, it is our duty to submit to the will of the Divine Disposer, since it has pleased Him to take her from us. It will be consoling to us to think that the loss will be temporary, and we know that "if earth has lost a mortal, Heaven gains an angel." Considering myself as one of your family, I can fully sympathize with you in this sad bereavement. It makes me very sad to think of poor little Emma. She is too young to lose her mother, and it will be a hard trial for the poor little thing to be without a companion. But it will be the widowed Dr. who will feel the deprivation more deeply.

If you are sufficiently composed, I would very much like to have you write all the particulars of her last moments and funeral. I suppose that she was buried by the side of your sister Harriette at the Laurel Hill Cemetery. How unfortunate it is that your letter did not come to me till Monday evening, for if the intelligence had not come too late, I should have gone to attend the funeral and take a last look at my dear departed friend. Will you do me the great favour to send me a lock of her hair, if you can spare it? It will be a precious memento of the dear friend I have lost.

I have recently communicated this sad intelligence to my father, and shall write to my mother also. I know they will partake in our sorrows, for they loved your mother as a sister. Hoping that we shall all have the pleasure of meeting again I must bring this letter to a close, and believe me ever yours affectionately

Th. Jefferson Trist.

R. J. Dunglison

Edgehill March 10, 1853

It was sad news indeed, my dear Willy, that your letter brought us yesterday. I had not felt as if we should lose our dear friend, she had so often been very ill before and recovered, and now what can I say to you? I can do nothing but say that you have all our sympathy for your heavy loss, for I know no loss greater than that of a mother, and such a mother as yours was too, and such a friend; we shall never have another so affectionate and kind. We had none out of our own family that we loved so much; she was indeed like a sister, and my heart melts with pity to think of you all. To you young people it is the beginning of the sorrows with which life is chequered. Why God sends them we do not know. We only know that we have to bear them, and that time takes the sting from grief; but we never forget the loved ones we have lost; they are always green and fresh in our hearts even when we have become again contented and happy. To your father the loss can never be repaired, but we old people feel that our friends have only gone before us, and that we too shall follow after some years, it may be fewer or it may be more, but death is nearer to us than to you young people, and we have more patience to bear. Our poor little Emma, we think of her much, and Mary joins me in sending you our sincerest and deepest sympathy, to each one individually for we think of you all, and believe us ever your most truly affectionate friends.

C. J. Randolph.

The following letter was written by Dr. John Staige Davis,[14] of the University of Virginia, teacher of Anatomy there. He is the son of Professor Davis my former colleague at the University of Virginia (see p. 42); and, after his graduation in medicine, spent sometime with us in Philadelphia. About two years before the death of my wife, he and Mrs. Davis—a most amiable lady—passed a few weeks with us in Philadelphia.

University of Virginia
March 9th 1853.
My dear Doctor,

The irregularity of the mail detained Will's letter on the road until this afternoon, and in the selfishness of my grief I could wish that it had never reached me, for it brings the afflicting intelligence that she who has long stood foremost in my admiration and warm affection, has gone forever. I shall not presume to suggest any topics of consolation, for the bereavement is bitter and the loss irreparable, but I trust that the memory of her surpassing loveliness and gentleness may sooth the intensity of a distress in which I have hastened to assure you of my earnest and lasting participation.

As I mentioned in my last letter, we had been looking forward with anxious pleasure to dear Mrs. Dunglison's spending the months of May and June with us, but it has pleased God to take her beyond the reach of the attentions and tenderness we had hoped to lavish upon her. If it would be grateful to you, we certainly will derive a mournful satisfaction from having you come at once to our house, where you shall be as secluded as you desire, and may be gratified by the tribute of our sincere and heartfelt sympathy. We are also solicitous that Emma should pay us a long visit. Can you not bring her, or in case your repugnance to leaving home is insuperable will you permit me to go on for her? I have always been able, my dear Doctor to subscribe myself your faithful and affectionate friend, but never before as faithfully or as affectionately as now.

J. S. Davis
Dr. Dunglison.

Although convinced of the accuracy of the diagnosis, I was exceedingly anxious, that a necroscopic examination should be made. This was accordingly done by Dr. Ellerslie Wallace, the Demonstrator of Anatomy in Jefferson Medical College, who kindly furnished me with the following Report of the appearances presented. These indicated, that the fatal malady was the Endo-pericarditis—the signs of the preceding and accompanying bronchitis, and also of the pneumonia, having mainly passed away. The evidences of previous attacks of endo-pericarditis were manifest; and it was a subject of astonishment to my medical friends who were present, how she could have lived so long in the

apparent possession of health, and so full of cheerfulness, with such lesions of both pericardium and endocardium. It is consolatory, too, to reflect, that if the amelioration, which took place in the disease had continued, nothing but a protracted agony could have been expected. Although the reports of the *post mortem* appearances in the case of my daughter and my wife were drawn up by two different individuals, the almost identity of language in describing the morbid phenomena is striking (see p. 158). They both died of the same affection, the former, however, of a first attack,—the latter of several.

Post-Mortem Examination. March 6, 1853.

Chest. All the cartilages of the Ribs were completely ossified.

Right Side. Pleuritic adhesions generally of recent character, uniting the lobes of the lung as well as the pulmonary and costal pleurae. The union was by narrow bands in great quantity—no broad adhesions. The anterior portion of the lung was splenified;—dark purple in colour and infiltrated with colored serum. The posterior part was nearly natural, slightly infiltrated. A free mucous effusion in the Bronchiae.— In the posterior part of the pleural cavity, there were about ten ounces of dark yellow serum.

Left Side. Pleuritic adhesions very firm and extensive; strongly uniting, by broad bands, the lobes of the lung and the two pleurae. The upper and back part of the pleural cavity was effaced. About four ounces of reddish serum in the cavity. Lung in nearly the same condition as that of the other side, rather less infiltration in the posterior part. No tubercles in either lung.

Heart. About ¼ larger than natural. The pericardial cavity was entirely effaced by old and recent deposits of lymph, so firm in parts that the opposing layers of the pericardium could not be separated. The *Right Auricle* was rather more than proportionately hypertrophied; was filled by a firm clot of "chicken fat" appearance. *Tricuspid valves* thickened by deposit between the layers of the endocardium forming the valves; the surfaces of the valves were smooth and free; slight cartilaginous deposit at the base of the valves. *Pulmonary valve* natural, a continuation of the clot in the Right auricle passed into the ventricle, wrapping around the posterior leaflet of the tricuspid valve, and thence passed into the pulmonary artery. The size of the clot in the Ventricle was very limited, small also in the pulmonary artery. The Mitral valve was much thickened by deposit between its layers, and the passage from the left auricle to the left ventricle was much diminished in size. *Chordae Tendineae* of mitral valve three times thicker than natural, and of nearly cartilaginous hardness. *Columnae Carneae* were in some instances tipped with cartilaginous deposit.

[14] John Staige Davis (1824–1885), M.D. 1841, University of Virginia, moved into one of the pavilions of the University of Virginia in 1830 when his father took the chair of law and thus grew up in association with the Dunglison children.

Aortic valves were thickened by internal deposit and also by commencing incrustation on their free margins. They did not quite close the aperture. There was a thick cartilaginous deposit passing from the mitral to the aortic valves at the base of the valves, so hard as nearly to resemble bone. There was a peculiar yellowish hue pervading the interior of the heart and the great vessels, very strongly marked in the *aorta*. The internal coat could be readily separated from the middle coat of that vessel. The middle coat was friable to a slight degree.

Stomach. Mucous membrane thickened and easily separated from the cellular coat. Of unusually red hue. Pyloric valve slightly thickened.

HONORARY DEGREE LL.D.

In consequence of the calamity, that had befallen me, I was unable to attend the public exercises at the commencement of the College held in March. I was gratified, however, to learn from Dr. Huston—the Dean—that the Board of Trustees, at their meeting on the preceding day, had determined to confer upon me the title of LL.D. Jefferson College at Canonsburg had already conferred the same degree of LL.D. on me. It had appeared to me to be doubtful, whether it was ever intended by the Legislature, that the Jefferson *Medical* College, when separated in 1838, I think, from the parent institution at Canonsburg, should have any other powers than those of a *Medical* college; but the Board of Trustees thought otherwise, and I could not but feel gratified at this mark of their kindness and confidence in me. Not having heard from Mr. Vogdes officially; after faiting a time I addressed him the following note.

Philad. Mar. 14, 1853

My dear Sir,

Dr. Huston has informed me of the kind intentions of the Board of Trustees, as expressed to him on the day of *Commencement*, in regard to myself; and I beg of you to convey to them my high sense of their kindness, and the great satisfaction I feel, that my course of conduct should have been such as to induce them to exhibit towards me any mark of their approbation.

> Be good enough to believe me
> with great respect and regard.
> Your friend and servant,
> Robley Dunglison.

At a subsequent period, and before the Meeting of the Board of Trustees, Mr. Vogdes—the Secretary—sent me an official notice of the action of the Board, when I begged him to substitute the following reply for the other.

Philadelphia, 18 Girard St.
Nov. 21, 1853.

My dear Sir,

I have the honour to acknowledge the receipt of your communication of the 8th of March last (then only just communicated to me) informing me, that the Board of Trustees of Jefferson Medical College of Philadelphia, had

unanimously *resolved,* that the degree of *Doctor of Laws* should be conferred upon me. May I beg of you to present to the Board my most grateful acknowledgments for this valued evidence of their goodness to me, and to believe me, with great respect,

> Yours truly,
> Robley Dunglison.

J. R. Vogdes Esq.
Secretary.

VOYAGE TO EUROPE

Before the death of my wife, it had been a cherished hope, that we should pay a visit to our friends on the other side of the Atlantic, but it was decreed otherwise. Her death, indeed, almost destroyed the expectation of pleasure from such a trip. My mother was, however, living, at the advanced age of 76, and I had not seen her, and some of my near relatives, for thirty six years. Her mind—as I was informed—had become greatly impaired—had been, indeed, so for years and although I felt, that the first meeting would be painful, I was desirous of seeing her, even in her present changed condition, once more before she died.

Soon after the termination of the Session of 1853-4 I prepared for my departure: took my passage in the Steamer *Pacific,* Captain Nye; and on the 31st of March 1854, left Philadelphia, accompanied by my son William, for New York, where we sojourned at the St. Nicholas. On the following day, we proceeded to the ship, along with my friend Mr. Trist, who came to see me off, and on the wharf we met with the Revd. Dr. Bethune, who had come down to the ship for the same purpose. So thick a fog, however, enveloped everything, that it was deemed imprudent to venture out of the dock. Mr. Collins, accordingly, announced, that the vessel would not sail until 9 o'clock the next morning (Sunday). My excellent friend Dr. Bethune insisted that Mr. Trist and my son should dine with me at his house in Brooklyn; which we did, and spent there a most agreeable and happy evening—Mrs. Bethune being, luckily, able to be present at table.

At a Stated Meeting of the Board of Managers of the Penna. Institution for the Instruction of the Blind held on Thursday aft. March 2nd 1854—

Judge Kane stated that the highly Esteemed Chairman of the Committee of Instruction [15] Dr. Dunglison, was about to visit Europe and offered the following resolutions which being considered were unanimously adopted.

Resolved That Dr. Dunglison the Chairman of the Committee of Instruction be requested during his absence from this country to Examine the different Institutions having Cognate objects with our own in such parts of Europe as he may visit and to report upon the same on his return recommending such changes if any as he may think it desirable to introduce into the organization and action of this Institution.

---

[15] Dr. Dunglison succeeded Dr. Patterson to this office in 1853.

Resolved that the President and Secretary do communicate this Resolution to Dr. Dunglison.

From the Minutes.
Theo Smyth
Secretary

Phila.   Arch Street, March 4th, 1954.
My dear Doctor,

As President of the Institution, I have the honor and the pleasure of Communicating the foregoing resolutions; and of sending with you in your voyage, about to take place, across the Atlantic, my cordial good wishes for your health and happiness, while abroad; every agreeable enjoyment of parental and filial love in your meeting with your aged Mother; and a safe return to your useful labors in your adopted Country.—

Saml. Breck
President.

At the meeting of the American Philosophical Society for March 17th 1854 (see Proceedings vol. VI. p. 20;) a resolution was passed "that Dr. Dunglison be requested to communicate on behalf of the Society with such of its foreign correspondents as he may have occasion to visit; and that the Secretary be instructed to write for him such aid as may conduce to the attainment of the object he has in view,—viz "the making himself familiar with the Scientific and Literary Institutions of Great Britain and the Nations of the Continent." He was likewise furnished with a letter from Governor Marcy,[16] Secretary of State, requesting diplomatic agents to extend to him every facility, and a letter from Professor Henry of the Smithsonian Institution, at Washington recommending him to the attention of all those bodies and persons with whom the Institution was in correspondence.

The following letters to my children form a kind of diary of my proceedings during my travels.

Liverpool
Waterloo Hotel
April 14th 1854.
My dear Willie,

I have just concluded my dinner, and as my luggage and paper have come up to the Hotel, I at once sit down to apprise you of my safe arrival after an agreeable voyage, somewhat agitated owing to head winds, but without any actual storm. I stood, my dear Willie, on the poop, waving my handkerchief towards you, but could not descry you and Mr. Trist. I wished you, however, to go away with the impression that I did so. That first day I enjoyed my meals, but on the day after, the wind was dead ahead, and the pitching of the ship soon disordered me greatly, so that for two or three days I did not venture down stairs to dinner or indeed to any meal. Gradually however, I became well, but immediately afterwards I was attacked with a violent fit of gout in the knee. On the Sunday I was confined to the State Room, and, for a day or two, was an object of commiseration to my fellow passengers. The joint rapidly swelled, however, so that the pain soon sub-

sided. I took my colchicum freely [17] and for the last few days, was able to take exercise on deck. Today I have been somewhat tired, for although we reached Liverpool at 8 A.M. we were not able to enter the dock until past 12; and I did not get away from the Custom house until past four. Not one of my articles was touched, tell Richard; and all his affairs went safe. I was called up as Professor Dunglison, and was most civilly received by the recording officer, who peeped in himself. The lid was soon closed down and I was let off. By the by, did you recollect that we left Harper's at the St. Nicholas. I hunted for it in vain in my carpet bag, and it all at once occurred to me that we had left it with the janitor at the Hotel. Our company on board was a very pleasant one. My room-mate turned out to be one of the best of Fellowes. He is a highly respectable man, the agent of the great Pulteney estate in New York embracing originally 1000000 acres. Geneva is situated in it. He was on his way to England on business connected with the estate and was all that I could desire as a room mate; exceedingly kind and attentive to me and anxious to anticipate all my wants,—another example of the fallacy of judging from first and imperfect impressions. Lord Mountcashel was most amiable, but not very profound. I had many communications with him on subjects connected with the United States, on which—so far as his ability had permitted he has gathered much interesting matter. The O'Sullivans were agreeable also. I did not touch on filibustering, and on other subjects he was rational and well-informed. She begged to be introduced to me, telling me that she had heard so much and so often of me that she wished to know me. His mother was on board. She is a sensible, well-mannered old lady. I shall not at present specify others, for all made themselves agreeable to me, and I endeavoured to return the compliment. The captain is an admirable seaman, and I like him altogether. He was very kind to me and told me the day after we set sail that somebody had spoken to him of me, but he could not recollect who it was. He strongly desired me to come to the Waterloo; I was going to the Adelphi, and I am not sorry that I followed the Captain's wishes. We had 164 passengers on board, and it is not possible to imagine more unanimity. Dr. Magoon officiated on the Sabbath, but I was too unwell to hear him; and on Wednesday night last, we had an amateur concert, very creditable to those who took part in it. On Friday the 7th inst. I was called out of bed to see an iceberg, and in the course of the day we saw many,—none at a less distance than ten or fifteen miles. On her last voyage home the Pacific had her copper rubbed off by the ice, and her cutwater injured, conditions which retarded us and made our voyage nearly twelve days. Everything during the voyage, with the exception of the inseparable concomitant sea sickness—and in my individual case, the gout, was pleasant, and although I sadly missed my own dear but now lamentable restricted circle, I experienced as much comfort as could be experienced under the circumstances.

I shall write to Aunt Mary and tell her that I shall remain here a day or two to recruit, and that I shall be at Keswick about the beginning of the week. This being Good Friday, I cannot learn whether there is any letter for me at the Post Office, and I cannot wait until tomorrow. The Cunarder too leaves to-morrow, by which this letter has to be conveyed to you. I am really fatigued

_____
[16] William Learned Marcy (1786–1857), Governor of New York 1833–1838, was in the cabinet of President Pierce (1853–1857).

[17] Dunglison often alluded to the personal use of colchicum in his medical writings (Practice of Medicine, 3rd ed., 2: p. 636, 1848; General Therapeutics and Materia Medica, 6th ed., 2: p. 205–210, 1857). It has been a specific for gout ever since it was first recommended by Alexander of Tralles in the sixth century.

with the trouble I have had in getting the luggage through the Custom House. My name was the 26th on a list of 164, and then the extortion and trouble to which you are exposed with porters and cabmen! The dinner I have just eaten was an excellent one. It consisted of mock turtle, Salmon, and steak with a couple of glasses of Sherry; and everything admirable of its kind. Lord Mouncashel, Mr. O'Sullivan, Captain Nye and many of our passengers are here. I am in a sitting-room fitted up as a bedroom; for the house is full, and am now writing in a bright light at seven o'clock P.M. But I must conclude, for the letter must be in the office before long.

May God bless and preserve you, dear Willie, dear Richard, dear Tom, and last but not least in my affections, dear Emma. You do not know how often I think of you all, and wish to have you around me. I shall indeed ever wish so, until the happy time comes when we shall be reunited. Tell Miss Dinnin everything was admirably arranged by her. I had no difficulty. I found, however, that I was losing our friend Kohler's buttons one after another, and shall need a restoration at Keswick. Only one or two of the shirts were as limber as the one you saw. I shall write to one of you, soon again, dear Willie, until which time and forever, believe me, with gushings of love to my dear children, my kind remembrances to Miss Dinnin and the servants,

Your affectionate father,
Robley Dunglison

P.S. I need scarcely ask you to present my remembrances to all my friends; who they are it is unnecessary to say to you. To-morrow (Saturday) I must attend to my letter of credit, etc. God bless you all.

The correspondence was, however, so voluminous, that instead of copying it farther, I determined to keep all the letters together for future reference. (In a case with these MS. volumes)

In one of the last of my letters, I refer to my having been in Edinburgh, where I received great kindness from Professors Simpson [18] and Bennett [19] more especially. I arrived there on Saturday the 12th August, at $\frac{1}{2}$ past 10 P.M. and went to the Queen's Hotel, Princess St. On the following morning I walked out towards the Calton Hill and fell in with Professor Flint of Louisville, the Revd. Dr. Meslar, whom I had met at Governor Vroom's in Berlin, and the Revd. Dr. Kipp of Poughkepsie. From Prof. Flint I was glad to learn, that he intended to accompany me back to the United States in the Pacific.

Professor Simpson was at his country seat, sick, and I did not see him until the next morning at 10, when he took me round with him to visit his patients. I lunched with him at 1: saw patients at his house until 3, and then accompanied him to see a patient at the Bridge of Allan, where we dined, and returned to Edinburgh, where we arrived at 10 P.M. After this, I went with him to Leith, and did not get to my hotel until midnight. No one could possibly be more atten-

tive to me. On the following day, I dined with him at his country seat where I met Dr. Flint, Dr. J. H. Bennett, Dr. Sellars,[20] Dr. Douglas Maclagan [21] and others; and, on the following morning, I visited the College of Surgeons, and the Materia Medica Museums of Drs. Christison [22] and Douglas Maclagan. I saw Mr. Syme [23] remove a bony tumour from the hand and was struck with the freedom, with which chloroform was used both by him, and Dr. Simpson.

On Wednesday, I breakfasted with Prof. Bennett, who, afterwards took me to his Museum at the University; and to the Medical Society. My interview with him was highly gratifying. He is a severe, but, I think, a judicious critic; and animadverted strongly on what he deemed unprofessional conduct.

My visit to Edinburgh—as I subsequently stated in my published introductory lecture—was very agreeable to me, and reminded me of scenes so familiar to me, nearly forty years before, when I was a student under eminent professors, all of whom are numbered with the dead.

On the 21st of August, I left Keswick, with my Sisters Mary Dunglison and Sara Ann Atkinson, for Liverpool, to sail in the Pacific, with my old Captain Nye, on the 23rd. Nothing—as it happened—could have been more fortunate than this determination. Prior to my mother's decease, I had always determined to sail from Liverpool on the 21st of September. I had partly engaged to spend a week with Sir John Forbes in London, and then to go to the annual meeting of the "British and Foreign Provincial Association" at Manchester, on the 14th of September. Her death altered all my plans however, and as my sister dreaded the September weather, I determined to abandon my intention to be in Manchester on the 14th of September; and to sail in the Pacific. If this circumstance, however, had not determined me, a message from Dr. Huston would have decided the matter. He begged of me, on account of the infirm state of his health, to return home early in September; as he thought some change would have to be made in the assignment of college duties, which he found himself unable to continue.

How providential was this change. The "Arctic," which sailed on the 21st of September was run into by an iron propeller; and almost all the passengers were

[18] James Young Simpson 1811-1870), Professor of Midwifery, who introduced chloroform as an anesthetic, was knighted in 1866.

[19] John Hughes Bennett (1812-1875), Professor of the Institutes of Medicine, instituted the practical study of histology, which does not seem to have impressed Dunglison.

[20] William Seller succeeded David Craigie as editor of the *Edinburgh Medical and Surgical Journal* in 1853.

[21] Andrew Douglas Maclagan (1812-1900), President of the Royal College of Surgeons of Edinburgh, wit and raconteur, author of *Nugae Canorae Medicae*, taught materia medica (John D. Comrie, *History of Scottish Medicine*, 2nd ed., 2: pp. 703-704, 1932).

[22] Sir Robert Christison (1797-1882), Professor of Materia Medica 1832-1877, was eminent in the field of medical jurisprudence.

[23] James Syme (1799-1870), famous surgeon lauded by Dr. John Brown in *Rab and His Friends*, was father-in-law of Joseph Lister (Comrie, Syme at Minto House in 1853 by Dr. Joseph Bell, *ibid.*, p. 253).

lost—including some acquaintances of my own—Professor Reed and Mr. Jacob G. Morris.

The first portion of our passage for five days was stormy. To employ the language of Captain Nye, "it was a January voyage saving the temperature." I suffered from seasickness for three days, but missed my attack of gout. My sister Mary suffered greatly from seasickness. My sister Sara Ann by no means so much. At the end of eleven days—that is, on Sunday *Nov. the third*, we arrived at New York on one of the hottest days of an unusually hot season, where I met my sons William and Richard; and, on the following day, I was in the bosom of my family; pleased with the long voyage I had made, and the numerous objects of interest I had seen; but delighted to be again with my dear children.

Dr. Huston I found to be in bad health, and very desirous to see me. He was anxious, that I should take the offices of Dean and Treasurer from him. A meeting of the Faculty was accordingly called; and although I had the greatest objection to assume the offices, I scarcely felt myself at liberty to refuse. I, accordingly, was appointed, at my desire, "Dean and Treasurer pro tempore." I had been so long the Dean's Counsel, that the duties would probably be easier to me than to anybody else; and one or two of my colleagues seemed to feel, that the School would be greatly benefited by my exertions in this direction. I made therefore no more difficulty.

Before I left America, I had been requested, by the Managers of the "Pennsylvania Institution for the Blind," to examine similar institutions abroad, and to report to them on my return. This I did early in October; and the report was directed to be printed for present distribution, and to accompany the Annual Report of the Institution.

My Introductory lecture to my class, delivered to them on the 9th of October, being an account of what I saw in Europe; and, therefore, entitled "Recollections of Europe in 1854" was requested of me for publication; and I did not withhold it from them. It contains an honest statement of the impressions made on me by what I saw and both it and the Report on the Blind are favorable to this country as regards the comparative views I took of it, and of continental Europe.

DEATH OF DR. PATTERSON

I arrived in Philadelphia the day before that on which my long valued friend Dr. Patterson died. He had, for some time, been dead to his family and friends. A premature subsidence of mental and corporeal power had occurred; so that, at the last, his decease was rather to be courted than dreaded. My excellent friend, Judge Kane, having been appointed to prepare an obituary notice of him, as a Preface to a volume of the Transactions of the American Philosophical Society, and having asked me to write him my opinion of Dr. Patterson, I gave him the following.

18 Girard St.
Sept. 26, 1854.

My dear Judge,

It is exceedingly gratifying to me, that the American Philosophical Society has requested you to prepare an obituary notice of our deceased friend, Dr. Patterson. No one knew him so intimately as you; and none can depict his virtues more eloquently.

It is now twenty six years since I first became acquainted with him, and my acquaintance with you—as you well know —dates from the same period. I cannot easily forget the hot summer's day on which your first visit was paid, together, to the University of Virginia, in consequence of a proposition, that had been made to him to fill the vacant chair of Natural Philosophy, and the favorable impression, that was made on the Professors with whom you were brought in contact. A general desire was felt, that Dr. Patterson should join us, and every effort was exerted, on their part, to effect the object. His acceptance of the chair was most pleasing to me, for a short intercourse only was necessary to impress me with his sterling worth. The excellence of his heart was, indeed, exhibited in a gratifying manner soon after he joined the University, under an afflicting dispensation which befell me, and at various times during my residence at the University, myself and family were indebted to him and his excellent wife for numerous kindnesses, in health and in sickness, as indescribable as they were grateful. For nearly the whole of the twenty six years, he and I were placed in greater or less correspondence with each other, and at all times, and on all occasions, I found him the same undeviating and firm friend. Throughout the whole of that period I do not recollect a solitary occasion in which the surface of our intercourse was, in the slightest degree, ruffled. At once, he obtained the confidence of his colleagues, and of the Board of Visitors, and was, on more than one occasion, appointed to the responsible office of Chairman or Presiding Officer of the Faculty. In that position, which involved the important matter of discipline of the first instance, he was mild and conciliating, when the nature of the offence would admit of his being so; but firm and decided, where sterner action was needed. He was decidedly popular amongst the students, not owing to misplaced leniency, which is so apt to engender good feeling amongst the governed; but to his uniformity of action. No where perhaps, is the remark, that certainty of punishment is of more importance than intensity, or severity more clearly seen than in the government of a college.

As a lecturer, Dr. Patterson was a great favorite with his class. Clear and eloquent without being gaudy or ostentatious, simple as every lecturer on science ought to be, with his various experiments always well arranged before hand, and certain to effect the elucidation he proposed, he never failed to interest his auditors, and to lead them onwards in his lectures on Physics, the department he taught in the University of Virginia from the elementary to the more abstruse, with progressively increasing interst. As a lecturer on science, he was one of the most successful I have ever heard.

Although eminently so as Professor of Physics in Virginia, he was desirous to occupy a chair of Chemistry in some great medical school; and it was the ardent wish of both of us, before I left the University of Virginia for that of Maryland, that we should again be colleagues (see p. 38) but this never happened.* (In 1835, he had accepted the situation of Director of the Mint in Philadelphia, and when in 1841 a vacancy occurred in the Chair of

---

* *This paragraph was omitted.*

Chemistry in Jefferson Medical College, with which I then was—as I still am—connected, it was my earnest desire, that he should be appointed to the same. It was necessary, however, to obtain first, the permission of the Secretary of the Treasury to his holding the Chair, if he should be appointed to it, along with the Directorship of the Mint. A reply which gave the necessary consent, did not come in time, however, and his name, consequently, was not placed before the Board of Trustees.)

Of the social qualities of our friend I need not say anything to you, who knew them so well. Frank, courteous, intelligent, he was the favorite of every circle. As a member of our "Five" (p. 149) we appreciated him as we ought to do. A more cheerful member there was not among us, and he possessed one enviable quality, where the circle is restricted, that of being apparently as much amused with a thrice-told tale as he was at the first narration of it. So liable were we to repetitions of this kind at the University of Virginia, that it should not be permitted for any one to tell the same anecdote more than *ten times*, unless a stranger was present.[24]

There is a melancholy pleasure in these reminiscences. The death of our valued friend makes the first breach in the ranks of the *"Five"*—the members of which have alas! and alas! been widely separated for some years. You and I are the only two, who have been left within a speaking distance as it were. All, however, are actively engaged in labours of honor and usefulness. That they may long continue to be so is the fervent prayer of, my dear Judge.

Your sincere friend,
Robley Dunglison.

I sent, also, the following article to Dr. Copland, of London, requesting him to procure its insertion into the *Gentleman's Magazine.*

ROBERT M. PATTERSON MD OF PHILADELPHIA.

Died, in Philadelphia, on the 5th of September, Dr. Robert M. Patterson, late Director of the Mint of the United States. Dr. Patterson was born in Philadelphia on the 23rd of March, 1787. He was the son of Professor Robert Patterson LL.D, who was, for many years, a Professor of the Faculty of Arts of the University of Pennsylvania, and for nearly twenty years, Director of the Mint, as his accomplished son was afterwards. He received his early education in the preparatory school of the University, and thence passed into the Academic Halls of that Institution, obtaining his first degree in the Arts in the year 1804. On leaving the University, he selected medicine for his profession, and pursued his studies under Dr. Benjamin S. Barton, a distinguished physician and teacher. In the year 1808, he received his degree of Doctor of Medicine. After this, he repaired to Paris, where he passed two years, and afterwards one in London to perfect his medical education, returning home in 1812. In 1813, he was appointed Professor of Natural Philosophy in the Medical department of the University, and, a year afterwards was elected to the chair of Mathematics and Natural Philosophy in the Faculty of Arts; and, soon afterwards, Vice-Provost. In 1838, he was chosen Professor of Natural Philosophy in the University of Virginia, and removed thither. Here he remained for seven years; when he was appointed, by

President Jackson, Director of the Mint of the United States in Philadelphia. This office he filled with so much ability, that, notwithstanding the different party changes which occurred, he was never disturbed and, after sixteen years of faithful service, retired from it, in July 1851, on account of his declining health.

In all these various positions, Dr. Patterson was eminently successful. As a lecturer, he was eloquent and clear, and as a physical experimenter uniformly happy. As a public officer, he was without reproach; and as a member of Society, he was honorable, upright, and esteemed by all. He was a zealous and able cultivator of physical science; and was one of the most active members of the American Philosophical Society, to which he was elected in 1809, when only twenty two years old; and passed, successively through the offices of Secretary, Vice-President, and President. He took a deep and abiding interest in the education of the Blind, and was a Vice-President of the excellent Institution for that purpose in Philadelphia. He was, likewise, one of the Founders of the Musical Fund Society of Philadelphia, and for a long time its President.[25]

I referred before (p. 169) to the loss of the Arctic steamer; and to Professor Henry Reed as being one of the ill fated passengers. In one of my letters, too, I mention having met him and Miss Bronson [26] in London. His death was the cause of deep regret in Philadelphia of which he was undoubtedly one of the most eminent literary characters. He was an ardent admirer of Wadsworth; and having conversed at Ambleside with Dr. John Davy, who was expecting Professor Reed there, I wrote the following letter to Professor Frazer, who had been appointed by the American Philosophical Society to prepare an obituary notice of him.

18 Girard St. Nov. 4, 1854.
Dear Sir,

On my visit to the Lake District of England, in April last, I had the pleasure of a conversation with Dr. John Davy, brother of Sir Humphry, who spoke to me, with much interest, of a visit he anticipated from Professor Reed, in the course of a few weeks; and informed me, that the people of Ambleside, and its vicinity, were much indebted to Professor Reed for a handsome subscription which he had transmitted from this side of the Atlantic, to be appropriated to some testimonial to his favorite poet, Wordsworth.

In May, on leaving the Royal Institution of London, I had the pleasure of meeting Professor Reed, accompanied by Miss Bronson, in Albemarle St. where he was then residing. He spoke to me of the pleasant visit he had paid to Ambleside and of his intention to return in the course of a fortnight on his way to Scotland.

His melancholy fate will, I am satisfied have caused profound regret in the minds of many accomplished persons, with whom, during his journey, he was thrown in contact, and who could not fail to have been strongly impressed with the rare and extensive information he pos-

---

[24] In his obituary of Patterson, his brother-in-law, Judge Kane, said: "Dr. Patterson's associates were among the scientific wits of the City, and also among the wits who were not scientific but merry instead. To Dr. Patterson I owe my admission into the Philosophical Society. At his suggestion, also, I became a member of the Wistar Club."

[25] Robley Dunglison was Vice-president (1850–1853, 1855–1856) and President (1853–1854). One son, William L. Dunglison, was Treasurer (1858–1872), another, Richard J., was President (1870). The Society still gives an annual concert.

[26] Henry Reed (1808–1854), lawyer and teacher of English literature at the University of Pennsylvania, visited Europe during the summer of 1854 with his sister-in-law, Miss Bronson, who also was lost when the Arctic sank.

sessed in the whole circle of literature, and with those striking intellectual and moral qualities, which were so well, and so feelingly depicted by you, at the last meeting of the American Philosophical Society.

I am, dear Sir,
very truly, your obed. Servant
Robley Dunglison

Professor Frazer.

Mr. Jacob G. Morris, who was lost in the same vessel, had been associated with me, for many years, as a Manager of the Penna. Institution for the Blind. He was a most useful man; an active philanthropist, and, I believe, strictly conscientious. On many occasions, our views did not harmonize in regard to the Institution; but at all times, I believe, he gave me the credit of acting in accordance with my convictions as to what were the best interests of the Institution as I most assuredly did him.

There was one other person in the Arctic, who sank with her, for whom my liveliest sympathies were felt. This was a young woman, Florance Hazard—who, after abandonment by her father, and the loss of her mother —was occupied in teaching dancing and had succeeded in supporting herself by her own skilful, and well directed exertions. She had gone to Europe for the purpose of learning any improvements there might be in her art, and was returning with her guardian, when the catastrophe occurred. Amidst all the notices, I observed none of her, and, I therefore, sent to the *Pennsylvania Inquirer* the following paragraph, which appeared in the number of the 17th of October (1854).

MISS FLORENCE HAZARD.

Amongst the passengers lost in the Arctic, we have seen no mention made of Miss Florence Hazard, well known of many of our citizens, who were greatly interested in her success. The writer of this paragraph knew nothing of her, excepting by name, as the teacher of his children in the art of dancing, until she called upon him on a matter of business, about a year ago, when he was struck with the charm of her manners, and not surprized at the warm feelings towards her entertained by her young pupils. She spoke most ardently in regard to a brother left with her, and under her counsel in some measure, owing to the recent death of her mother; and so impressed the writer, that although not in business, he interested himself to obtain employment for her brother.

As one whose children were benefited by her accomplishments, and all of whom deeply deplore her fate, he cannot resist this feeble tribute to her memory.

The alteration in all my plans, to which I have before referred (p. 169) prevented me from again seeing my valued friends—Dr. Copland, and Sir John Forbes, either in London or elsewhere. For a time, however, I indulged a hope, that I might be able to go for a short time, to London, as I had left some matters of interest to me unseen, under the expectation, that I should stay a short time in that city on my return from the Continent. The meeting of the "Provincial Medical and Surgical Association" was to be held at Manchester on the 14th of September, and I hoped, for a time, that I

should be able to meet Sir John Forbes at it, but I fortunately did not go. If I had, I should have been a passenger in the unfortunate Arctic.

On writing to my friends above mentioned, that I should not be able to see them again in London, I received the following replies.

Old Burlington St. July 29

Dear Dr. Dunglison,

I hope you will yet be able to pay me a visit before you leave England. I expect to remain in town for three or four weeks, and, after that, to go to my son's in the country, where I can receive you quite as well as in London.

The meeting of the "Provincial Medical Association" takes place at Manchester on the 13th or 14th of Sept. and of the "British Association" at Liverpool, on the 20th of Sepember and following days. I expect to be at both. I am afraid these meetings may be too late for you, if not, I shall be delighted to see you there.

Yours faithfully,
John Forbes.

Pray come to me at anytime. I need no preparation to receive you as (alas!) I live quite as a Bachelor.

J.F.

The following is the reply of my valued friend, Dr. Copland, who received me with all the affection of a brother, and with whom, in the month of May, I passed ten days of real enjoyment. Nothing could exceed the attention he paid me, and the manifestations of desire that I should pass as long a time as possible with him, under his roof.

5 Old Burlington St.
18 August 1854

My Dear Dunglison,

I am sorry at hearing of the loss you have sustained in the death of your Mother, and of your decision to proceed forthwith, to the States, without again visiting London. I was in hopes of having the pleasure of a visit for a few days on your return from the Continent. My sister, nieces, and friends, whom I have seen, regret that you have thus determined, and that the opportunity of seeing you again, this Autumn, will be denied them. We too may meet again, and we may never; but, at all events, the happiness I have received by your friendly visit to me will live in my remembrance to the last, and the hopes of our meeting on this or the other side of the Atlantic, or even on both sides, at some future time, will not soon be relinquished by me. I wish you, my dear Friend, to make my best regards acceptable to all the members of your family, and I beg of you to cause them, when they visit the "old country" and come to London, to make their way with their luggage to my house, where I, and those who belong to my family, will feel happy in receiving them, and treating them as the children of an old and most esteemed friend ought to be treated.

My niece (she was laboring under consumption) continues much the same as when you saw her, perhaps a little better. My sister requests to be kindly remembered to you, and with sincere wishes for a prosperous voyage, and for health, happiness and prosperity to yourself and to all connected with you, I remain,

My dear Dunglison,
your affectionate friend,
James Copland.

Prof. Dunglison MD.
etc. etc. Keswick

APPOINTED DEAN OF JEFFERSON MEDICAL COLLEGE

I have before (p. 170) remarked, that I had assumed the office of Dean and Treasurer of the Faculty of Jefferson Medical College *pro tempore* with great reluctance. I experienced the same reluctance at the annual meeting in March 1855 to be appointed to the office of Dean and Treasurer for the year 1855–56. I could not, however, well escape from it. Its duties were disagreeable to me, and it appeared proper, that they should not devolve upon the oldest (by appointment) Member of the Faculty. Dr. Huston, however, positively declined it; indeed, the state of his health precluded all expectation, that he could, if appointed, execute the numerous exigencies of the office. No other member of the Faculty was proposed for it, and therefore, when unanimously chosen, I determined not to decline it. My repugnance to it is not, however, in the slightest degree diminished, and my experience of upwards of half a year has shewn me, that much of the success of the College is dependant upon the mode of conduct of the Dean in his official and private capacity. Still, I thought I might be aided—as I have been—by my son Richard,[27] and that the assistance he rendered me might give him a knowledge—a *savoir faire* of business matters, which he might not otherwise readily attain. These considerations determined me to hold on for at least one year more.

When I was sojourning with my brother William in London, I was attacked, whilst shaving, with giddiness to such an extent as to compel me to throw myself on the bed. It soon passed away. I felt, however, that in my contemplated journey on the Continent of Europe I might be attacked with some fatal malady; and I therefore wrote in the front blank pages at the commencement of a small *Daily Pocket Remembrances for 1844*, which I carried in my pocket, a memorandum of which the following is a copy.

1854 May 9. In consequence of a sudden attack of giddiness today, I deem it right to request, that should I die during my absence in Europe, my body may be deposited where I die; and that no useless expenditure be indulged in. I direct this, in order, that my dear family may receive all that belongs to me, as directed in my Will, which is in the possession of my Son Willie. I should desire, however, that, at their discretion, a plain monument should be erected at Laurel Hill, on which should be inscribed my name, the name of her most cherished by me, and whose loss I shall ever deplore, although satisfied, that she must now be amongst the blessed, also the names of my dear deceased children Harriette and Robley John, with spaces left for names of those of my family, who may die hereafter,—but after, I trust, long and virtuous lives. Such

---

[27] Richard James Dunglison (1834–1901), A.B. 1852, A.M. 1855, University of Pennsylvania, M.D. 1856, Jefferson Medical College thus became assistant to the Dean while still a medical student. His brother, Thomas, graduated from Jefferson Medical College in 1859.

are my desires, although I hope the time may be distant when they have to be carried into effect.

Robley Dunglison
Now at 18 Camden Road Villas.

The Memorandum was not necessary with the exception of repeated attacks of my old malady—the gout—some most severe; others less aggravated but still sufficiently annoying, I completed my tour in Europe, returned to this country, and until the present time (September 24, 1855) have no return of the Vertigo, to any extent at least.

End of Volume VIII

SUPPLEMENTARY ANA

AMERICAN PHILOSOPHICAL SOCIETY

In the third volume of these *Ana,* my election as a member of this Society was recorded. This was in the year 1832, and on my removal to Philadelphia, I became a regular attendant at the meetings as its "Proceedings" will show. In 1840, I was selected *Secretary,* and was so until I was made Vice President in 1853.

In the office of Secretary, I first gave more full abstracts of the remarks made by members and urged the publication of the *Proceedings.* A committee was appointed on this subject on the motion of Dr. Bache, who, as senior secretary, served as Chairman. Dr. Wood, in his obituary of Dr. Bache, states, that the publication of the *Proceedings* was owing to Dr. Bache. To a certain extent only this is true, as he was Chairman of the Committee, but that the act was greatly owing to my movements is sufficiently shewn by the fact, that, at the Supper of the American Philosophical held not long afterwards, Judge Kane proposed my health as one of the Secretaries and as the originator of the *Proceedings.* He told me afterward, that more than one person had claimed the credit of it. It was a small matter, and is only mentioned here, owing to the published observations of Dr. Wood.

After having served as Secretary for many years, until 1849, I found, owing to my repeated attacks of gout, and other private reasons, that it was inconvenient to me to attend the meetings. I wrote to Judge Kane, then Senior Secretary, expressing my desire to withdraw from office. His note in reply to me is appended to this. I, accordingly, held on, after having become Vice President, and on the occasion of his death when President and I Senior Vice President, I sent in a communication dated June 8, 1858, informing the Society, that, "on account of the attacks of an oft-recurring and disabling malady, it would not be in my favor, after that year, to undertake the duties, that appertain to any official appointment by the Society."

The causes, that gave occasion to my declining office, existed with greater force subsequently, so that I have rarely been able to attend its meetings. The last time I was there was when I delivered obituary notices on Professor Tucker and the Revd. Dr. Bethune. (October 3d 1862.)

My dear doctor,

No one can gainsay it. You have served the Society so well and so long, that you have all the rights of a veteran; and I believe all the members know it, though none so fully as myself. Yet, I am *fashed* that you should think of retiring when the half pay list is full. In one year more there will be a vacancy, as I happen to know. Now, why not let your name stand a little longer at the head of the list of Secretaries! I feel that your juniors will not quarrel with you about the distribution of duties, if you will continue to allow them to call you their associate. If you are willing, I will sound them on the point; and as I think that you must be willing, I shall delay any other communication to them till I hear from you again.

Excuse me for giving you the trouble of a second note:— it is the only subject on which you can ask of me a favour twice.

Truly yours
J. K. Kane

Rensselaer
17 Dec. 1849

HUGH BLAIR GRIGSBY LL.D

Soon after my return from Europe, in 1854, I made the acquaintance of Hugh B. Grigsby, LLD, of Virginia who was sojourning in Philadelphia. He was a resident of Norfolk, and likewise of (Charlotte) Halifax, Va. and had married later than usual in life. He was a friend of Professor Tucker, and corresponded with Mr. Randall, whilst the latter was preparing his life of Jefferson.[28] Mr. Grigsby was a well educated and well informed man, fond of historical research, especially as regarded men and things of his own country and I was led to form a most favorable opinion of him as a literary man, and a statesman; who, with feelings distinctly Southern, was disposed to render ample justice to everyone. After his return to Virginia, I was surprized and gratified to find that he had given the following too flattering notice of me in a Southern paper. (See Grigsby, Hugh Blair, in "Allibone's Dictionary of Authors")

Mr. Grigsby's notice of Dr. Dunglison:

THE SOUTHERN ARGUS
NORFOLK:
FRIDAY MORNING, December 3, 1858.
Professor Robley Dunglison.

If any one of the States is especially entitled to the honor of giving this distinguished man to the Union, Virginia is the State. There are many, we hope, among our readers, who remember the founding of the University of Virginia under the auspices of Thomas Jefferson, and the hopes and fears which the institution of so grand a scheme of education excited at home and abroad. Not among the least sources of interest throughout the Union was the appointment of the men who were to fill the respective chairs.

This was a subject environed with inumerable difficulties.— At that day we could not have filled the chairs with Southern men; because, although we had many able and learned men among us, they were either in existing institutions, or were indisposed to embark in a new scheme in another State, or were engaged as the pastors of religious associations. To have appointed one of the last mentioned would have been to raise a hue and cry of sectarianism which might have materially hazarded the success of the University.

Nor could the Visitors look to the Northern States. There, too, the most learned men were engaged either in teaching or preaching; and, as is unfortunately too common in New England, had borne a prominent part in the discussions of a great political question which had then recently shaken the Union itself to its centre. A Missouri restrictionist was no fit teacher for Southern youth. To bring men among us whose first lesson in Moral and Political Philosophy, would be to announce the right and duty of the slave to cut the throat of his master, would have been as imprudent then as it would be now, and equally at war with the interests and the self respect of the Southern people. The only alternative was to go abroad, and select most of the professors among the promising young men who had distinguished themselves at the British Universities, and who might be willing to exchange the uncertain and distant prospects of success in an old and staid society for the immediate rewards which await solid abilities and real worth in the new. An accomplished agent was sent over to England by Mr. Jefferson, and was charged with that most delicate office, which he executed with consummate ability. In a short time the professors were embarked on their voyage to Virginia; but the voyage was long and tedious, and Mr. Jefferson at one time feared that the ship and her precious burden had gone to the bottom. But she arrived at last; and after a fatigueing land journey the professors were cordially welcomed at Monticello, and were duly installed in their respective positions.

The professors were undoubtedly among the ablest literary and scientific men of the day.—Still there was a murmur of dissatisfaction loud and deep among certain disappointed persons. It was said that the foreign gentlemen were at best but deists; some went so far as to say that they were atheists, that they would soon put forth an Encyclopaedia Methodique of their own, and get up another French Revolution. Unfortunately for these prophets of evil, the foreign gentlemen were, we believe with but one exception, connected with some one of our orthodox churches; and the only survivor of them all now living in this country is a member of St. Stephen's, Philadelphia.— Finally the sound common sense of the people prevailed; students crowded to Charlottesville; and our noble University has flourished with a freshness and a grandeur unparalleled in the history of American Colleges. It has become the Alma Mater of statesmen, governors, senators, doctors, divines and civilians, is one of the purest gems of the Commonwealth, and is winning the generous love of the whole South. Among those foreign professors who a third of a century ago gave eclat to the infant institution, the only survivor residing in this country is Dr. Robley Dunglison, of Philadelphia. Having filled a professorship for eight or ten years, he accepted a chair in the Jefferson College of Pennsylvania, which has established its medical branch in the city of Philadelphia, and has numbered among his pupils hundreds and thousands of young men from every part of the American continent. When thirty odd years ago he first set his foot on our soil, he did not know probably a single human being from Maine to Florida. Now he cannot pass through the remotest villages of the Union without being recognized by men who have either received

28 See Appendix I.

his instructions or have seen him in his chair.—His career has been crowned with fame and wealth. With a mind eminently philosophical he ever keeps abreast, if not ahead, of his age, in the great departments which it is his peculiar province to teach. Nor does he confine himself to the mere technical range of his office. One of the lectures of his course that is looked for with delight and received with applause is on the philosophy of medicine.

It might be presumed that a professorship in the Jefferson University would be too absorbing to allow much time for authorship; and it is certainly true that from various causes our popular medical professors have not written much more than a digest of their course of lectures. But very different has been the case with Dr. Dunglison. He is probably the most voluminous medical writer of the age; or, if he should be surpassed in this respect, on the score of separate works, it is certain that more copies of his works have been issued from the press and diffused through this country and Europe than those of almost all the medical writers of the Union put together. In the number and popularity of the volumes of his works no single European writer in ancient or in modern times can vie with him. His volumes are large octavos. One of them now before us contains a thousand closely printed pages. Now of these large works more than one hundred thousand volumes have been sold and distributed abroad. The general reader may form some notion of the mass of paper consumed in their composition from the fact that if the volumes were placed on a single shelf, they would reach more than half a mile, or on shelves they would more than fill the largest edifice in this city. This immense popularity was the amazement of the literary men in England. They have never seen or heard any that approached it, in the department of medicine.

The secret of this unwonted success is to be found in the works themselves. The thorough scholarship of the doctor, his untiring assiduity, his wonderful system of arrangement which enables him to conquer details that affright the most industrious writers, his conscientious revision and judicious amendment of each succeeding edition, in which old matter is retrenched and the new substituted in its place, so that the latest edition is the real reflection of the science of the times, impart to his books their peculiar value. The latest edition of his Medical Dictionary, just published by Blanchard & Lea, illustrates very forcibly what we have said. When a former edition was published, the British and Foreign Medico-Chirurgical Review (July 1853, page 205) spoke of it in glowing terms, declared that the labor of its preparation "was something prodigious," and added "that the work, however, has been done and we are happy in the thought that no human being will have again to undertake the same gigantic task." Yet in the edition just published no less than six thousand subjects and terms not to be found in that so highly praised by the Review, have been added. The amendments alone of the new edition of the Dictionary would have consumed one third of the life of an ordinary writer in their preparation, and could only have been accomplished by our author by his consummate skill in arranging his stores, and his thorough mastery of all the topics of the sciences embraced by medicine.

The doctor is between sixty and sixty-five years of age, stout built, and, unless his days are shortened by the sudden attacks of a disease which for a thousand years has delighted to fasten on the thews and sinews of Englishmen, is likely to live many years. A huge pile of forehead trending majestically upward afford a fit rampart for a brain ample enough to hold the facts of every theory true or false which medical wisdom or folly has put forth from Apollo to Hippocrates, and from Galen to the late Dr. Mitchell of

cetaceous fame.[28a] He has a capacious chest, breathes freely, enjoys a flow of spirits rarely allotted to lexicographers and throws a charm over any circle in which he happens to be. For he is so well informed on all matters disconnected from his profession that a stranger meeting him from home might converse with him for hours on general subjects of literature or science, without dreaming that he was in the company of the most popular medical writer in the annals of the profession. Such is the man who aided in imparting its early lustre to our University, and who, called to Virginia by her authorities in his youth, has enriched by his genius the reputation of our common country.

						Hugh B. Grigsby, LLD

HON. EDWIN M. STANTON [29]

In August of the year 1859, I met Mr. Stanton at Cape May, where he was making a temporary sojourn with his wife and family. I was there as an invalid with my son, and, soon after my arrival, was introduced to him by an excellent gentleman, who, in a complimentary manner remarked, that we ought to be known to each other. During the few days I spent at Congress Hall with him, we were often together, and I had ample opportunity for appreciating the character and extent of his intellectual powers. We belonged to different professions, and I soon found, that he had not restricted his attention to the ordinary routine of his own; but was well read in the literature of State and other trials, in which interesting and important medicolegal and other questions were involved, on all of which he had brought his cultivated mind to bear, with signal success. I have rarely, indeed, met with one as generally well informed and as accurate in his conclusions.

It was pleasing, also, to witness the suavity of disposition which he exhibited towards his own family, and the unaffected fondness for his infant, as he playfully bore it along, when, in company with his accomplished wife, we took an evening stroll along the beach.

The cares and anxieties of office, and the perpetual, and often untimely intrusions to which he must have been subjected in his official career, may have led to occasional abruptness of manner towards some with whom he may have been thrown in contact, but it is difficult to conceive, that one so full of the kindlier sensibilities, prior to his entrance in office, could be so changed as to be guilty of the wanton or intentional official rudeness, which has been ascribed to him. His whole career—it appears to me—has been characterized by eminent ability, and by an integrity of purpose, which has rarely been assailed or questioned.

[28a] Samuel Latham *Mitchill* (1764–1831), M.D. 1786, Edinburgh, whose extensive bibliography contains many articles on marine animals (Courtney Robert Hall, *A Scientist in the Early Republic*, pp. 141–150, 1934).

[29] Edwin McMasters Stanton (1814–1869), Attorney General in President Buchanan's cabinet at this time, became Lincoln's Secretary of War during the Civil War.

My dear Sir:

Accept my thanks for a copy of your Introductory lecture.

Mrs. Stanton and myself have read it with much interest, and I hope, on some occasion this winter, if it be permitted by the rules of your institution, to *hear* a lecture delivered so as to enjoy the charm that is ascribed not only to the matter but the manner of your lectures. Mrs. Stanton joins in compliments and great wishes to yourself and son. I expect to visit Philadelphia soon, and one of the pleasures anticipated on that occasion is the happiness of meeting you.

Yours truly,
/s/ Edwin M. Stanton
Washington, 7 Nov. '60

Professor Dunglison

SIR HENRY HOLLAND

In vol. VII page 840 (see p. 153) of the *Ana*, I have referred to the visits made to this country by Dr.—afterwards Sir Henry Holland. These were subsequently repeated more than once. The last he paid me was on the day before the arrival of Dr. Acland [29a] with the Prince of Wales, in October 1860. Dr. A. was the Prince's medical attendant and it was not wholly agreeable—and I think the proceeding was of doubtful propriety—that Sir Henry should have joined the party in America, as it naturally gave rise to the surmise, which had no foundation, that Queen Victoria had sent over her own physician to supersede Dr. Acland.

In his visit to America, four years before, I had called with him on Lea & Blanchard, who agreed to reprint the then recent edition of his *Notes & Reflexions* which I saw through the press for him. His correspondence with me on the subject is prefixed to the American edition in my Library. On the return of Dr. Kane from his last Arctic expedition, he went to England for the benefit of his health, and, amongst the physicians whom he consulted, was Sir Henry Holland. By desire of Judge Kane, the father of Dr. Kane, I wrote to ask of Sir Henry his opinion of the Doctor's case, and the letters of both the gentlemen are appended.

I may add, that Dr. Acland, who was Regius Professor of medicine at the University of Oxford impressed me favorably, the short time that I had an opportunity of seeing him. He is, evidently, an amiable and intelligent gentleman, anxious to obtain information on all subjects, and especially on those connected with his own profession.

My dear doctor,

Our news of Elisha by the last steamer make us very anxious about him. They come through Lady Franklin, who had solicited an interview with Sir Henry Holland to make inquiries as to the dangerous symptoms of his complaint. She promises to let me know the result; but

[29a] Sir Henry Wentworth Acland (1815–1900), M.D., F.R.S., on his return from this trip was appointed Honorary Physician to His Royal Highness Albert Edward, who became Edward VII of England.

it has occurred to me that a letter from you to Sir Henry, might not improbably draw from him more definite information. Elisha was to sail for Cuba on the 10th, and it may be right perhaps that one of his brothers or myself should go there to meet and attend on him.

Can you with entire propriety write such a letter to Sir Henry?

Very faithfully yours
J. K. Kane
Fern Rock.
19 Nov. 1856.

Dr. Dunglison

London
December 19th

My dear Sir,

I received your letter of 21st three days ago, and hasten to answer it by the earliest steamer.

First let me again thank you for your kindness touching this Philadelphia Edition of my book. It is earlier through the press than I had expected—you will have judged before this time whether it was worth while to give insertion to the note I sent, in some other place. I shall not be at all disappointed if you have omitted it altogether. Your judgement I should much prefer to my own on this point.

I wish I could satisfy more completely your inquiry respecting that excellent man, Dr. Kane. I am led to hope that by the time you receive this letter there may be tidings of his arrival at Havannah. I shall rejoice if they be tidings of good import.

I saw him few times during his stay in England. Had he not been 5 miles out of London, I should have gladly made my intercourse with him much more frequent.

I was made aware of the severe attack of acute rheumatism he had undergone in earlier life, and was led to infer from his accounts, (though without any exact details) that there had been some affection of the Heart in connexion with, or in sequel to it. The symptoms affecting him while here were as follows—Rheumatic swellings of different joints, and shifting from one place to another. Symptomatic fever, attested by a pulse generally above 100, chills and heats during the day, with profuse perspirations at night.—The action of the Heart indicating a tendency to become affected in its irritable and jarring beat, but without any active evidence of endocarditis, or affection of the pericardium. Dry cough, with great susceptibility of all the membranes of the air passages—Much reduction of flesh and strength, as might be expected.

Under the circumstances thus briefly noted, I considered it impossible for him to make the journey to the S. of Europe, and unfit that he should remain through the early part of the winter in England. I therefore willingly (and I trust, rightly) seconded his own desire of proceeding to Cuba; where I learnt that a comfortable home and friends awaited him. The latter (friends) I considered on every account essential to him in his present state of health and spirits. It was very satisfactory to me to know that Mr. Grinnell would be with him in his voyage to Cuba.

You will judge what part of the information I have given you should be communicated to his Father. Probably indeed what I have related may be superseded by later accounts from the Havannah. I shall be earnestly solicitous to hear good accounts of him then. Little though I saw of him here, it was quite enough to give me a high regard for his eminent qualities; to which indeed the narrative of his voyage gives ample testimony.

Farewell, and believe me, my dear Sir
ever yours faithfully,
H. Holland.

Richmond　7 Oct. 1860
Hon. W. C. Rives
Dr. Acland
My dear Dr. Dunglison,

I feel great pleasure in being permitted to be the instrument of making you acquainted with my friend Dr. Henry W. Acland, Regius Professor of Medicine in the University of Oxford, who accompanies the Prince of Wales in his American tour. Dr. Acland is, I am sure, already well known to you by his reputation, both social and professional. I am persuaded that there are no two gentleman who would be likely to find more pleasure in an acquaintance with each other. I will, therefore, only add how much pride I feel in being the friend of both, in which character I subscribe myself most truly and faithfully yours.

W. C. Rives
Dr. Robley Dunglison
Philadelphia

Richmond Va. Oct. 7 1860
My dear Sir,

I venture to forward this letter by post because I am most anxious not to miss you.

I am half ashamed half vexed to say we have only one day in Philadelphia. There are three things as at present advised that I very specially wish to accomplish in that day. The one some quiet conversation with yourself—2d some time in Dr. Morton's collection—3rd to visit the Asylum for the Insane with Dr. Kirkbright.

The latter I am to do at 8 AM on Wednesday our only day.

I would wait on you on *Tuesday Evening,* after our arrival, and fix a time for seeing you and ask your advice as to the hours and places I should attend to next day, if you would kindly send a note to our Hotel to say when and where I may have the favour of meeting you.

I am thus depending on Mr. Rives's assurance that I may throw myself on your kindness unreservedly.

I am my dear sir
most faithfully and respectfully yours
Hny. W. Acland

PROFESSOR TUCKER

I have elsewhere (see pp. 24 and 29) spoken of my old friend Professor Tucker, who was the first Chairman of the Faculty in the University of Virginia. He removed, subsequently, to Philadelphia, where I saw a good deal of him. His "Life of Jefferson" was an able work, and he endeavoured to be correct and impartial and on that account it was not so popular with the uncompromising admirers of that illustrious man as was that of Mr. Randall. At his death, which occurred in Virginia, in April 1862, in consequence of an accident, which he received at Mobile, on the appointment of the American Philosophical Society, I delivered an obituary notice, which is hereto appended.

AN OBITUARY NOTICE
OF
PROFESSOR GEORGE TUCKER
READ, ACCORDING TO APPOINTMENT,
BEFORE THE
AMERICAN PHILOSOPHICAL SOCIETY,
October 3d, 1862.
BY ROBLEY DUNGLISON, M.D., LL.D.
(From the "Proceedings" of the Society)

Professor George Tucker was born in Bermuda in the year 1775. He came to this country when about twelve years of age, to be educated under the superintendence of his relative, Judge St. George Tucker, who was Professor of Law in the College of William and Mary in Virginia, and was the father of Judge Beverly Tucker, afterwards Professor of Law in the same college, and of Judge Henry St. George Tucker, Professor of Law in the University of Virginia, and author of Commentaries on the Laws of Virginia. Professor Tucker's collegiate education was at the College of William and Mary, after which he studied law, and practised his profession in Richmond, and afterwards at Pittsylvania and in Lynchburg, and for a considerable distance around, with great success. He was elected to the Legislature of Virginia from Pittsylvania, and in 1819, whilst a resident of Lynchburg, was chosen member of Congress to represent the district composed of the counties of Pittsylvania, Halifax, and Campbell. He was in Richmond at the time of the terrible sacrifice of life by the burning of the Theatre in 1811, and from a falling beam, received a severe wound, which resulted in a permanent scar over one eye.

Whilst in Richmond, he contributed to the *British Spy,* edited by Mr. Wirt, and wrote amongst other communications, in the year 1800, on the Conspiracy of the Slaves in Virginia, and in 1811, on the Roanoke Navigation, which were printed. In the State Legislature, and in Congress, he was most distinguished as chairman or member of important committees, in which his services were highly valued, and he was twice re-elected to Congress. In the year 1822, he published *Essays on various subjects of Taste, Morals, and National Policy, by a Citizen of Virginia,* which were so favorably thought of, as was, indeed, his whole course in the Legislature of Virginia, and in Congress, by President Madison, that he urged and obtained his appointment to the Chair of Moral Philosophy and Political Economy in the nascent University of Virginia.

In the year 1819, after the death of a daughter at an early age, who had given promise of varied excellence, he wrote in Lynchburg, *Recollections of Eleanor Rosalie Tucker.* In 1824 appeared *The Valley of Shenandoah,* a novel, intended to illustrate the manners of the Old Dominion, which was republished, the writer has been informed, in London in 1825, and in Germany the year after.

In consequence of the protracted voyage—of fourteen weeks—from England of the vessel in which were the writer of this notice, and two of the professors, the opening of the University of Virginia, which was to have been on the 1st of February, did not take place until April, 1825, when Professor Tucker, the oldest of the professors, and the one most familiar with the habits of the country, was chosen Chairman of the Faculty for the first session.

During his residence at the University, he engaged in many literary labors. In 1827, he published a work of fiction entitled *A Voyage to the Moon,* the evident aim of which was to fulfil for the existing age, what Swift had so successfully accomplished for that which had passed by; to attack, by the weapons of ridicule, those votaries of knowledge, who may have sought to avail themselves of the universal love of novelty amongst mankind to acquire celebrity, or who may have been misled by their own ill-regulated imaginations to obtrude upon the world their crude and imperfect theories and systems, to the manifest retardation of knowledge. It was reviewed by the writer in the *American Quarterly Review* for March, 1828.

In 1837, Professor Tucker published *The Laws of Wages, Profits, and Rent Investigated,* and in the same year, his *Life of Thomas Jefferson,* in two large volumes, which received high commendation in the *Edinburgh Review* from Lord Brougham, as "a very valuable addition to the stock of our political and historical knowledge." In it, Professor Tucker does not always accord with the illustrious subject of his biography. The work, indeed, manifests a laudable desire to do justice, and to decide impartially on contested topics; and hence, perhaps, it failed to give satisfaction to the ardent supporters, as well as to the bitter opponents of Mr. Jefferson.

In December, 1837, he delivered before the Charlottesville Lyceum, "A Public Discourse on the Literature of the United States," which was published in the *Southern Literary Messenger* for February, 1838; and in which he enumerates many of the contributions made in this country to the domains of science and literature, concluding with glowing auguries of their future "progressive brightness."

In 1839 appeared a small volume, entitled *Theory of Money and Banks,* the copyright of which Professor Tucker was unable to dispose of in Philadelphia or New York, and which was published in Boston, and soon passed to a second edition. His *Progress of the United States in Population and Wealth in Fifty Years, as exhibited by the Decennial Census from 1790 to 1840,* was a valuable contribution to statistics and political economy. It was a thorough analysis of the census for the period mentioned, and led its author to important inferences on the subjects of the probabilities of life, the proportion between the sexes, emigration, the diversities between the two races which compose our population, the progress of slavery, and of productive industry, &c. To this he added an appendix in 1855, when eighty years of age, containing an abstract of the census of 1850, in the preface to which he expresses the patriotic hope "that these authentic exhibitions of our growth and improvement, so gratifying to the pride and love of country, will lead our citizens to greater party forbearance, and give them new incentives to cherish that Union to which, under heaven, they owe the blessings they enjoy." Impelled by the same sentiments, he gave "A Public Discourse on the Dangers most Threatening to the United States;" (Washington, 1843.)

Professor Tucker's last production at the University of Virginia, was a *Memoir of the Life and Character of Dr. John P. Emmet,* the accomplished Professor of Chemistry and Materia Medica in the University, who died in 1842.

During the whole of this period of his life, he had been a prolific contributor to the public journals, and to the more imposing periodicals, as the *North American,* the *American Quarterly,* the *Southern,* and the *Democratic Reviews,* and at an earlier period, to the *Portfolio* of Philadelphia; and when his colleague, Professor George Long, left the University of Virginia, to occupy a professorship in the University of London, and became editor of the *London Journal of Education,* and of the *Penny Cyclopaedia,* Professor Tucker was, at his request, the author of various educational articles in the former, and in the latter, of sundry biographical notices, as of Presidents Jefferson and Madison, and of geographical contributions in regard to the United States.

From the first opening of the University of Virginia, it had been thought by many of its most intelligent friends, that it presented a favorable occasion for the establishment of a literary journal. It was presumed that eight or nine professors, who were daily occupied in communicating the fruits of their studies to others, would be qualified to make such a work at once useful and interesting to the public. It was known that the plan of the Institution was principally the work of Mr. Jefferson, and that important innovations had been made in its discipline and course of instruction, whence it was inferred that a lively curiosity would be felt to learn the progress of an experiment, made by one of the most popular and most philosophical statesmen of his age. It was not, however, until the year 1829, after the University had been visited by an endemic disease, from which no locality, however healthy, is exempt, and the feeling of the faculty, that if such a medium of communication had been in existence, they might have been able to allay popular apprehension, and prove from unquestionable evidence the general salubrity of the place, that they determined on the establishment of a weekly periodical, entitled *The Virginia Literary Museum, and Journal of Belles-Lettres, Arts, Sciences, &c.,* the editorial charge of which was assigned to Professor Tucker and the writer. The first number appeared on the 17th of June, 1829; but although its contents were diversified and interesting, it was discontinued at the end of the year, and mainly for causes which have proved fatal to so many undertakings of the kind,—the failure of the contributors to afford the aid they had profusely promised, and hence the editors found, that to furnish the requisite materials from their own resources, demanded more of their time than was consistent with their other duties and engagements. The contributions of Professor Tucker were numerous and varied, but were, generally, popular essays on the subjects that appertained directly or indirectly to the chair he held in the University.

In the year 1845, at the age of seventy, with his mental powers undimmed, he resigned his Chair in the University of Virginia, and decided to spend the remainder of his days in comparative leisure. At all times fond of social intercourse with the enlightened, he had never failed to pass his vacations away from the University, and generally spent a portion of the time at the summer resorts of the refined and intellectual. Philadelphia was his choice for a permanent residence, both on account of its intelligence, and the opportunities afforded by its libraries to the seeker after knowledge. He was chosen a member of this Society in 1837, and was likewise, a member of the Historical Society.

From the time Professor Tucker took up his residence in Philadelphia until his death, with brief intervals of relaxation, he adhered to his student life, and continued his contributions to various literary periodicals, and especially to those which were devoted to the elucidation of great questions of politics and political economy.

His undiminished intellectual activity is signally shown by his having commenced about the year 1850, or when seventy-five years of age, the herculean task of collecting materials for a political history of the United States. To aid him in the execution of his work, as he himself remarks, it had been his good fortune to have a personal knowledge of many, who bore a conspicuous part in the Revolution, and of nearly all those who were the principal actors in the political dramas which succeeded. The history extends to the elevation of General Harrison to the Presidency, in 1841. This seemed to Professor Tucker as far as he could prudently go, at least, without obtaining some testimony from public sentiment of his fairness to his contemporaries. The work was comprised in four volumes, the first of which appeared in 1856, and the last in 1857. The first chapter is devoted to colonial history prior to the Declaration of Independence, and the remainder to the Confederation and the United States.

Nor was this elaborate work the last production of its venerable and indefatigable author. In 1859, he printed, and was his own publisher of *Political Economy for the People,* being in substance a compendium of the lectures on Political Economy, delivered by him in the University of Virginia, with such alterations and additions as his farther experience and reflection had suggested; and lastly, in 1860, when eighty-five years of age, he issued on his own account, *Essays, Moral and Metaphysical,* some of which had been already published anonymously or separately, but were now republished, and added to the series. These essays were respectively, On our Belief of an External World; On

Cause and Effect, read before this Society; On Simplicity in Ornament; On Sympathy; On the Association of Ideas; On Dreams; On Beauty; On Sublimity; On the Ludicrous; On Classical Education; On the Siamese Twins, read before this Society; and On the Love of Fame.

Professor Tucker's protracted and useful existence was now verging to a close. The death—in the summer of 1859 —of his wife, the constant and faithful participator in his joys and his sorrows for upwards of thirty years, gave occasion to a thorough revolution in his domestic arrangements, and in place of wisely determining

"To husband out life's taper at the close,
And keep the flame from wasting by repose,"

he undertook extensive and harassing journeys. In the early portion of the summer of 1860, he visited Baltimore, Washington, Norfolk, and the Eastern Shore of Virginia; and in the middle of June, in company with his son-in-law, Mr. George Rives, of Virginia, travelled as far as Chicago, to look after property which he had there. He did not suffer from the long journey he took on this occasion, and subsequently in Virginia, and returned to Philadelphia in the early part of the winter, with the intention of escaping the severity of the northern winter, from which he had suffered greatly the previous year, by a sojourn in the South. In December, he left Philadelphia, and in company with a friend proceeded from Richmond, in Virginia, to Columbia, in South Carolina; and afterwards to Charleston, Savannah, and other Southern cities. The last letter the writer received from him was dated Savannah, in February, 1861. In it he feelingly and deploringly depicts the condition of Southern sentiment as exhibited there. "The state of public affairs," he remarks, "is indeed gloomy, even to heart-sickening. People seem to be crazed in the fancies of imaginary evils, and of their strange remedies."

Some weeks after the date of this letter, the writer was pained to learn from Mrs. Rives, the eldest daughter of Professor Tucker, that while landing at Mobile from a steamboat from Montgomery, her father had been struck down by a bale of cotton, which was being removed from the vessel; and that the shock to his system was so great, that for two or three days he was insensible, or more or less incoherent. Under a most hospitable roof, he remained at Mobile, until his son-in-law reached the place, when he was removed to Sherwood, in Albermarle County, Virginia, the residence of Mr. Rives, where, surrounded by his estimable relatives, he gradually sank, and died on the 10th of April, at the advanced age of eighty-six.

Few persons have contributed more to the literature of the period than Professor Tucker. He himself estimated the amount of his more fugitive productions,—about one-half of which were anonymous and gratuitous,—at ten thousand pages. His talents were at one period directed greatly towards the composition of works of fiction, and he occasionally wooed the muse. When at the White Sulphur Springs of Virginia, in his extensive journeyings in the summer before his death, he composed measured lines, upwards of one hundred in number, entitled Life's Latest Pleasures, the manuscript of which he gave to the writer, before setting out on his last journey to the South, in which, to use his own language, he casts a look on the future,

"And midst old age's cares and pains,
Asks what enjoyment yet remains."

His forte was not, however, the imaginative. It is as a successful and equitable writer on great questions of politics and political economy, and of intellectual philosophy, that he will take his place. His Biography of Jefferson, and his History of the United States may, indeed, be regarded less as narratives of occurrences than views of great na-

tional and political questions, as they from time to time arose, logically discussed, and conveyed in language which has usually the merit of great terseness and perspicuity.

During his residence in Philadelphia, Professor Tucker was a frequent attendant on the meetings of this Society, and at the time of his death was a member of the Board of Officers and Council.

REVD. DR. BETHUNE

I have elsewhere, in these *Ana* (see p. 158) spoken of my excellent friend; who died in Florence, Italy, in 1862. The American Philosophical Society appointed me to deliver an obituary memoir of him, which I did. Appended to this sheet are some of the notices taken of it by friends, who heard, or received a copy of, it. It contains much of what I had to say of him, but it is impossible for me to depict, adequately the loss I sustained by his decease. He bequeathed to me his Greek and Latin works relating to medicine. Mrs. Bethune has remained in Europe since his decease; and is now in Vevey.[30] I often hear from her (Sept. 1865, Jan. 1868). She was very desirous that I should prepare a lengthened memoir of her beloved husband, with his letters and other matters, but I was compelled to decline. It is said to be in the hands of the Revd. Dr. A. Van Nest Jr., who is, I hope, a competent person.

My son William and myself were amongst the mourners at his funeral in New York.

* * *

Jan. 1868 the memoir has appeared as edited by Dr. Van Nest. I regret I did not see a proof of what concerned me as the errors are numerous and some of them not a little provoking.

New York
December 4, 1862.

My Dear Sir

Thanks to your kind remembrance, I have received the pamphlet you sent and it has afforded me great gratification by its clear and graphic narrative of a friend whom I revered. I framed a plan of a little memorial of Dr. Bethune collecting the action of different ecclesiastical and literary bodies and adding the funeral services and my own sermon. This has been delayed by Mr. Bancroft who has not yet furnished his remarks before the New York Historical Society—I had hoped that your obituary might be included and now regret it the more when it is complete and satisfactory. Still I heartily thank you for publication. I have done what was in my power to supply Mrs. Bethune with photographs. As to the biography, I have sent the best advice according to my judgment—first recommending her to assume the nominal direction and then asking different friends to assist her—as one might know more of Dr. Bethunes early life—another would understand better his ministerial position—another could describe his oratory etc. I was led to this by finding that she was disposed to arrange the papers and hoping that it would give her an object to live for, while it probably would present the best portrait of his diversified character. Afterwards at her request I sent the names of several eminent clergymen, friends of Dr. Bethune and added names eminent in litera-

---

[30] Switzerland.

ture who might be good biographers—holding this as a first principle that the person selected should be one having a warm devotion to and thorough appreciation of his subject. Whatever lies in my power I will do for her in the life of sorrow and loneliness which overhangs the future. Always delighted to receive any communication from your pen I am

Your much obliged servant
Abm. R. Van Nest Jr.
No. 149 Madison Avenue

Dr. Dunglison

P. S. Change of residence caused a delay in my reception of your note.

Washington,
Dec. 6, 1862.

Dear Doctor,

I thank you for the copy of the beautiful notices of Prof. Tucker, and Dr. Bethune that you have been kind enough to send me.

One of the bright spots of my autumn was hearing the reading of those papers before the Amer. Philos. Soc.

Yours truly,
A. D. Bache

Dr. Robley Dunglison.

My Dear Sir,

I have to return you my sincere thanks for copies of your obituary notices of Prof. Tucker and Rev. Dr. Bethune. A further examination has only increased the admiration with which I heard them. There is an easy style and earnest spirit, pervading them—which give them a *natural* character, which we know—however—to be the very highest attainment of *art*.

Very truly yours
Geo. Sharswood
Dec. 15, 1862.

Dr. Dunglison

Dear Doctor Dunglison:

Let me hasten to thank you for your courtesy, in sending me a copy of your scholarly paper on the life and character of our friend Dr. Bethune. I have read it with great satisfaction.

It is a perfect pen photograph of that noble hearted and noble minded man.

In purity of style, perspicuity of thought and beautiful analysis of character, your Obituary notice must be accounted a model for such productions.

With sincere regard,
Very truly yours
W. H. Odenheimer
Riverside,
Dec. 19, 1862.

Germantown Dec. 23, 1862.

My Dear Dr. Dunglison

Yes—your inscription is the proper one, and I shall see that it is adopted, and early in the season ask you to inspect the whole. Mr. Ingersoll must be appeased somewhat! (This alludes to the epitaph on the tomb of Mr. William B. Wood.)

Yes—again—I read your pamphlets with particular pleasure. It is quite refreshing to find your pen indulging in such a theme and such reminiscences, just now, when everything and everybody is absorbed in politics and war fevers. Enjoying as I did Dr. Bethune's talents and conversational powers, I regretted his removal from Philadelphia where he was so highly appreciated. We want just such men in

our city and find too few. Alas—"the five," now either *gone* or removed; I had the pleasure of knowing and esteeming them all, but of the gone, I feel most the loss of Dr. Patterson, whose goodness and bonhommie were so delightful and not again to be matched.

I sincerely hope you may find time and inclination to comply with Mrs. Bethune's request and edit the work as you only can do it. Such a production would be creditable to you, and useful to the world. It should be done by a layman not by a dogmatist.

Ever most sincerely yours
Jno Jay Smith

My Dear Sir,

Accept I pray you my thanks for the obituary notice of our beloved and sainted Dr. Bethune which you were so kind as to send me. Your intimacy with him and your affection for him have given you the power to speak of him as no man could who knew him less.

I feel it to be no slight honor that you have chosen to incorporate in your choice and beautiful tribute the extract from my sermon. He was my friend and my father's friend. I had revered him in childhood and admired and loved him in riper years. It gave me peculiar pleasure that my transfer to New York to be his colleague in the pastorate of 28th St Church was at his suggestion and ministered comfort to what proved to be his last moments.

I take the liberty of sending by this same mail with this a copy of the sermon of which I beg your acceptance.

With very great respect
I am my Dear Sir
Truly Yours
Alex. R. Thompson

105 E. 18 St.
New York
9 Jany. 1863.

Dr. R. Dunglison
Philadelphia

New York Jan. 26, 1863.

Dr. Robley Dunglison
My Dear Sir,

Accept my sincere thanks for the copy you have sent me of your genial and eloquent tribute to the memory of our mutual friend the late Dr. Bethune, and also for your very kind references to my own discourse on the same grateful theme. I was not aware until informed of it by Mr. D. S. Jones in a recent letter that you had not received a copy of my sermon—but thought that one had been sent you among the first that were distributed. It is not too late however to repair the omission and by this mail you will I trust receive one. I venture also to send you a copy of another discourse on the late President Frelinghuysen whom possibly you knew as I did only to admire and love him.

I received a cheerful and long letter a few days ago from Mrs. Bethune. It is truly wonderful how she has borne her great sorrow added to all of her previous afflictions. She requests reminiscences of the Doctor's life in Philadelphia and I hope to be able to furnish her with a few at least which could not have been put forth in public discourse, but which will illustrate some of his noble traits. Reflection upon his memory only deepens our sense of the great loss sustained by his lamented departure. But doubtless the triumphant faith in which he lived and died has been crowned with glory unspeakable.

With great personal regard I remain.

Yours very respectfully
William G. R. Taylor

127 S. 7th St.
Jan. 29/63

My dear Doctor,

Tis my fixed habit, sleepy or wakeful, to go to bed as the clock strikes 12 but last night my rule was interfered with by your notice of our common friend Bethune in the transactions of the Philosophic Soc. which I picked up a few minutes before 12 and positively could not lay down till I had finished it. Of your many happy efforts, Doctor, in this way this is the happiest and this result, doubtless, was attained from it having been done con amore.

very *sincerely* and cordially
Jno. Penington.

Dr. Dunglison

Philadelphia
248 So. 8th St.
Sept. 5, 1865.

R. Dunglison M.D. L.L.D.
Respected Sir,

I can scarcely refrain from giving some expression to my gratitude for the noble tribute you have paid to one whom not only his choicest friends held dear, but whom the whole country cherished. His death was a national loss. I first heard Dr. Bethune in my early life, when he was in Utica, and though at that time belonging to another parish and another communion I was attracted to his service and have ever since admired and joyed in his noble service for the good of man and the glory of God. And it is no common gratification to me to find myself in the presence of a people with whom his memory is still fresh and fragrant.

I am still so much a stranger to the church, that I feel a delicacy in making any suggestions, but I should earnestly hope that the love which so yearns to perpetuate the memory of such a husband and such a minister, in the church of his toils and his successes, would find an instant response and complement.

With many thanks,
I remain, Dear Sir
Most truly yours
E. R. Beadle

Castle Hill, Cobham
P.O. 1 July 1867

My dear Sir,

I am greatly mortified when I contemplate the wide gap between the date of your letter of 15th April, which I found here on my recent return from New York, where I was kept a prisoner for three months by a long and dangerous illness, and the date of this my acknowledgment. I had gone to New York about the middle of March, to attend a meeting of the Peabody Trustees, of whom I am one; and my exposure while there and on the journey, brought on an attack of nervous, intermittent fever, to which I was already predisposed, that pressed very sorely on the spring of life in a septuagenarian patient. Mrs. Rives came on to assist in nursing me, and it is only within a week or two, that we have been re-installed in our home.

Your letter was most gratifying to us both, as a souvenir of your kind feelings and of our ancient friendship. I regretted only to learn from it that you, like myself, had become a sufferer from disease during the long period of our suspended intercourse. I trust, however, that as we were contemporaneous sufferers, we shall be contemporaneous convalescents, in which category I am happy to be able to report myself to you at present.

I am sorry to say that neither of the Discourses mentioned by you, which you had the kindness to transmit to me has ever come to my hands. If you should still have any copies left, I should be most happy to possess one of each, especially that on the character and life of our friend Mr. Tucker, which would have the double attraction of the author and the subject for me.

Mrs. Rives joins me in kindest regards to you, and with the hearty reciprocation of your cordial and friendly sentiments, believe me, my dear Sir, most truly and faithfully yours.

W. C. Rives

Dr. Dunglison

Castle Hill, Cobham
6 July 1867.

My dear sir,

I am very much obliged to you for the prompt transmission of additional copies of your obituary discourses on our friend Mr. Tucker and the Revd. Dr. Bethune. I have read them both with very great interest. The notice of Mr. Tucker, though but a sketch, I found very interesting, as recalling the leading events of a long and varied career. He was, as you say, undoubtedly deficient in imaginative conception, while clear and chaste in exposition and unadorned narrative. His writings display the talent of a literary photographist but not that of a painter. His earlier inductions written in the freshness of vigorous manhood, have always appeared to me to be his best.

Your Discourse on Dr. Bethune is a finished composition, executed in the very best taste and presents in an easy but graphic style, the traits of a man of remarkable genius, and of most attractive moral and social qualities. His reputation was not unknown to me; but I had no idea, until I read your Discourse, of his eminent gifts and very varied powers. I must thank you for giving me further conception of the gifts and adornments which confer honor on the American name.

With my thanks, accept renewed assurances of my cordial esteem and regard.

Very truly and faithfully yours
W. C. Rives.

Dr. Robley Dunglison,
Philadelphia

JUDGE KANE

Of my valued friend Judge Kane, I have spoken more than once in these *Ana* (see pp. 35 and 149). Our friendship continued uninterruptedly until his death, which occurred on the 21st of February, 1858. He was, at the time of his death, President of the American Philosophical Society; and, as Senior Vice President, it was my sad duty to call a meeting of the Society to take action thereon; when I made remarks, which are published in the *Proceedings* (see 6: 389) briefly expressive of my feelings toward him as an officer and member of the Society; and as my friend. At the meeting of the Society, Mr. Thomas Dunlap, an old friend of the family, was requested to prepare an obituary notice of Judge Kane to be read before the Society and to be perpetuated among its records.

The resolutions were brought forward by Mr. Fraley, and, when they were first read, my name was in the place of Mr. Dunlap's. I suggested to Mr. Fraley, that Mr. Dunlap would be the proper person. He said he had declined. Mr. Dunlap just then came into the Hall of the Society, and I suggested to Mr. Fraley to ask him if he was still of the opinion. He then stated that he would undertake the office. I was mortified, however, to find, that, year after year no progress was made; and drew the attention of members of the Society to the circumstance. Mr. Dunlap promised, however, to them, and to the family, that he would, certainly, fulfill his engagement. At his death, however, in 1869, matters remained in *statu quo.* Anxious as I still was that the notice of the deceased President, who had served the Society so faithfully, should have appeared in the usual form of an obituary notice, it was difficult for me to introduce one after so long a period had elapsed, since his death and this difficulty impressed some of my friends, so that I did not prepare a notice and, moreover, the Society had not appointed any one in the place of Mr. Dunlap. This apparent—but I do not think intentional—neglect or forgetfulness made me declare to some of the members of the Society, that I could not take my place in its deliberations until the omission was supplied. As I have elsewhere remarked in these *Ana* (see p. 173), the state of my health excluded me from the meetings except on one or two occasions after I resigned as an officer. The accompanying notice of Judge Kane was delivered by me before the Pennsylvania Institution for the Blind.

A BRIEF MEMORIAL
of the late
JUDGE JOHN K. KANE,
Senior Vice-President of the Pennsylvania Institution for the Instruction of the Blind.
PREPARED BY DESIRE OF THE BOARD OF MANAGERS
By
ROBLEY DUNGLISON, M.D.,
A Member of the Board.
MEMORIAL

Their late distinguished vice-president—JOHN K. KANE—from the very first, took profound interest in the success of the Pennsylvania Institution for the Instruction of the Blind; but it was not until the Fifth Annual Meeting of the Board of Managers, held on the 6th of March, 1837, in the hall of the American Philosophical Society, that he was elected into the Board. From the moment of his admission he became one of its most active, energetic, and able members. The first motion, made by him on the 6th of June, was to determine what measures should be adopted to testify the respect of the Board for Mr. Birch, a great benefactor of the Institution, who had recently died; and the result was the tablet erected to his memory in the exhibition-room, and the purchase of the burial-lot at Laurel Hill. On the 3d of July, following, the first Committee of Instruction was appointed, of which he was a member; as he subsequently was of the Executive Committee, the Committee of Admission and Discharge, the Committee of Finance, and, lastly, of the present Committee of Instruction, of which he still formed part at the time of his decease. In the year 1847, he was appointed one of the vice-presidents; and, on the resignation, by Dr. Robert M. Patterson, of the office of vice-president, in 1854, became senior vice-president.

In addition to the various official positions which he occupied in the Institution, he usually formed an important part of committees on special subjects; and, when it became necessary to enlarge the Institution, he was, on the 20th of January, 1842, made chairman of a committee to inquire into the expediency of remodelling the whole; and on the 14th of February, an organization, on a more extensive scale, was recommended by the committee, in conformity with a general plan suggested by the Board, the details of which were left subject to modification by the committee. The Board, at the same time, resolved, that it was expedient to erect a "Birch Retreat," and that the same committee should be charged with its organization. This "Retreat" was designed for those pupils whose terms had expired, and was the prototype of the present "Home." On the 21st of April, 1842, the Committee of Organization reported a system of general arrangement and management, with a series of revised by-laws, such as exist essentially at the present day. The last meeting of the Board at which he appeared, was on the 7th of January, 1858.

A reference to the Minutes of the Board of Managers exhibits the many important proceedings of the Board, in which Judge Kane was personally engaged for the promotion of an institution for which he felt so profound and abiding an interest; and yet none but those with whom he was associated can know how zealously and ably he co-operated with them. There was not, it may be safely affirmed, a single question of magnitude in which he did not participate, and on which his cultivated intellect did not shed light, and facilitate a satisfactory determination. Although absorbed in matters of the greatest moment, first in the active exercise of his profession as an advocate, and afterwards in the elevated and responsible position of Judge of the District Court of the United States, he was generally present at the meetings of the Board; and it may be said with truth, that the presence of no member was hailed with more satisfaction. At all times, and under all circumstances, courteous, gallant, and conciliating, he was as open as he was fearless in the exposition of his views; ever solicitous to avoid giving pain, and never permitting himself to be unduly ruffled at the honestly-entertained and frankly-expressed convictions of others. The Board have, indeed, lost in him an able, energetic, conscientious, and exemplary associate; the blind, one who could not be surpassed in intensity of devotion to their best interests; and it is to commemorate the sad bereavement, which every one connected with the Institution has sustained by his death, that the Board of Managers have directed this brief memorial to be preserved in the archives of the Institution.

At a meeting of the Board of Managers, held on the 4th day of March, 1858, it was resolved, that a copy of the above memorial be appended to the next Annual Report of the Board of Managers, and that it be printed in raised characters for distribution among the pupils of the Institution.

DR. ELISHA K. KANE

This gentleman, who was the son of my valued friend, Judge K. Kane, I knew from an early age. He was, for some time, in the Philadelphia Almshouse where I encouraged him to undertake some investigations on Kyesteine, which formed the subject of his inaugural dissertation, when he graduated in the University of Pennsylvania, and which were published in the *American Journal of the Medical Sciences* for July 1842, and are referred to in *Elder's Life of Kane*, p. 45.[31]

Dr. Elder's book gives an account of Dr. Kane's service in the Navy etc., during which he wrote to me a letter hereto appended; and of his adventures in India, Mexico, etc. until he became engaged in Arctic voyaging. His own account of his voyages has been ably given in the interesting volumes he published. In the first of these, he was senior medical officer, and, in the second, commander. On his return from the first, he used every exertion to get up the second; made numerous addresses in different cities and excited an interest: which was liberally responded to, especially by Mssrs. Grinnel and Peabody, without whose aid the expedition could scarcely have been undertaken. He was anxious to obtain the powerful aid of the Hon. John P. Kennedy—the then admirable Secretary of the Navy, to whom I gave him a letter of introduction of which the following is a copy:

My Dear Mr. Kennedy:

I beg to present to you my friend Dr. E. K. Kane of the United States Navy—the son of my valued friend Judge Kane, and himself, doubtless, known to you in connection with the Grinnel expedition. You will find him full of intelligence, and in all respects, an estimable gentleman. He is now engaged in preparing for publication an illustrated narrative of that arduous expedition; the details of which cannot fail to prove, in his hands, most interesting to the public, whilst they will convey valuable information to his own profession, and to those in particular, who may be, hereafter, engaged in similar undertakings. In introducing him to you I know, that he will become personally acquainted with one, who will feel entire sympathy with him, as an officer connected with the department over which you preside, and as a contributor to the literature of his country, which has already received so many admirable contributions from your pen.

> Believe me, dear Mr. Kennedy
> faithfully your friend,
> Robley Dunglison

As elsewhere remarked (Supplementary Ana on Hon. John P. Kennedy), he obtained the powerful and invaluable aid and support of the Secretary in his second expedition, from which he unfortunately returned in bad health. He was advised to have recourse to change of climate, and proceeded to England, where he did not improve. His friends were, indeed, so anxious about him, that, as elsewhere remarked (Supplementary Ana, Art. Sir Henry Holland) that I wrote, by desire of his father, to Sir Henry Holland whom he had consulted. By his medical friends in

England he was advised to go to Havannah, where he died on the 16th of February 1857.[32] Of the honors paid to his memory, Dr. Elder has given the most ample details in the volume already referred to. Amongst these, he alludes to the meeting of citizens held in Philadelphia, of which I was one of the Vice Presidents. On the same evening, March 27, 1857, I presided—as Vice President of the American Philosophical Society—at a special meeting, which I had called

under the conviction, that the Society would be desirous to pay their tribute of respect to the memory of a distinguished member, whose services to science and humanity are appreciated by the whole civilized world, whose life had been one of adventurous daring and of genuine philanthropy; and whose enlightened efforts had greatly enlarged the boundaries of geographical and general Knowledge.

The Society passed sundry resolutions, amongst them one directing, that an obituary notice should be prepared of him. The Revd. Dr. Boardman was appointed to this office, who declined; and Professor A. D. Bache was substituted. The memorial was, however, never presented. (See *Proceedings* of the Amer. Phil. Soc. VII: 241).

Dr. Kane named a Cape on Smiths Straits, Smith Sound, after me. It lies between Cape Isabella and Cape Sabine. (See Kane's *Arctic Explorations* ii, 385) & *Lippincotts Gazetteer*, art. Cape Dunglison.

> Washington Jany 2, 1853

My dear Dunglison

I have had several favors from you, and, most acceptable of all, the recent resolutions of The American Philosophical Society, the receipt of which I have acknowledged in the letter which accompanies this. I would have written before, but they leave me no time for correspondence with friends. I am really ashamed to look at the immense file of letters unanswered upon my table—but I can't help it, and must wait for day of enfranchisement—the 4th of March—to make my excuses. Meantime, it is very pleasant to me to find that my administration of the Navy Dept. is so well approved. I rejoice to see these public notices of Dr. Kane, for he truly deserves all the honor they bestow upon him. Tell him, when you see him, I shall write to him in a few days. I have a letter from Lady Franklin praising him to the skies. How glorious it would be for him—and for this generation—if this Second Expedition should bring back either the live Sir John, or the certain vestiges which may tell his fate—Lady Franklin has hopes—and so, I hope.

Can't you find a week to come and look at us this winter? I can give you a bed, and a laugh, as often as night comes, —but hardly in day time as I work like a horse.

Although I have not even *acknowledged* your letters— slave that I am—I believe I have done everything that you have asked—which I hope will show you that a friend in deeds is worth at least a friend in word.

A happy New Year to you, my dear Dunglison—and as many as you intend to spend on this vale!

> Yours ever & truly
> John P. Kennedy

---

[31] Kiesteine, announced by Nauche of Paris in 1831 as a new urinary substance which was an indubitable test in cases of suspected utero-gestation, had been studied in 1840 by Mc-Pheeters and Perry at Philadelphia Hospital at the request of Dr. Dunglison, and their results were published in the *Medical Intelligencer* in March 1841. Further studies then made by Kane formed the basis of his inaugural thesis in 1842 (William Elder, *Biography of Elisha Kent Kane*, p. 45, 1858).

[32] He suffered from rheumatic heart disease from the age of sixteen (*ibid.*, p. 31).

Frigate United States
Cape Palmas

My dear Doctor Dunglison,

With the exception of L'Ile au Prince, a beautiful wooded little island near Fernando Po, hardly a green thing has greeted me since leaving home. Sand and Sunshine, Sky and water are my only features and these so completely epitomise *Guinea* and the *Gulf of Guinea* as to spare you details of a description. Look then upon this letter with more charity than we generally extend to these one sided conversations and if my addled brain hatches nothing attribute it to a faulty process of thermal incubation.—Even Dr. Dunglison could not lecture from a bake oven.

I see that General Diathesis as modified by place of nativity and its accompanying influences is beginning to excite attention from its influence upon the character of recuperation after severe injuries and surgical operations. Our surgical successes with the Chinese are certainly remarkable and the cases repeatedly published by the surgeons of British India seem to establish a similar success for the natives of Hindostan. Since leaving you, however, several intelligent medical officers of H. B. M. Squadron on this coast have alluded to the operation of these influences upon the native African claiming for him extreme want of constitutional impressibility as well as rapid recovery after gun shot wounds, & extensive mutilations from shark bite. On this account I am led to send you a couple of cases—the results of my vagrant experience—in the hope that they may have something more than a merely surgical interest.

One is a lascar malay—a rice eating animal—the other a Gola Prince in the family of Governor Roberts at Cape Palmas. The cases are so extreme that I should have hesitated in presenting them, were it not that I have for one the voucher of the Officer of the Indiamen and for the other the association of our mutual friend Dr. Dillard. Should you find them worthy of notice they are at your kind disposal, either for the scissors and Dr. Hay's,[33] or the scissors and Mrs. Dunglisons Allumette box.[34] Unlike Garrick's "nor tragedy nor comedy man" my genius must befit them for one or the other direction.

As regards the Climatorial Fever I am busily collecting cases: for, remembering your advice in the old days of Kyestine and Blockley Hospital, I am about to renew that plodding and statistical method which connects one's conclusions with his facts.—What a delightful thing it must be to be "an authority," a something whose opinions are matters of reference, whose facts are matters of moonshine. It is a hard thing this reduction of one's enthusiasm to arithmetic.

I hope, my dear doctor, now that my Father has settled down into the quiet gentility of a Judgeship—that your kindly intercourse continues unabated. A long existing source of excitement has been suddenly closed upon him and I cannot help feeling that he will now need the stimulus of well selected friends.

The Junta,—that respected combination of Hock and Science, and Theology and oysters,[85]—I hope it still retains its quintuple sociality.

Present my respectful regards to my friend Mrs. Dunglison, and if, by any chance, my brass buttons should have made their proper impression upon that dear little Emma, include her in my kindly messages. This done, my dear Sir, let me, as a sort of "Carthago delenda" episode, assure you that of all miserable blanks in one's existence the most

miserable is an African Cruise, and then secure of your sympathy, remain

Very faithfully
Your friend
E. K. Kane.
U. S. Navy

Professor Dunglison
Tenth Street

Office of the Corr. Sec. of the Kane Monument Association
New York July 12, 1858.

Dear Sir,

I have been directed to inform you that at a meeting of the Kane Monument Association held at Clinton Place Hotel on the Evening of the 7th instant the committee on permanent organization unanimously reported your name as one of the Honorary Members of the Association and you were thus elected.

The object of the Association is to erect in the City of New York, an enduring monument to commemorate the worth and services of Dr. Kane and to honor his memory.

You will please signify to the undersigned at your earliest convenience your acceptance or non acceptance of the post.

Very Resp. yours
Sidney Kopman

Dr. Robley Dunglison
Phila.

Article 5 of the Constitution.

The Legislative assembly shall have power to elect as many Honorary Members of the Association, as to them seem desirable, but no active duties shall be imposed on Honorary members; any efforts therefore on their part will be voluntary under the dictation of their own feelings of admiration for the man, and the interest they may feel in the completion of the work before us.

Article 7

Any person of any nation, Male or Female, may become a member of the association by calling upon such person or persons as may be hereafter designated signing his or her name to the list of members and paying the sum of one dollar.

Among the Honorary members who have accepted Posts I will name a few. Hon Rufus Choate, Edward Everett, W. H. Prescott, Boston, Prof Silliman, K. Rayner, Dr. H. L. Haukes, Allen Pike.

RESOLUTIONS OF THE FACULTY OF JEFFERSON MEDICAL
COLLEGE OF PHILADELPHIA.

April 17, 1868.

To Robley Dunglison M.D.

late Professor of "Institutes of Medicine & Medical Jurisprudence" and "Dean of Faculty"—in Jefferson Medical College.

April 17th 1868.

Dear Sir,

We, the remaining Members of the Faculty of Jefferson Medical College address you, recently their esteemed colleague and honoured head, with a few sad words of leave-taking. We cannot part from you with the customary and merely formal expression of regret. It is altogether unsatisfactory to us, as it would be superfluous, to place upon our minutes, the trite and usual record of respect for a Co-Professor retiring from his position or snatched away

---

[33] As editor of the *American Journal of the Medical Sciences.*
[34] Match box.
[85] See the Club of Five, page 149.

by the fate common to all men. The entire history of the College for a third of a century, a generation of our race, is itself, from beginning to end, little else than just such a record. Its progress, rise, and well deserved reputation and prosperity are well known to be closely connected with and dependent upon your energy, your zeal, your learned labours, your eloquent prelections, your sagacious and popular administration of its affairs.

Throughout all our land and far beyond its limits you have made for yourself a name surrounded with bright and noble recollections; and a host of grateful Pupils have testified widely to your distinguished ability as a Teacher and your estimable qualities as a gentleman, while they have illustrated by their usefulness and success the school over which you have so long presided.

Your resignation—for which your obviously progressive infirmities, borne with manly courage and philosophic patience had reluctantly prepared us—has nevertheless filled us with the sincerest sorrow, not unmingled with some degree of apprehension concerning its influence on our cherished Institution, thus deprived and impaired.

If any consolatory reflection presents itself to us under the contingency we so much lament; if any compensation can be found for the loss we thus suffer, it is in the hope that the leisure which nature compels you to seek, and the rest from toil which has become necessary to you, may restore in some gratifying measure your health and strength worn out in the struggle to endure, & fulfil all duty incumbent on you.

That your valuable life may be prolonged through many years of tranquil ease and that you may enjoy in all comfort an honoured and beloved old age, is the earnest prayer—Dear Dr. Dunglison

Of Your Friends and Associates
Saml. H. Dickson
S. D. Gross
Ellerslie Wallace
B. Howard Rand
J. B. Biddle
Joseph Pancoast

APPENDIX

I. DUNGLISON–RANDALL–GRIGSBY CORRESPONDENCE PERTAINING TO THE ANA

Robley Dunglison began to write his *Ana* in 1852. During the 1850's Henry Stephens Randall was writing the *Life of Thomas Jefferson* which, since its publication in 1858, has been a basic source of reference to all succeeding historians.

In the summer of 1855 Randall, having learned of Dunglison's *Ana*, requested to see them and on August 29, 1855, Dunglison wrote the following letter to Randall in reply:

I am very apprehensive from the tenor of your letter that you place too high an estimate on the information, which I can offer you. My memoranda were not made at the time, but a few years ago, for the information of one who alas! no longer exists. I have no doubt, however, that they would be suggestive to you and that I may have made some remarks worthy of notice. A short time ago, I surprised my friend Professor Tucker by reading over to him some of them, which were full of interest to him. I know of no way in which they can be made available to you, except to your inspecting them under my own eye; by which I mean, that many explanations and amplifications might be made. So perfectly unreserved are the *Ana*, that I could not let them pass into any other hands, than my own; and I have no idea, that they could be worth a journey such as you mention, to see. It may happen, however, that business connected with your publishing arrangements, may bring you in this direction, when I shall have great pleasure in an interview with you, and in farthering the object you have in view:—not only on account of the illustrious object of the biography but of the biographer. The statements you get from me could easily be dovetailed into your own manuscript. As an old writer, I know that such dovetailing is entirely practicable.

I mentioned to my friend, Mr. Trist, who visited me last evening, that I had received a communication from you and from Mr. Randolph and that I would aid you as far as was practicable.

By the way, you have no doubt noticed, in your reading, the work of Von Raumer[1] on this country. He told me, when he was here, that he had come over to witness the result of the "experiment of Jefferson." I saw him last year in Berlin; and in an introductory Lecture to my class —entitled "Recollections of Europe in 1854," I refer to this remark of the Baron.

I regret that I cannot go farther than I have mentioned, and am, dear Sir,

> with great respect,
> Your obed. Servt.
> Robley Dunglison.

Henry S. Randall, Esq.

Randall may have suggested in his reply that he would come from his home in Courtland, N.Y. to Philadelphia to see the *Ana*. On September 7, 1855, Dunglison then wrote to him:

Dear Sir: I still feel that I have nothing so important to communicate to you as to be worthy of a special journey

[1] See page 150.

to Philadelphia. Yet should you be nearer me, there may be points which I may be able to elucidate *in conversation* with you; and I may be able to *read* matters to you, that may be eminently suggestive. I have received an invitation to be present in New York, on the 27th of this month, at a dinner to be given by a publishing association to publishers and authors; but I shall not be able to be there; otherwise, we might, perhaps, have been able to meet at less inconvenience to you. In regard to the last scene in the life of the philosopher and patriot, I have not had a great deal to say; and I am sure what I have need not subject you to much inconvenience in your MS. hereafter. I would advise you, therefore, to proceed; and should you, hereafter, be able to see me, I will certainly give you what aid I properly can.

Every note I have made would ·require—as it would be desirable, that it should have—explanation which can only be given you in person.

The Duke of Saxe Weimar visited Monticello; and he is the subject of one of my notes.[2] His book was translated— I think by Dr. Godman, although not so stated in the title page—and published in Philad. in 1828. A gossipy paragraph is at page 51, on hearsay; and an account of his visit you will find at page 197.

Levasseur—the Secretary of Lafayette—published on his return to Paris. His book was translated, and entitled *Lafayette in America*. Philad. 1829. He refers to Monticello and to Mr. Jefferson. I have made him, also the subject of comment in my *Ana*.[3] I do not at present recollect any special notice of Mr. Jefferson elsewhere.

> I am, dear Sir,
> Very truly yours,
> Robley Dunglison.

This letter, as well as others by Dunglison to Randall between the years 1855 and 1861 which are subsequently quoted in this appendix, are in possession of the Alderman Library of the University of Virginia.[3a]

Randall while writing his biography of Jefferson also came into corresponding contact with Hugh Blair Grigsby of Norfolk, Virginia. Grigsby, by profession a lawyer and journalist, was a bibliophile, historian, and scholar of distinction with a network of friendships of like-minded people and membership in the leading historical societies such as the Historical Society of Pennsylvania and the American Philosophical Society.

Late in 1855 he came to Philadelphia for several months on account of the health of his wife. Having an intimate knowledge of Virginia men and events, he carried on an extensive correspondence with Randall concerning the biography of Jefferson, much of which is in the Huntington Library of California and has been published.[3b] On the seventh of February, 1856, Randall inquired of Grigsby: "Have you picked up no anec-

[2] See page 31.
[3] See page 28.
[3a] Ms. 38–600.
[3b] Frank J. Klingberg and Frank W. Klingberg, The Correspondence Between Henry Stephens Randall and Hugh Blair Grigsby, 1856–1861, *University of California Publications in History* 43, 1952.

dotes this winter in Philadelphia, which would give zest to my volume?" Replying two days later, Grigsby says: "I will probably see this evening at the Wistar Party of Dr. Pancoast, Dr. Dunglison and probably Mr. Trist." [3c] Further on in the same letter he wrote:

Let me suggest to you that the information which you will have it in your power to obtain from Dr. Dunglison will be of deepest interest, and will tend to illustrate not only the latter part of Mr. Jefferson's life, but incidentally the earlier also. You should by all means see his intelligence before you model even your early chapters. I should say that you could not see him too soon; as he is advanced in life, though as hale as a man of forty. But his information oral and documental should be secured at all events promptly and thoroughly.[3d]

Dunglison had just passed his fifty-eighth birthday when this was written. It is interesting to learn that he was thought to be "advanced in life, though as hale as a man of forty." His health did not really break down before another decade had gone by and he lived to the age of seventy-one, his death occurring in 1869.

Dunglison and Randall were also writing to each other at this time. February 9, 1856, the former wrote:

Professor Tucker is now convinced, if he ever doubted it. In the preparation of his life of Mr. Jefferson—as Col. Randolph[4] said—he did not ask many questions. He certainly did not apply to me for any details. Hence, much can be said, which he has omitted; and I am glad to find so many scenes have been opened out to you.

I think the family all pleased with the side view of Mr. Jefferson by Stewart. It is good; but I am not so much impressed with it as with the protrait by Sully, which I mentioned to you, and which—as I remarked—Mrs. Coolidge,[5] when I took her to see it, thought admirable.

I was glad to find, that Mr. Grigsby, of Virginia, had entered into correspondence with you. He is full of zeal and intelligence, and I have been much gratified to make his acquaintance. He seems to have an idea, that you ought to have at once a coup d'œil of the whole of the object (individual) of your biography; and, I fear, places too high an estimate on any information I could give you. I think the course mentioned in your last letter, and which I hinted at once before to you, is the true one. Write a full draught of your biography for which you have ample materials; and leave the details of your picture to be filled up afterwards. You ask if I shall be in Philadelphia in the Spring and Summer. This is very doubtful. I have it in contemplation to visit the South, I mean South of Virginia; perhaps as far as New Orleans, and if I carry my desire into effect, I shall go in Spring; and certainly not be there in Summer. Whether I shall get off is a matter of doubt. In October, I went to press with a new edition of one of my works; in January, I commenced another; and when it is completed, I expect to be called on for a new edition of a third; so that it is not improbable that before next October, I shall have seen 8000 pages through the press. I may not be able, for this reason, if for no other, to travel early. If you could visit Philadelphia before the end of March, the city would offer more attractions to you than afterwards.

Our Wistar parties, at which you would see everybody, once a week, would, I am sure, be agreeable to you. It was at one of them tonight that I had a conversation respecting you with Mr. Grigsby. He tells me you and your family have suffered from the rigours of this winter; but I hope not materially. We have had more protracted severe weather than has ever been known by the "oldest inhabitants"; and the ice and snow still hold possession of us, so that when there is a thaw, owing to the quantity in the streets the temperature cannot rise much above that of melting ice; and the consequence is that it disappears slowly.

I am, dear Sir,
with great respect,
Yours truly,
Robley Dunglison.

P.S. I understand Mr. Rives is to write the life of Mr. Madison(?).[5a]

About a week later (February 15, 1856), Randall wrote a long and plaintive letter to Grigsby. His lust for the Dunglison memoranda, as well as for word of mouth information from Dunglison and others in Philadelphia who had known Jefferson on a personal and intimate basis, had been whipped up to fever pitch, but he was weather bound in up-state New York. In this letter to Mr. Grigsby he wrote.[5b]

Yours and a long letter from Dr. Dunglison reached me by the same post—both giving me most gratifying evidence of the kindly interest which you take in my life of Mr. Jefferson. . . . —The opinion which you expressed to Dr. Dunglison (as he writes me) that I ought to have at once, and at the outset, a coup d'œil of my whole subject, is, in my judgement, undoubtedly the correct one. No painter or sculptor ever wrought a great work, without having a perfect and vivid ideal before him.

There can be no patching or dovetailing in high art.— There can be no growth in the conception after the execution is in part completed. It must come from the brain, as Minerva came from Jupiter's—full grown and armed! I pressed Dr. Dunglison, so far as propriety permitted, to let me take and use his work (his Ana) and I would have been glad to have stipulated to show him, before publication, every passage of his I had used, and every allusion to a passage, and then make him the final arbiter whether they should stand or be stricken out.—But, personally, he was a stranger to me,—and I could not and did not blame him for his disinclination to place his manuscript in my possession.—I had several reasons for not making an immediate journey to Philadelphia.—It was excessively inconvenient for me to leave home before Spring—and excessively disagreeable to do it, during the cold weather, I having been much out of health for a year or two. . . . I am not a very old man (forty-four) but owing to my exhaustion and ill health, I have contracted a morbid aversion to going abroad, unless driven by an absolute necessity.—Then I have found by experience, that breaking off the main thread of my story, to follow out very thoroughly incidental or side points, puts me out, as children say. I confuse dates, and jumble up incidents, forgetting whether I have referred to them before or not. The even, flowing, gradually widening and deepening current is broke.—The spirit—the confident decisive tone, is lost.—and then again, you cannot well tell how to take the facts, traits, allusions and the like, from the writings of another, before you have been over the

3c Ibid., p. 27.
3d Ibid., p. 28.
4 Thomas Jefferson Randolph, who, like his father, was a Colonel.
5 Granddaughter of Jefferson.

5a University of Virginia, Alderman Library, Ms. 38–600.
5b Klingberg and Klingberg, ibid., p. 29.

ground and know how to apply them, unless you copy everything. This last I desired to do, through a Secretary, but Dr. Dunglison has matter in his Ana, I suppose, which he does not choose to submit to a secretary—and indeed, he expects to be consulted in regard to all that is used. I could not, I feared, accomplish much *thus,* without first writing through my work, and then going back and *filling in* what he permitted me to use.—Dr. D's rule is one which he has every right to make;—but had he known me, and could he have relied on my discretion and my honor, with the distinct understanding that I should *show him every passage in any way referring to the contents of his Ana,* it would have been inexpressibly gratifying and *beneficial* to me. Then, unhurried, with attention undivided, with the *whole* subject calmly and clearly in my mind's eye, I could have completed my ideal down to the minutest particular and when its counterpart was given to the world it would have showed that perfect *unity* of expression, without which every effort of the mind or hand, is but secondary. It would be no substitute for *this,* for me to go to Philadelphia tomorrow, and spend two weeks in studying the Dr.'s Ana. I know my own ways, my own habits of composition, when and under what circumstances, the spell comes over me. I cannot mix the charm by rote and by rule, and hampered by regulations, and broken in upon by clubs and dinners. Clubs to me are dullness, and dinners are literal abominations, when I have any engrossing object to accomplish.— No, as you say, I want the *whole* before me (not a part here and a part in Philadelphia). I want time—*my own time.* I want the deep, eloquent solitude of the country— and not the hot, hateful life fever of the city about me. I want my own chair—my own table. I want to hear the winter wind moaning through ancestral trees, or the voices of countless birds pealing down from them in the fragrant and sparkling morning, or when the solemn twilight is deepening around. I want to look upon ancestral faces, and hear their mute lips whispering to me "up—onward—fear not—flag not!"

It is most unfortunate to me that circumstances take just the Shape that they do. Dr. Dunglison writes me very probably, most probably, he may leave Philadelphia by the first of April and that he will be gone so long that I cannot avail myself of his Ana at all, for my work must go to press this Summer. Yet it is almost madness in me to tempt another return of my long, dreary prostration by getting away from my own chimney corner during this terrific winter. The thought of it is *torture.* Tell me not of the benefit of changes of scene.—I have drank out all that is good in city life, to the bottom of the goblet. Until time and solitude bring restoration, I detest it. A new face is a dread to me. I am like the knocked hunter, knee-sprung, gravelled, backstrained and what not! A years run at grass (with shoes off) *may* recover me, for I naturally had the endurance of two common men.

And the worst of all this is that I cannot do half as well what ought to be done, by going to Philadelphia! Oh, that Dr. Dunglison fully knew me! What difference *can* there be, among men of Education and breeding and *experience* as to what it is delicate and proper and *expedient* to use, in a free journal, and what not?—I have no doubt I shall be more fastidious than the Dr. himself in committing *him* in any disagreeable way. I this day have the custody of an ancient and voluminous family correspondence deeply affecting the feelings and honor of the most distinguished family this state ever had, and that custody is continued by the wish of the present head of that family! Two days would bring the Dr.'s papers *perfectly safely* here by express. They could all be sent back, with a citation of every line used or referred to before his southern journey, and the citations be submitted to his decision. Is it not a pity

that the thing stands as it does? I have gone into these details, on account of the great interest you have manifested in my work, and because Dr. D. in addition to speaking of you as a friend, himself mentioned that you and he had discussed the same topic.

Yours cordially
Henry S. Randall

Confidential

My dear Sir,

I commenced going into great details in the enclosed letter to show *you* why I did not go promptly to Philadelphia to see Dr. D., because I desired to vindicate myself to you, who have so kindly enlisted yourself for me, from what *without* explanation, might seem a very strange omission.—I never have any *half* confidences. Before closing my letter (*just* as I was closing it) it occurred to me that if Dr. D. *knew precisely how things stood* he *might change his mind.* To write to him as I have to you, would have been a degree of importunity transcending the limits of good taste. Besides, when he decided the point, we had arrived at none of our present freedom of correspondence.— A thought has occurred to me, and I submit it for your consideration. Dr. D. spoke with *high respect* of you in his last letter to me. *He,* without any remark on my part, voluntarily reopened the subject of my having the use of his "Ana" in advance, by telling me *your* (perfectly correct) views on that subject, and then saying, that he thought the other course would do which "*I* suggested." (I *did* suggest it rather than not see the papers at all.) The Dr.'s tone I *think,* is that of a man *shaken* in his opinion and who is looking round for a backer. I think he *felt* the force of your position. Now I shall very quietly tell him that I agree with you, but I shall enter into no argument, and, of course shall not ask him to revoke what *he* has *settled.* I don't know him, nor what his texture is,—but it does seem to me that if he understood the whole, he would *send me his Ana.* You can judge what effect a full knowledge of the facts would have on him (I could, too, if I had had one look at him, and heard him speak three sentences.) If you make up your mind that he could be *carried*—and that reading to him what I have written *you* on the subject would be likely to have any effect on him, his re-introduction of the topic would be a full excuse for *my writing you,* and for you (as my letter to you *is not a confidential one*) disclosing to him what it contains. But entirely, I pray, you be governed by your own judgement and knowledge of parties in the matter.—The thing looks like a little *stratagetic* at first blush. If I thought it savored a particle, in *spirit,* of trick or even of pious fraud, it might be damned for all of me! The way it presents itself to my mind, at first view, is that of opening a convenient way for a gentleman who has put his foot down in an unfortunate place (for another) taking it up again gracefully, if the spirit moves him so to do. . . . I shall write Dr. D. herewith—. . . .

Cordially yours,
H. S. Randall.

Unshaken, nevertheless, in his resolve not to risk transmitting his Ana out of his own control, Dunglison and his friend Grigsby seem to have been somewhat perturbed by the importunate letters from Randall, and we find Mr. Grigsby promptly replying.[5c]

[5c] *Ibid.,* p. 34.

Philadelphia, February 18, 1856.
At Mrs. Allen's, 309 Chestnut.

Dear Sir:

Your letters of the 15th came safely to hand this morning, and, although I will take an early opportunity of seeing Dr. Dunglison, and conversing with him on the subject of the Ana, I have thought it best to chat with you at once, as I know more *now* about the Ana than I did when I wrote you. I have since seen the duodecimo pocket books which contain them, some four or six, and heard Dr. D. read passages from them. This is the nature of his Ana. He evidently prepared them as memoranda of his personal life for the eye of his wife now dead and of his children; and they contain a mass of matter which no other eye during his life save that of a descendant might be allowed to peruse. So far as I could see and judge, there is no connected account of Mr. Jefferson, but simply here and there an allusion to him intermixed with his own private history. Much that he read me I, who am familiar with all that is printed about Mr. Jefferson and Mr. Madison, knew before; although for the purpose for which the memoranda were compiled the facts stated are of much interest. I remember a conversation of the Dr. with Mr. Jefferson, in which Mr. Jefferson expressed an utter disregard for posthumous fame —a mere flash of the moment evidently, as you remember in one of his more solemn letters to Mr. Madison he alludes to the fact that Mr. Madison was preparing a history of their joint administrations, and says: You have been to me a column of support through life—*take care of me when dead:* And in his last will he bequeathes "his daughter to his country," which bequest could be justified only on the ground that his services would make a favorable lodgment in the bosoms of posterity.

Since I have had a glimpse of the Ana of Dr. D., I do not think it so imperative as I did before that you should consult them at once—but I am free to say that the concluding chapters of Mr. Jefferson's life would be greatly enhanced by a free conversation with Dr. D. as well as by reference to his notes. Should you come on by the middle of March, I think it probable that you will meet with Dr. D., who, let me say, is a gentleman in every aspect of the term, eminently learned, liberal, courteous, and full of that chivalry of literature which leads him to think that the greater wealth he can bestow upon his literary compeers the richer he is himself. He was a confidential friend and physician of both Mr. Jefferson and Mr. Madison, and has a mind capable of comprehending the characters of the men. You and I may not embrace his conclusions, but we must highly value the facts which lead him to draw them. Could you get the Dr. to prepare an account of the last illness of Mr. J., I presume that, besides its authentiticy as coming from the attending physician, himself a man of the greatest reputation, it would fill up the closing scene of Mr. J. with graphic effect, though, of course, you will prepare the account of his last illness with literal intelligence from other quarters.[6]

Gov. Coles, my friend, knew both Mr. J. and Mr. Madison intimately. He told me the other day an amusing story of Mr. J. It should seem that Mr. J., on a return from some distant jaunt had a cutaneous eruption which his physician Dr. D. pronounced to be the itch. Jefferson thought differently, but as his physician believed it to be the itch, he consented to use mercury which salivated him. This annoyed him excessively—and he complained to Gov.

Coles of the unpleasant effects of the salivation. The Governor observed: "Mr. Jefferson, you are the last man in the world, from your hatred of physic, whom I would have expected any excess in taking physic." Mr. J. added with warmth: "Yes, but I would rather have *the devil* than the itch." So it seemed that, though he hated doctors, he hated the devil more.

Dr. Dunglison, by the way, read me a paragraph in which there was an allusion to the itch which Jefferson had caught on one of his jaunts.[7]

My impression is that a few hours in Dr. D's study would enable you to get through his memoranda, and that you can obtain more from him orally than from his written words. . . .[7a] Memo: I consider the testimony of Dr. D. as valuable, not only in itself, but as coming from an eminent foreigner. By the way, he quoted to me the remark of Von Raumer, I think, about coming to this country to see "the experiment of Jefferson"; showing that the fine minds of Europe regard republicanism and Jeffersonianism as convertible terms.[7b]

February 29 Grigsby wrote again to Randall:

I was with Professor Dunglison yesterday and was sorry to learn from him that he had fallen on the icy pavement the day before, spraining his right wrist. He, however, shows a brave spirit, attends to his pressing duties at this season, without seeming inconvenience, and expects to hold his "Wistar" tomorrow evening. . . .

On the subject of the manuscripts of Professor D., I have not pressed matters since I saw you, and I am inclined to think that you will agree with me that there is not that importance to be attached to an early perusal of them as I anticipated before I had a glance at them. They will be valuable indeed; but their value will be available as well, or nearly so, towards the close of your work as at an earlier period. To converse with him and with Gov. Coles[8] will be most desirable for you; you will thus obtain information which neither of those gentlemen would put to paper, but not the less valuable on that account. . . .

Early in the spring Randall finally determined to come to Philadelphia as is evident in his letter to Grigsby, March 3d., 1856, in which he says:

Dr. D. writes me that it is very doubtful whether he gets away from Phila. by the first of April; but I must come before that for the purpose of making your acquaintance. Unless something unforseen prevents, I will start for there about the 17th inst.

Are you "posted up" enough in regard to Phila. to tell me where I can find a quiet hotel convenient to Dr. D. I know about as much of Philadelphia (as is generally the case with New Yorkers) as I do of Pekin. . . .

Grigsby then recommended "The St. Lawrence Hotel" to Randall. "It is new, quiet, elegant, and distant fifty yards from my residence and one hundred and fifty from Dr. Dunglison's. It is on Chestnut between 10th and 11th."[8a] By the end of March, Randall finally

---

[6] In his *Life of Thomas Jefferson,* Randall presented Dunglison's memoranda concerning Jefferson's last illness in several installments (3: pp. 512–519; 547–549), as well as a letter from Dunglison with further recollections (3: pp. 670–671).

[7] See p. 32 and footnote 51, Chapter II.
[7a] Klingberg and Klingberg, *ibid.,* p. 36.
[7b] *Ibid.,* p. 37.
[8] Edward Coles, cousin of Dolly Madison and private secretary to James Madison, married Sarah Logan Roberts, of Philadelphia, moved from Virginia to Illinois in order to manumit his slaves and was Governor of Illinois 1822–1826.
[8a] Klingberg and Klingberg, *ibid.,* p. 42.

arrived in Philadelphia, his letters bearing testimony to the pleasures he was deriving from the convivial company of Grigsby, Coles, Dunglison and others. He stayed through most of April, and on his return to Cortland, N. Y., wrote (May 2nd 1856) to Grigsby: ". . . Oh, my friend, I have *paid*, if not *dearly*, yet *paid* for my flow of health and spirits in Phila.—drawn out by the noble genial men I met there. Dr. Dunglison, Mr. Gilpin, yourself, Judge Woodward, good old Govr. Coles, etc., etc.—" [8b] A month later Grigsby wrote to Randall: "of our excellent friend Dr. Dunglison I have not heard directly since I left the Quaker City. I never think of his goodness and his rich stores of learning without regretting my destinies have removed me so far from him." [8c]

However, Randall and Dunglison were henceforth in intercommunication with each other. In the middle of May, 1856, Randall received the following letter from Dunglison: [8d]

My dear Sir:

I have for some days—indeed before I received your last letter—been determined to write to you, but that proverbial "thief of time," procrastination, has impelled me to postpone my good intentions. Your letter has acted as an additional stimulus—if, indeed, I had needed any—and I now thank you for it. I regretted much to learn, that you had not been as well since your return home; and therefore regret, that the "high pressure" system—as it agreed with you—was not continued longer with us. I do not learn, that your malady was the gout; and, therefore, your generalization may have been not perfectly logical as to the "drinking largely" sobering you again. I have no doubt, however, that such a change of your accustomed associations and habits was as beneficial to you as I have found it to be in my own case; whatever may have been the cause of the indisposition you suffered subsequently. For myself—since I saw you—I have remained unusually well. Although this is my accustomed time for gout, I have not been laid up a day; and this very day, at table, I asked my boys if they had ever known me so little affected as I have been this winter and spring; and they unanimously said, No!

We have had but little change since you left. My family are in *statu quo*, and all are thankful for your kind remembrance of them. Richard and I have been talking of taking a trip, by sea, to New York, for the benefit to us of the sea air; and we shall probably stay there a day or two, starting perhaps the week after next.

In the summer I shall probably take my daughter, and as many of the boys as will accompany me to Niagara; and perhaps, we may cross over into her Brittanic Majesty's dominions. I have seen the Governor (Coles) but rarely since you were here. My friend Mrs. Coolidge has been in town for about three weeks, and I met her at his house, and also saw him and his family to meet her at Mr. Tuckers. I am sorry Mrs. Coolidge was not here when you were. She returns tomorrow to Boston. Mr. Tucker still pursues his historical career; and has nearly—he tells me—completed his first volume. What a task to be yet executed for a man of his age! Mr. Grigsby returned to Virginia soon after you quitted us; and I sent your letter after him.

We have not heard from him—any of us, I believe—since his departure. I spent an evening, a week ago, at Mr. Gilpin's to meet Mr. & Mrs. Astor of New York. I found him a plain unassuming man, with apparently good sense; and no affectation. He evidently takes great interest in the Astor Library, and will not, I think, be niggardly on anything that will be demanded of him to extend his sphere of usefulness. My friend Trist is evidently much occupied with his agency, which, I trust, will bring him profit, if not much honor. What a pity that one who is so capable of better things, cannot find an appropriate sphere of action. His family—it is probable—will not remain in Philadelphia. A position has been offered them in Florida of which they may, perhaps, avail themselves. His occupation will, however, retain him in the North.

I have not forgotten your letter to me. It is one, however, very difficult of answer on most points. They are so well and so closely put by you, that they will require reflection, and some of them I may not be capable of giving any definite reply to. I do not intend to neglect—but to postpone. When do you expect it to be in press? Your work will be most interesting. Take care that it does not deserve the remark on a "big book" in general—a great evil. The children desire their kind regards. Faithfully yours,

Robley Dunglison.

A little later upon returning from the New York trip, Dunglison wrote another long letter to Randall:

Philadelphia
June 1, 1856.

My dear Sir:

I did not receive your letter of the 25th ult. until on my arrival from New York yesterday. Richard and I went by sea on Tuesday last, anticipating an agreeable trip at this season. Nor were we greatly disappointed. The sea, however, was rough and the latter part of the voyage somewhat uncomfortable from the motion of the vessel. I retired early, and so did Richard. I escaped sea sickness; he suffered much—so much, that he will not take the outside voyage again. Nor do I think, that I shall. Before I left, Mr. Tucker told me he had heard from you on the matter on which you ask my recollections. I have not the slightest reason for believing that Mr. Jefferson was, in any respect, guided in his selection of Professors for the University of Virginia, by religious considerations. The question was certainly never asked of me by Mr. Gilmer, who chose some of the Professors in England—myself amongst the rest;—and in all my conversations with Mr. Jefferson, no reference was made to the subject. I was an Episcopalian, so were Mr. Tucker, Mr. Long, Mr. Key, Mr. Bonnycastle and Dr. Emmet. Dr. Blätterman, I think, was a Lutheran; but I do not know so much about his religion as I do about that of the rest. There certainly was not a Unitarian amongst us.

You have Mr. Robinson's Christian name spelled properly.

In regard to the difficult subjects you placed before me in a former letter I will endeavour to make a running commentary. It would not be easy indeed, to do more; and even that I make with much hesitation, so difficult is it to speak of the intellectual and moral powers of anyone *distinctively*.

Mr. Jefferson was not a man who could be regarded as an eminent conversationalist. He was rather reserved; and did not often enter into great questions—political or moral—in my presence. I should say, that his views were generally based on principles, and legitimate theories on great matters. On those of a minor character he might not

8b *Ibid.*, p. 46.
8c *Ibid.*, p. 56.
8d University of Virginia, Alderman Library, *ibid.*

be so free from partialities, on the contrary. He was fond of mathematics; and would not admit of variations which a wide expansive mind might have received, perhaps, more readily. In architectural details this was strikingly evinced. Palladius must be followed by line and rule; and if a deviation would have added comfort and convenience, it must not be adopted; because unsanctioned by high architectural authority. It was said—and said truly—that the outside of the *pavilions*—as the habitations of the professors were called—received primary attention from him, whilst the insides must shift for themselves.

This want of expansion applied, I think more or less, to his views on some other subjects. Undoubtedly, he could not be considered verily as an *adroit* man. He conceived for himself; acted upon his conceptions, and succeeded in inducing others to embrace them; had a powerful influence, therefore, upon his fellowmen; and was justly entitled to the epithet *"great."*

His philanthropy was actual and active. It embraced, I believe the whole globe. His desire was to see all people prosperous and happy—all *peoples,* I may say. He did not like the *government* of England; was careful to separate it from the people. He certainly had no objection to Englishmen, as such;—on the contrary, his kind feelings toward them were exhibited with frankness and sincerity. I never knew him express a hatred towards political opponents of distinction. He would deplore what he considered the malign influence they were exerting. *His correspondence,* and excellent feeling to the last, with the elder Adams, sufficiently exhibited that differences in political sentiment did not preclude a warm appreciation of the man.

You ask me, what were his private virtues that appeared conspicuous to all acquaintances? I do not think there were any, that would call for comment. He was kind, courteous, hospitable to all; sincerely attached to the excellent family that were clustered around him; sympathizing with them in their pleasures; deeply distressed in their afflictions. I mentioned to you the scene I witnessed on the approaching death of a grand-daughter Mrs. Bankhead.

I knew nothing of any private vice of any kind, never heard from him a loose or indecorous speech. I would say in your language, that he was always in my observation "peculiarly decorous, modest and 'decent' in all things."

As to his "personal characteristics";—he was of commanding aspect, dignified, and would have been striking to anyone not knowing in whose presence and company he was. He was, as I before remarked, courteous. His expression—as I recollect it—was pleasing, intellectual, contemplative. He was tall and thin;—nothing, as far as I recollect, marked about the head. I do not speak *phrenologically,* the results of my observation having shewn, that no satisfactory inference can be drawn from its *details.* Of the minute expression of eyes, mouth etc. I could speak well from recollections; but, as a whole, I liked his countenance much.

In his general knowledge he appeared to me to be accurate and precise. His examination of any subject that engaged his attention for the time was full. I *never* knew him loose and inaccurate; but I am writing—you know—of him as he was to the best of my recollection; and after a lapse of upwards of thirty years. I ought, therefore, to give my recollections with becoming caution.

As a University officer he was always pleasant to transact business with, was invariably kind and respectful; but had generally formed his own opinions on questions, and did not abandon them easily. The first regulations of the University, which were mainly, I believe, his work, were the results of his reflections, but did not act well; and had to be abandoned—some of them, I know, with great reluctance on his part. He had a great respect for men of

science and letters, and was always glad to do them honour; had a horror of superficial knowledge, as seen in his desire to get the best informed men for Professors, no matter from what country they came; and would have delighted to patronize talent and learning united with worth wherever he found it.

To sum up. I had the most exalted opinion of him. I believe him essentially a philanthropist, anxious for the greatest good to the greatest number, a distinguished patriot, whose love of country was not dimmed by any consideration of self; who was eminently virtuous, with fixed and honorable principles of action not to be trammelled by any unworthy consideration; and whose reputation must shine brighter and brighter, as he is more and more justly judged and estimated.

To conclude, my children desire their kindest regards to you. We often speak of you, as we did today at table with my old friends Messrs. Wood and Garesche. I am glad to hear you are proceeding with your book. Do not be too long in letting it appear. The *"nine years"* of thrall are, in such cases simply an absurdity.

my dear Sir,
faithfully yours,
Robley Dunglison[8e]

H. S. Randall

After Randall returned to his home in Cortland Village, N. Y., the correspondence between him and Dunglison continued at a decelerating rate. The following letters from Dunglison are still in existence at the University of Virginia. Dunglison and his Philadelphia friends were immensely useful to Randall in the final preparation of his book. Their influence is gratefully acknowledged in his preface to the work.

Philadelphia
Mar. 31, 1857.

My dear Sir:

It has often been a remark in our family circle, "I wonder how Mr. Randall proceeds with his interesting labours," and when we shall have an opportunity of perusing the results, and glad was I to observe from an article in the *New York Commercial Advertiser,* whilst I was staying for a few days with my valued friend the Revd. Dr. Bethune of Brooklyn, that, before long, ourselves and the public will be gratified. As an author I know well the amount of time and reflection which has to be bestowed upon such undertakings, and congratulate you on the prospect of your coming to a close. The article to which I have referred, and which I presume, has met your eye, is stated to be from the pen of "one of the most eminent literary men of the country"; and one who is evidently a most ardent admirer of the illustrious subject of your biography; who considers, that Jefferson could do no wrong, and that everything relating to him may and ought to be given to the world. On this he is aware that I do not agree with him; but he does me injustice when he affirms—for I presume the "Dr. D." alludes to me—that "the literary friend" as well as myself recommended, in opposition to your views, that you should make *"a short life"* of it. The only advice I gave you, and I repeat it now with the same conviction I had then, was, that you should not let it be *"too long."* Nothing is better established in my mind than this;—that if a biographer desires to have the character and history of the subject of his work extensively disseminated and appreciated, that work should

[8e] A large part of this letter was published by Randall (*Life of Thomas Jefferson* 3: p. 670, 1858).

not occupy too large a space. I never attempted to give the number of pages; but I felt, and still feel, that the work should not be "*too long.*" When the *Correspondence* of Jefferson was published by his grandson, it was believed, that nothing which he could say, or rather have said, was unfit for publication. I did not coincide with this; and I well recollect my objections to the publication of the letter in which he ascribed to Lafayette "a *canine* appetite for popularity," at the time when Lafayette was still living. The letter was published, however.

After enumerating many interesting topics to be expatiated on in the work, the author of the article in the newspaper facetiously asks "on which of these departments would Dr. D. *bring down* his amputation saw?" I am not in the habit of operating in surgery—even metaphorically; nor have I seen the departments in detail to which he refers; but would answer—perhaps on none of them. They may be all executed in the way that would appear to me most desirable. No elision whatever may be needed; and yet I must express a hope, that they will not render the book "*too long.*" The 'Journal' to which he refers of Mr. Jefferson's last illness and death I never kept.

You are aware that the Ms. to which I gave you free access, consisted of reminiscences penned at an after period.

I do not know, of course, the author of the article; but the tenor of his remarks savour more of zeal than of discretion; and if he had left me out of the question, I perhaps might not have noticed them; but if he is inaccurate as regards me, he maybe so as respects others; whilst his arguments are far from impressing me with the sentiments which he inculcates.

The whole matter, however, rests with you, and I have more confidence in your judgments than I have in his. His remarks could fail, however, to exert a good effect on your forthcoming publication, which I hope on your account, as well as for the sake of the illustrious subject, and the *whole cause*, may not circulate only amongst the wealthy, in which class is not to be found the greatest proportion of good democrats; but that its price—which is regulated by its dimensions—maybe such as to enable it to be diffused—spread broadcast—amongst those whose means would not enable them to obtain a highly expensive work. This, it is true, might be effected by an abridgement published at a subsequent period; but I should greatly prefer, that all should be able to procure the original.

Your friends here are much as when you visited us. We have, at this time, two Ex-Presidents with us—Van Buren and Pierce. Mr. Tucker dined with me yesterday, to meet my friend Dr. Bethune. Mr. Grigsby I have not heard from lately. I have had unusual wear and tear mentally during the winter, for in addition to all my other duties, I have had to see two new editions through the press; and one of them—my formidable *Dictionary.* Fortunately the same labor will not recur for the next six years.

My own family circle often speak of you; and now desire their kindest remembrances to you. Mr. Trist proposes to take up his residence in Illinois, whilst his family go to Virginia. They have been doomed to constant separation; and I do not see any strong prospect of their speedily coming together. It is sad to think that those who are so well fitted to render each other happy should be so constantly divided owing to the *res augustae domi.*

I am, my dear Sir
very truly yours,
Robley Dunglison

H. S. Randall Esq.

Philadelphia
Apl. 6, 1857.
My dear Sir:

I regret greatly that you should have troubled yourself to write to me—or rather to dictate anything—whilst you are on a bed of sickness. I am glad, however, to find, that your indisposition will probably be but temporary. I should not think, that fresh *bile* enter with your indisposition, and can readily understand that it might be dispersed by the little expended in your biography. I doubt not, that some will feel themselves *galled*, but they will deserve it.

The matter on which I wrote to you disturbed me but little; and yet I determined to make that little known to you. I have got rid of it entirely in the communication I sent you; and whilst, I repeat, I regret that it may have subjected you to inconvenience. I have been gratified, to have it answered by so accomplished an amanuensis, to whom I beg of you to present my warm acknowledgments. I hope, my dear Sir, that you may be soon at your labors again; and that before long, the world may be able to hail the appearance of the interesting results. "Hope deferred" —too long deferred—has undoubtedly its disadvantages.

I am, my dear Sir
faithfully yours,
Robley Dunglison

Henry S. Randall

Philadelphia
July 26, 1857.
Dear Mr. Randall:

Your letter of the 21st reached me at a time, when I was suffering greatly from the most atrocious (*dolor atrox*) attack of gout, that I have experienced for many years; and I am, now, writing to you one foot from a lower bed from which I have not been able to move farther for the last week. It is now one month since I have been out of the house and upwards of three weeks that I have been confined to my chamber. I sent you a newspaper on a slight matter of history the day after your letter reached me. Today (Sunday) I am as crippled as ever, and I fear I shall be confined here for many days yet. Luckily, I could not have had a confining attack, when it could have subjected me to less inconvenience. I am delighted to hear, that you are at last in press; and as I have said to you all that I could or ought on the subject of the size of your work, I will only remark, that, I doubt not, what you have added will render it much more valuable to the most intelligent part of the community, and I only wish—for your sake—that they were in larger proportion. I saw from the very first communication I had with you, that you could find it difficult to restrict yourself in a matter on which you felt so much zeal and enthusiasm, and it was on that account that I urged you not to make the work "too large"; although doubtless, it will be beyond the means of many. I hope you have been careful in your bookseller, for I should be sorry to learn, hereafter, that you have had any pecuniary difficulty with him. Look well after this, for times have been hard recently with the craft, and Putnam, whom you mention—has—I see by the newspaper—been compelled to make an assignment. I have had some little experience with booksellers and my preference is decidedly for the old established houses; which, if they should move less quickly, are apt to do it more evenly. It takes some time to establish a connection, which is extensive and known to be trustworthy.

You will have a troublesome business with the proof reading; but I hope you will not shrink from it. It is most important, that the proofs should be carefully read; and all important errors avoided; Do not, therefore, let the printer push you beyond your capabilities. You are kind enough to express a desire, that you had me near you. It would have given me great pleasure to afford you any assistance I am able; but certainly, I am now at too great a distance. You will have to exercise your calm judgment on the questions to which you alluded; and especially where your feelings are—as you say—"perhaps too deeply enlisted." It will be easy for you, however, by very slight modifications to soften down any expression which may appear to you to be of a doubtful character. I shall look forward with the most lively interest to the appearance of the work; and feel satisfied I shall be pleased and instructed by it. Mr. Grigsby has been for a few days in Philadelphia, to read a paper before the "'Historical Society" in opposition to remarks made by Dr. Hawkey in North Carolina. He spoke of you in the kindest terms. Mr. Tucker is at Long Branch. He is preparing the Index for his "History," what I hope may be extensively read; and prove profitable to him; but I fear it will not. The children and myself are thankful for your kind interest. Willie has just returned from a short visit to Mt. Carbon near Pottsville—Dr. Richard is on his return from England whither he went in one of Coxe's packet ships from this port. Tom is by me; and Emma has been my faithful attendant during my long indisposition, which does not prevent me from reading or—as you see—from writing, but cripples me in *toto*.

Present me in the kindest terms to your family, and believe me, my dear sir

faithfully
Robley Dunglison

Philadelphia
Feb. 19, 1858.

Dear Mr. Randall:

I have just received, through Judge Woodward, from your publishers Messrs. Derby & Jackson, and doubtless, by your direction, the first volume of your biography of Jefferson and I hasten to say to you how much pleasure I have had in running over its pages. It will be a most valuable work; and will shed bright light over matters on which obscurity previously rested. Every page indicates the trouble, time and thought you must have bestowed upon it; and I cannot doubt, that it will receive great attention from everyone who desires to be informed on the real history of the country. The style too is nervous and the language flows agreeably, and is entirely free from obscurity.

I sincerely hope that the work is destined to have a wide dissemination. One or two notices, which appeared, some time ago, in our newspapers—long before I had seen the volume—I sent to you. I do not know from whence they proceeded. Mr. Grigsby sent me his second notice, which you have doubtless seen.

I shall devour with avidity your remaining volumes; and beg of you to accept my thanks for the one you have sent me, and to believe me

faithfully yours
Robley Dunglison.

PS. I notice a few literal errors of which I shall make a note, so that hereafter should you think it worth while, you may alter the stereotype plates.

RD

PS. I have just heard, that Judge Kane is alarmingly ill of pneumonia at his residence, about eight miles from the city.

Philadelphia
Apl. 8, 1858

My dear Mr. Randall,

I hasten to say to you, that you have copied my *Ana* inaccurately. The word is *"consciousness,"* not *"anxiousness"*; and I hope this will reach you in time to rectify your error. Dick will give you a certificate of the very words.[10]

I am more and more pleased with your work and ask of you to believe me

faithfully yours,
Robley Dunglison.

Dear Mr. Randall,

The sentence of which you speak reads as follows:—

"In the course of the day and night of the 2d of July, he was affected with stupor; with intervals of wakefulness and consciousness; but on the third, the stupor became almost permanent."

Truly yours—
Richard J. Dunglison.

Dr. Robley Dunglison—
his opinion on my Life of Jeff.
June 17, 1858.[9]

Philadelphia
June 17, 1858

Dear Mr. Randall,

I have had the great satisfaction to receive, from your publishers, the last volume of your excellent work, and hasten to congratulate you on the termination of years of anxious but successful labour. The materials you have added to the history of the country, and to the proper appreciation of one of its most illustrious sons are invaluable. I have looked over the volumes with interest as deep as it will be abiding and have derived both pleasure and instruction from this perusal. I do not think, also, that you could have adopted a better plan for your index or have carried it out more happily.

The strong ground you have taken on many contested topics will give occasion, I doubt not, to animadversions, which may try your equanimity; but, now that you have had your say, based on conviction, I hope you will not permit yourself to be drawn into angry controversy; except to subserve the great cause of historic truth.

In the transcription of my memoranda, and the printing from your Ms. I do not observe many errors. One or two I will ask you to have modified now:—the others when you are revising the plates; and should I notice any other typographical inaccuracy, you will pardon me for drawing your attention to it, for, unless an author be a practised proofreader and even then—literal errors will escape him.

My friend, Judge King, my son Richard (the Doctor) and myself have been waiting for fine weather to take a trip to the coal region of Pennsylvania; we had determined to make a tour by Mauch Chunk, Wilkesbarre, Delaware Water Gap, Easton and home, and to leave Philadelphia on Monday last; but the northeasterly wind and damp weather caused my friend and the Judge to hesitate under the fear, as he said—"that the gout might be driven from my extremities to the "vitals"—of which I have had no apprehension. We shall probably set out on Monday next, and be away until the following Saturday.

Mr. Tucker has been in Virginia for some weeks. Our other friends, except Mr. Trist, whom I see frequently, I

---

[10] Dr. Dunglison was apparently having difficulty in writing with his arthritic hands at this time and made use of his son, on occasion, as his amanuensis.

[9] This endorsement was written at the top of this page by Mr. Randall.

have not met with for some time. My children are all well and with me. They bear you in kind remembrance, and this is refreshed by your interesting volumes, which they are all eager to peruse.

They desire me to present their kind regards to you, and have the goodness to believe me

faithfully yours,
Robley Dunglison.

Henry S. Randall

Philadelphia
Aug. 20, 1861.

My dear Sir,

A friend this morning, placed in my hands the number of Littells *Living Age* for August 31, published in advance, and which contains an article on the Private Character of Thomas Jefferson, from the *New Englander*, a publication, which I never see. This circumstance drew vividly to my mind the remembrance of our, to me, agreeable intercourse and correspondence, and determined me to resume the latter, chiefly with the view of learning something of your condition and occupations, and of stating to you the little of interest that appertains to my own. Since we last communicated with each other I have lost my old friend Professor Tucker. He spent three weeks of the early part of last winter with me in Girard Street;—he had lost his wife in the preceding summer;—and proceeded to the South with the view of escaping the winter. At Mobile he was knocked down by a bale of cotton and so much shaken by the accident, that he never afterwards recovered. His son-in-law went to Mobile, and succeeded in conveying him to Virginia where he died. The American Philosophical Society requested me to prepare an obituary notice of him, which I shall do, as soon as I am able to obtain some facts from Virginia from his family; but there is no sign of when this can be accomplished, in the present unhappy condition of the country. And, here, let me mention the great satisfaction I felt in perusing this evening the admirable observations of your friend Mr. Dickinson. Intensely union as I am they suit me precisely; and are a thousand times more valuable, as proceeding from him than if they had emanated from the believers in a different political creed. I accord with him *in toto*.

Living, as you do, in the country, you have but little desire, I presume, for change of residence during the summer. My own taste has always been civic; and this year, I have been almost wholly in town. The only exception has been a short visit to my friend the Revd Dr. Bethune at Catskill, where he resides during the summer, and, on that occasion, I went because he was unwell, and had a desire to converse with me. I may still go to Atlantic City for two or three days, but I cannot do this, until after the death of the son of my friend Dr. Pancoast, who cannot live many days. Most of my friends are temporarily absent: but few, however, have gone far away:—all anxious to be near their homes in the existing condition of affairs. Reverting to the article in the *New Englander*, what bitterness of spirit is exhibited in the allusion to oft refuted calumnies in regard to Mr. Jefferson. "In the region where he lived, the traveller now can hardly fail to hear the most unfavorable reports touching his private history narrated in detail and specifically as to persons and circumstances. And as truth sometimes invades the region of romance, we are not surprised if the representations go beyond the reality, at least in the surrounding scenery of the principal figure *when his residence is fitted up with chambers, casements and passages of mysterious import, with all the inventions of sensual art to hide itself from intrusion or indulge its desires in pictures, statuary, and*

"The lascivious of the tinklings of hitting instruments, the softening voices of women, and of beings less than women." *The palace of Sardanapalus* could not have contained *more incentives* to unlawful gratifications *than the dwellings where fancy has built for the voluptuary of Monticello*. All these tales cannot pass for fables springing out of the brains of his political enemies!!!

The whole article is written in the worst spirit, and is totally unworthy of notice. He classes me amongst "interested witnesses,"—to which I do not object in one sense. I certainly was "interested" in testifying to the truth.

I have not seen Mr. Trist for some time. His excellent wife and sister come to see me frequently, but she has been recently on a visit to her daughter in Alexandria; who is intensely union but united to an equally intense secessionist. My daughter is still in England: and I have my sons Richard and Tom with me. Willie has been for some weeks absent, in consequence of the delicate health of his wife, to whom I recommended a change of air, society and scenery, which has greatly benefited her.

I am, my dear Sir,
faithfully yours
Robley Dunglison.

II. DUNGLISON'S LIST OF INSTITUTIONS TO WHICH HE BELONGED

Before leaving England, to the

Royal College of Surgeons, of England
Society of Apothecaries, of London Licentiate and afterwards member.[1]
Associated Apothecaries and Surgeon-Apothecaries of England and Wales (member of council)
Hunterian Society of London (member, and member of council)[2]
Medical Society of London (member and Secretary of Foreign Correspondence)[3]
University of Erlangen (Doctor of medicine by written examination in the Latin language).
(Original diploma at the bottom of square red box)[4]
Linnean Society of Paris
Royal Society of Arts, Letters etc. of Nancy

[1] The archives of the Society of Apothecaries are in the Guildhall Library, London. The original certificate of R. Dunglison reads: Qualified as L.S.A. (Country Practice), 1st January, 1818. Apprentice to John Edmondson of Keswick for four years; also to Charles T. Haden of Sloan Street, London for sixteen months. Presented testimonials from both as well as proof of lecture attendance:

2 courses Anatomy
2 " Theory and Practice of Medicine
2 " chemistry
1 course Materia Medica.

Hospital attendance as probationer at Royal Infirmary Edinburgh—12 months. Examined and approved by Mr. Johnson (Guildhall Library Ms. 8241/1, p. 22). (Courtesy of Mr. A. J. Dickson, Sub-Librarian, Wellcome Historical Medical Library, London, England.)

[2] For complete list of officers in 1824, see *London Medical and Physical Journal* 51: 261, 1824. Dunglison's father-in-law, John Leadam, was elected a member of council the same year as Dunglison (*London Medical Repository* 17: 261, 1822).

[3] For an account of the fifty-first meeting of the society with a list of officers, see *London Medical and Physical Journal* 51: 261, 1824.

[4] Now in possession of the College of Physicians of Philadelphia.

Société de Médecine of Paris
Royal Academy of Marseilles
Society of Pharmacy of Paris
Physico-Medical Society of Erlangen
Academic Society of Medicine of Marseilles
The Royal Humane Society of London (member of the medical committee)
Eastern Dispensary of London (Physician Accoucheur)

Since Coming to America

1825 Yale College (Doctor of Medicine—Honoris Causa)
1825 University of Virginia (Professor appointed in 1824 in London)
1832 American Philosophical Society (member)
1833 Medical and Chirurgical Faculty of Maryland (Licentiate)
1833 University of Maryland (Professor)
1833 Baltimore Infirmary (one of the physicians)
1834 Medico-Chirurgical Society of Baltimore   ·
1834 St. George's Society of Maryland [St. Andrew's]
1835 Philocretan Society of Baltimore
1836 Jefferson Medical College of Philadelphia (Professor and clinical teacher)
1836 Philadelphia College of Pharmacy. (Hon. member)
1836 Maryland Academy of Science and Literature (Baltimore)[4a]
1836 Massachusetts Medical Society
1836 Medical Society of the State of New York
1837 Athenian Institute (Phila; member and counselor from the commencement)
1837 Franklin Institute
1837 Society of the Sons of St. George, Philadelphia. (Member and physician)
1837 Medical Society of Philadelphia[5] (Hon. member)
1838 Provincial Medical and Surgical Association of England [now British Medical Association]
1838 Philadelphia Hospital (Physician and Lecturer on Clinical Medicine)
1838 Medical Missionary Society of China
1838 Meteorological Committee of Franklin Institute (Chairman)
1838 Member of the Vestry of St. Stephen's.
1839 College of Physicians of Philadelphia
1839 Pathological Society of Philadelphia.
1839 Member of the Wistar Party.
1840 American Statistical Association (Boston)[6]
1840 American Philosophical Society (Secretary)
1840 Member of the Publication Committee of the U. S. Pharmacopoeia (with Professors Wood and F. Bache)
1840 Diagnothean Literary Society of Marshall College, Mercersburg[7]
1840 Musical Fund Society (manager)
1840 Frankford Lyceum of Science (Pa.)

1840 National Institution of Washington[8]
1841 Philomathean Society of St. Mary's (Emmitsburg)
1841 Asylum for the Insane Poor of Pennsylvania (trustee)
1841 Medical Society of Hamburg
1841 Historical Society of Pennsylvania
1842 Northern Academy of Arts and Sciences, Hanover, N. H.[9]
1843 Phi Beta Kappa Society of Union Coll., Schenectady.
1844 Young Men's Literary Soc., Knoxville, Tenn.
1844 Central Commission of Statistics of Belgium (correspondent)
1844 Pennsylvania Institution for the Blind (manager)
1844 Sydenham Society of London (hon. local Sec., Phila.)
1844 Royal Society of Northern Antiquaries (recommended by President and Council to be member but declined in consequence of having $30 to pay for the honor!)
1845 Medico-Chirurgical Society of Toronto, Canada
1845 Medico-Chirurgical Society of Montreal
1846 Société Française de Statistique Universelle, Paris
1846 Council of Society of the Protestant Epicsopal Church for the advancement of Chistianity in Pennsylvania
     Delegate on two or three occasions to the Episcopal Convention.
1848 Commissioner to attend the Assay at the Mint.
1849 Society of Sons of St. George, Phil. (honorary member)
1849 Aesculapian Society of the Univ. of Virginia
1849 Sydenham Society of London (Vice-President)
1850 Musical Fund Society of Philada. (Vice-President)
1850 Hart Institute of the Central High School (Philad.)
1851 Society of Apothecaries, London. (liveryman)
1851 Member of a Permanent Committee for the erection of a monument to Dr. Jenner in London.
1852 Jefferson College, Canonsburg;—the degree of LL.D.
1853 American Philosophical Society—Vice-President
1853 Phrenakosmian Society of Pennsylvania College, Gettysburg.[10]
1853 Pennsylvania Institution for the Blind. Chairman of the committee of instruction.
1853 Academy of Natural Sciences of Philadelphia.
1853 Jefferson Medical College of Philadelphia, the degree of LL.D.
1853 Philadelphia County Medical Society
1853 Director of the Pennsylvania Training School for Idiotic and Feebleminded Children[11]
1853 Musical Fund Society—President
1853 Vice-President of the Pennsylvania Training School for Idiotic and Feebleminded Children
1854 Iowa Lyceum and Museum of Natural History (Fort DesMoines) member
1854 American Association for the Advancement of Science.
1854 Dean of the Faculty of Jefferson Medical College (pro tem.)

[4a] Formed about 1822 and dissolved in 1844.
[5] Philadelphia Medical Society. See Samuel X. Radbill, The Philadelphia Medical Society, 1789–1868, *Tr. & Studies Coll. Phys. Phila.*, 4 ser. 20: 103–123, 1953.
[6] Founded in 1839.
[7] The Diagnothian Society, a student literary group founded in 1835, which, when Franklin and Marshall College were consolidated in 1853, moved to Lancaster, Penna., with its fine library (William J. Rhees, *Manual of Public Libraries, Institutions and Societies*, p. 359, Phila., 1859). The College of Physicians of Philadelphia is in possession of the letter Dunglison wrote to W. P. Schell, Esq., Cor. Sec., acknowledging enrollment as an honorary member.

[8] The National Institute for the Promotion of Science was founded in 1840 (William J. Rhees, *ibid.*, 513, 1859).
[9] The Northern Academy of Arts and Sciences was founded in 1841 and composed mainly of professors of Dartmouth College (William J. Rhees, *ibid.*, 212, 1859).
[10] Founded 1831 (William J. Rhees, *ibid.*, 350, 1859).
[11] Now the Elwyn Training School at Elwyn, near Media, Delaware County, Pa., founded February 10, 1853, through the efforts of Dr. Alfred L. Elwyn (James Clark Fifield, *American and Canadian Hospitals*, p. 1060, Minneapolis, 1933).

1855 Dean of the Faculty of do. do.
1855 Vice-President of the Musical Fund Society (declined)
1856 President of the Musical Fund Society
1856 Academy of Science of St. Louis (corresp. member)
1857 Member of the American Medical Association.
1857 Hon. member of the Natural History Society of Montreal
1858 Declined office in the American Philosophical Society after this year.
1858 Hon. member of the Kane Monument Assoc. of N. Y.
1859 State Historical Society of Iowa. Hon. member
1859 Burlington County Lyceum, Mount Holly, corr. member
1861 Associate member of the U. S. Sanitary Commission. (resigned, or rather, declined)
1862 Chief surgeon of the City of Philadelphia to act with the commissioner to superintend drafting (declined)
1862 Pennsylvania Institution for the Blind (Vice-President)
1866 Philomathean Society, Illinois State Univ. (Hon. member)[12]
1867 Euterpean Literary Society. (Muhlenberg College, Allentown, Pa.)
1867 Offered his resignation to his colleagues of his professorship and Deanship, but was induced to remain.
1868 Resigned these situations (see resolutions of faculty and board of Trustees)[13]

III. ORIGINAL WORKS ISSUED BY DUNGLISON

1823 *Dissertatio Inauguralis de Neuralgia.* Erlang. 1823.
1824 *Commentaries on Diseases of the Stomach and Bowels of Children.* Lond. 8vo. 1824.[1]
1827 *Syllabus of the Lectures on Medical Jurisprudence, and on the Treatment of Poisoning and Suspended Animation delivered in the University of Virginia.* 8vo. University of Virginia, 1827.
1829 *An Introduction to the Study of Grecian and Roman Geography* by Geo. Long, Esq. late of the University of Virginia, now of the University of London, and Robley Dunglison, M. D. of the University of Virginia. 8vo. Charlottesville, 1829.
1832 *Human Physiology.* 2 Vols. 8vo. Philad. 1832. 2nd edit. 1836. 3rd edit. 1838. 4th edit. 1841. 5th edit. 1844. 6th edit. 1846, 7th edit. 1850; 8th edit. 1856.
1833 *A New Dictionary of Medical Science and Literature, Containing a Concise Account* of the Various Subjects and Terms with the Synonymes [*sic*] in Different Languages and formulae for various officinal and empirical preparations etc. etc. 2 Vols. 8vo. Boston. 1833.
The second edition was published in Philad. in one volume in 1839 with the biographical and bibliographical notices omitted and also the German synonymes.
The third edition. 1842: 4th edit. 1844: 5th edit. 1845: 6th edit. 1846: 7th edit. 1848: 8th edit. 1851: 9th edit. 1852.[2]

---

[12] At Springfield, Illinois.
[13] In the Library historical collections, College of Physicians of Philadelphia.
[1] Critical Analysis in the *London Medical and Physical Journal* **52**: 493–504, 1824.
[2] 10th edit. 1853; 11th edit. 1854; 14th edit. 1856; 15th edit. 1857; 16th edit. 1860; 17th edit. 1865; 18th edit. 1866; 19th

1835 *Elements of Hygiene.* Under the title *On the Influence of Atmosphere and Locality; Change of Air and Climate; Seasons; Food, Clothing, etc. on Human Health,* constituting elements of hygiene. 8vo., Philad. 1835.
The second edition was published under the title *Human Health, or the Influence of Atmosphere and Locality; Change of Air and Climate; Seasons; Food; Clothing; Bathing and Mineral Springs; Exercise; Sleep; Corporeal and Intellectual Pursuits etc. etc. on Healthy Man; Constituting Elements of Hygiene.* 8vo. Philad. 1844.
1836 *General Therapeutics or Principles of Medical Practice, with tables of the chief remedial agents and their preparations, and of the different poisons and their antidotes.* Philad. 1836.
To the second edition, published in 1843, Materia Medica was added, and it appeared under the title *General Therapeutics and Materia Medica, adapted for a medical textbook.* 2 Vols. 8vo. Philad. 1843.
The third edition appeared in 1846; the 4th in 1850; to these two editions numerous illustrations were added; 5th edit. 1853; 6th edit. 1857.
1837 *The Medical Student or Aids to the Study of Medicine including a Glossary of the Terms of the Science, and of the Mode of prescribing; bibliographical notices of medical works; the regulations of different Medical Colleges of the Union;* etc. etc. 8vo. Philad. 1837.
A second edition appeared in 1844 under the title: *The Medical Student, or aids to the Study of Medicine, a modified and revised edition.* 12mo. Philad. 1844. [It was reprinted in Phila. in 1887.]
1839 *New Remedies: The Method of preparing and administering them;* their effects on the healthy and diseased economy. 8vo. Philad. 1839.
This work appeared first in the *American Medical Library* [3] of which the author was editor. The second edition—a transcript of the first—was published by Messrs Lea and Blanchard, the same year. The third edition appeared in 1841; the fourth edition in 1843, under the title *New Remedies, pharmaceutically and therapeutically considered.* The fifth edition appeared in 1846, under the title *New Remedies;* the sixth in 1851, under the title *New Remedies with formulae for their administration.* [A 7th edition was published in 1856.]
1842 *The Practice of Medicine, or a Treatise on Special Pathology and Therapeutics.* 8vo. 2 Vols. Philad. 1842. The second edition appeared in 1844 and the third in 1848.
1860 *A Dictionary for the Blind in tangible type on the basis of Worcester's small English dictionary,* prepared by W. Chapin, Principal, under the supervision of Dr. D: in three huge volumes, folio.

---

edit. 1868; 20th edit. 1874 edited by Richard J. Dunglison; 21st ed. (by Richard J.) 1893; 23rd edit. edited by Thomas Lathrop Stedman (1853–1938). In 1911 Stedman then published his own *Practical Medical Dictionary* with a new publisher, starting a complete new progression of editions, of which the 20th has just appeared in 1961, twenty-three years after Stedman's death.
[3] *Dunglison's American Medical Library* published by A. Waldie, Philadelphia, 1838, under the title of *Medical and Surgical Monographs,* contains (pp. 93–104) a "Formulary of New Medicines," which Dunglison states is an epitome from the *British Medical Almanac* for 1836 (by W. Farr?) containing all the most valuable practical matter from the later versions of Magendie's Formulary (see also Chapter I, note 48).

IV. WORKS EDITED OR TRANSLATED OR BOTH BY DUNGLISON

1822 *On the Use of the Moxa as a Therapeutical Agent.* By Baron D. J. Larrey etc. etc., translated from the French, with notes, and an introduction containing a history of the substance, by Robley Dunglison, etc. etc. 8vo. Lond. 1822.[1]

1824 *Formulary for the preparation and Mode of employing several new remedies; namely, Morphine, Iodine, Quinine, Cinchonine, Hydrocyanic Acid, Narcotine, Strychnine, Nux Vomica, Emetine, Atropine, Picrotoxine, Brucine, Lupuline, etc. with an introduction and copious notes* by the late Charles Thomas Haden, Esq. translated from the French of the third edition of Magendie's *Formulaire.* 2d. edit. with numerous alterations and additions by Robley Dunglison, M. D. etc. 12mo. Lond. 1824.

1824. *Appendix to the Formulary for the preparation and Mode of employing several new remedies; containing the pharmaceutical and therapeutical properties of the Hydriodates of Potass and Soda, the Ioduret of Mercury, the Cyanurets of Potassium and Zinc, the Oil of the Croton Tiglium, Piperine, Jalapine,* etc. by Robley Dunglison, M. D. etc., etc., translated from the French of the fourth edition of Magendie's *Formulaire* published in July. 12mo. Lond. 1824. The *Formulary* was reprinted in Philadelphia. 1825.

1824 *The Surgeons Vade-mecum Containing the Symptoms, Causes, Diagnosis, Prognosis, and Treatment of Surgical Diseases. Accompanied by Engravings, to illustrate the modern and approved methods of operating; also Select Formulae of Prescriptions and a Glossary of Terms.* 3d. edition. greatly enlarged. London. 12mo. 1824.
     This edition of the *Surgeons Vade-mecum* of Dr. Hooper was edited by Dr. Dunglison, but his name did not appear on the title page.

1830 Translated the portion of Eichhorn's *Weltgeschichte,* which relates to ancient history, and entered into an arrangement with Gray and Bowen of Boston to publish it; but for reasons mentioned elsewhere, it never appeared. (See p. 47.)

1837 *Medical Clinics of the Hospital Necker: or Researches and Observations on the Nature, Treatment, and Physical Causes of Diseases.* By I. Bricheteau etc. Translated from French for the American Medical Library. 8vo. Philad. 1837.
     This translation was made by Dr. Dunglison but it is not so stated on the title page.

1838 *Medical and Surgical Monographs.* By Messrs. Andral, B. Babington, C. J. Beck, Bright, B. C. Brodie, Burne, Carmichael, Clutterbuck, Cormack, Dubois, W. Farr, Itard, Louis, Maunoir and A. S. Taylor. with occasional comments, by the editor of the Library and others. 8vo. Philad. 1838.

1839 *Outlines of Physiology, with an Appendix on Phrenology,* by P. M. Roget, M.D. etc. etc. First American edition ; revised with numerous notes. 8vo. Philad. 1839.
     This edition was by Dr. Dunglison, but not so stated on the title page.

1841 *Outlines of a Course of Lectures on Medical Jurisprudence.* By Thomas Stewart Traill, M.D., F.R.S.E. etc. First American from the second American edition.[2] Revised with numerous notes. 8vo. Philad. 1841.

---

[1] 8vo. pp. 148. Critical Analysis in the *London Medical and Physical Journal* **48**: 511–526, 1822, also in the *London Medical Repository* **18**: 242–248, 1822.

[2] I.e., First American from the second Edinburgh edition.

This edition was by Dr. Dunglison, but not so stated on the title page.

1845 *The Cyclopaedia of Practical Medicine, Comprising treatises on the Nature and Treatment of Diseases, Materia Medica and Therapeutics, Medical Jurisprudence* etc. edited by John Forbes M.D., F.R.S., etc. etc., Alex. Tweedie, M.D., F.R.S., etc. etc. and John Conolly, M.D., etc. etc. Thoroughly revised with numerous additions by Robley Dunglison, M.D. in 4 Vols. 8vo. Philad. 1845.
     Besides the interstitial additions, the following articles were written by Dr. Dunglison: *Acrodynia; Anaemia* (with Dr. Marshall Hall) ; *Anthracein; Aphthae* (with Dr. A. Robertson) ; *Asphyxia of the Newborn; Blood, Morbid States of* (with Dr. M. Hall) ; *Bronchial Glands (diseases of the)* ; *Bronchitis, summer; Cachexia; Cholera infantum; Cirrhosis of the lung; Colon, torpor of the; Croup of the adult; Delirium tremens; Dengue; Diarrhoea, adipous; Disinfectant; Enteritis* (with Dr. Stokes) ; *Eutrophie; Fever, Malignant remittent; Galvanism* (with Dr. Apjohn) ; *Glanders; Heart, Polypus of the; Intussusception; Jaundice of the Infant; Laryngitis, chronic; Milk sickness; Mind, soundness and unsoundness of* (with Dr. Prichard) ; *Molluscum; Nauseants; Ophthalmia* (with Dr. Jacob) ; *Palpitation* (with Dr. Hope) ; [3] *Parturients; Ptyalism; Rheumatism* (with Dr. Barlow) ; *Sedatives* (with Dr. A. T. Thomson) ; *Spermatorrhoea; Spleen, diseases of the* (with Dr. Bigsby) ; *Statistics, medical* (with Dr. Bisset Hawkins) ; *Stomatitis; Strophulus; Throat, diseases of the* (with Dr. Tweedie) ; *Toothache; Toxicology* (with Dr. Apjohn).

V. JOURNALS EDITED BY DUNGLISON

1823 and 1824 *The London Medical Repository, Monthly Journal and Review.* Edited by James Copland, M.D. and Robley Dunglison. 8vo. Lond. Vol. XIX, XX (1823) and new series, Vol. 1 (1824) till October 1824.

1823 *The Medical Intelligencer or Monthly Compendium of Medical Chirurgical and Scientific Knowledge, being a review of the Contents of the various Transactions of learned Societies, and of the Monthly and Quarterly Journals, English and Foreign,* forming a concentrated record of Medical Literature. 8vo. Vol. IV. Lond. 1823.
     The editor's name was not on the title page. The earlier volumes were edited by Dr. Armstrong, Mr. Alcock, Mr. Haden.

1830 *The Virginia Literary Museum and Journal of Belles Lettres, Arts, Sciences etc.* Edited at the University of Virginia. 8vo. Charlottesville, 1830.
     This periodical was edited by Professor Tucker and Dr. Dunglison, although not so stated on the title page.

1837 to 1842 *The American Medical Library and Intelligencer, a concentrated record of Medical Science and Literature.* By Robley Dunglison, M.D. etc. etc.
     The *Medical Intelligencer* portion is in five volumes. 8vo. Philad. 1837 to 1842 inclusive.
     It was published once a month; with a reprint of some valuable foreign work on medicine, which appeared in detached portions monthly. Some of the monographs were united together to form a volume.

---

[3] This was James Hope; it must have been written before his death in 1841; or else Dunglison merely revised what Hope had previously written.

VI. ORIGINAL ESSAYS AND CONTRIBUTIONS TO PERIODICAL AND OTHER LITERATURE [1]

1817–1818 *Monthly Magazine* for 1817–1818. Various articles; the first contribution he ever made, that was printed, was on a "Floating Island in Derwentwater Lake," Cumberland, in the number for May 1817, p. 304, and other articles on the "Wind of a Ball," "Anthropophagi," "Collectanea Dietetica" etc. under the signature *Philos.*

1819 *London Medical Repository*, 12: 210–212, Case of considerable disease of some of the thoracic and abdominal viscera.

1820 *Annals of Philosophy*, Vol. X, p. 432.
A communication on certain phenomena of vision produced by dilating the pupil with the extract of Belladonna.

1822 *London Medical Repository* 18: 286. Case of Arachnitis cerebelli (mentioned in the *London Medical and Physical Journal* 49: 26, 1823).

1824 *Quarterly Journal of Science and the Arts.* Vol. XVII. p. 335. A review of Daniell's Meteorology.

1823 *London Quarterly Review;* edited by W. Gifford; for October 1823, p. 133.
An article on "Malaria" partly in answer to one by Dr. McCulloch in the 72d number of the *Edinburgh Review.*

1823 & 1824 *The London Medical Repository, Monthly Journal and Review.*[2] Edited by James Copland, M.D. and afterwards by him and Robley Dunglison, M.D. (sundry articles and reviews).

1823 *De Neuralgia*—Dissertation Inauguralis Medica etc. etc. pro' gradu doctoris summisque medicina et chirurgia honoribus, juribus et privilegiis rite obtinendis. Erlangae. 1823 (vide p. 8).[3]

1823–4 *Eclectic Review:* different articles.

1824 *The Universal Review or Chronicle of the Literature of All Nations.* (Edited by the Reverend G. Croly, D.D.). Sundry articles in the first and second volumes.

1827 *American Quarterly Review;* edited in Philadelphia by Robert Walsh. Sundry articles, as *Gastronomy of the Romans,* Vol. 2, p. 422. A review of Professor Tucker's *Voyage to the Moon* Vol. 3 p. 61.[4]
Popular Superstitions Vol. 3. p. 423.
Longevity Vol. 8. p. 380; *College Instruction and Discipline,* Vol. 9, p. 283; *Infirmities of Men of Genius,* Vol. 15. p. 214.

1830 *Virginia Literary Museum*
Sundry articles; Signed Zy, *, D, Z.Y., Ψ, X.Y., W.Y., YY, β, Δ, J, ④, Z, φ, Y, Σ, Zephaniah Stump.

1833 *A Lecture delivered to the Medical Class of the University of Virginia at the conclusion of the course, and on the occasion of his leaving the Institution.* July 9, 1833; published by the Class. 8vo. Charlottesville, 1833.

1834 *An Address delivered to the Graduates in Medicine at the Annual Commencement of the University of Maryland* on Wednesday, March 19th, 1834. Published by the Graduates and the Students. 8vo. Baltimore, 1834.

1834 *Baltimore Medical and Surgical Journal and Review;* edited by E. Geddings, M.D. etc. etc. Nos. 3 and 4. Baltimore 1834.
Reviews etc.

1834 & 1835 *North American Archives of Medical and Surgical Science.* Edited by E. Geddings, M.D. 2 Vols. Baltimore 1834–5.
Sundry articles.

1834 & 1835 *American Cyclopaedia of Medicine and Surgery.* Edited by Isaac Hays, M.D. Vol. 2. Articles— *Asphyxia* and *Atmosphere* (therapeutically and hygienically regarded).
The work ended with the letter A. The two volumes were subsequently issued under the title *Essays on Practical Medicine and Surgery* by Drs. Chapman, Bache, Coates, Geddings, Griffith, Mitchell, Wood, Dunglison, Condie, Dewees, Hays, Jackson, Patterson, Hodge, Horner etc. etc. 8vo. Philad. 1841.

1833 to 1836 *American Journal of the Medical Sciences.* Edited by Isaac Hays, M.D. Nos. 22 to 35. Philadelphia 1833–1836.
. Sundry reviews.

1836 (Oct.) *British and Foreign Medical Review.* Edited by John Forbes, M.D. etc. etc. No. IV. p. 583 (on the present state of medicine in the United States) ; and No. V. p. 273 (the subject continued. These articles contain an account of the institutions, associations, medical education etc. of the United States at the dates referred to).[5]
No. 43. p. 245. A letter marked (C) and dated America, Feb. 28, 1846, containing the Author's opinion on "Homeopathy, Allopathy and Young Physic of Dr. Forbes"; published by Dr. Forbes, but not objected to by Dr. Dunglison. The letter has no signature. The other articles are signed "Robley Dunglison."
These are the only contributions by Dr. Dunglison to the *British and Foreign Medical Review* excepting, I believe, a notice of Dr. Harlan's "Medical and Physical Researches" in No. IV.

1837 *Address to the Medical Graduates of the Jefferson Medical College; delivered March 11th, 1837.* Philad. 1837.

1837 *A Lecture on Animal and Vegetable Instinct,* delivered before the Athenian Institute of Philadelphia (published in the *Young People's Book or Magazine of useful and Entertaining Knowledge* edited by John Frost A.M. Nov. 1841 to Jan. 1842 inclusive.

1838 *The Medical Examiner and Record of Medical Science.* Edited successively by Drs. Clymer, Biddle, Huston, Francis Gurney Smith.
Numerous Reviews of Books, etc. etc.

1838 *An Appeal to the People of Pennsylvania on the Subject of an Asylum for the Insane Poor of the Commonwealth.* Philad. 1838. (Published by the Committee.)

[1839 *Liability of Vaccine Virus to Deterioration.* Published in the Proceedings of the American Philosophical Society, January 4, 1839, vol. I, p. 68.
A Report of Estlin's virus sent to Dunglison. See pp. 109–111.][6]

1839 *Introductory Lecture to the Course of Institutes of Medicine and Materia Medica in Jefferson Medical College of Philadelphia for the Session of 1839–40.* Philad. 1839.

[1840 "A Worm in a Horse's Eye." Published in the *Proceedings* of the American Philosophical Society, April 17, 1840, Vol. I, p. 200–201.

---

[1] For additional bibliography of the writings of Dr. Dunglison, see John R. Quinan, *Medical Annals of Baltimore,* p. 95, Baltimore 1884.

[2] Volumes 19, 20, 21, and 22.

[3] Also Appendix III, p. 196.

[4] Tucker's "Voyage to the Moon" appeared in the *American Quarterly Review* in 1827, and Dunglison's "Review" of this article in the March issue, 1828.

[5] See chapter I, fn. 65.

[6] Items in brackets are not in original manuscript but have been added by the editor.

A case of Filaria reported to him in a letter from Dr. Joshua I. Cohen, of Baltimore.][7]

1840   A Second Appeal to the People of Pennsylvania on the Subject of an Asylum for the Insane Poor of the Commonwealth. (Published by the Committee.) Philad. 1840.

1841   Introductory Lecture to the Course of Institutes of Medicine etc. delivered in Jefferson Medical College. November 1, 1841. Philad. 1841.

1842   On Certain Medical Delusions. An Introductory Lecture to the Course of Lectures on Institutes of Medicine etc. in Jefferson Medical College of Philadelphia. Delivered Nov. 4, 1842. Philad. 1842.

1843   An Introductory Lecture to the Course of Institutes of Medicine in Jefferson Medical College delivered November 8, 1843. Philad. 1843.

1844   Introductory Lecture to the Course of Institutes of Medicine etc. in Jefferson Medical College. Delivered Nov. 4, 1844. Philad. 1844.

1844   A Public Discourse in Commemoration of Peter S. Duponceau, LL.D., late president of the American Philosophical Society, delivered before the Society pursuant to appointment on the 25th of October, 1844 (published by order of the society) 8vo. Philad. 1844.

1845   An Introductory Lecture Delivered before the Class of Institutes of Medicine in Jefferson Medical College, Nov. 3, 1845. 8vo. Philad. 1845.

1847   A Charge to the Graduates of Jefferson Medical College of Philadelphia, delivered Mar. 25, 1847. 8vo. Philad.

1847   An Introductory Lecture delivered to the Class of Institutes of Medicine in Jefferson Medical College, Nov. 4, 1847. 8vo. Philad. 1847.

1848   An introductory Lecture delivered to the class of Institutes of Medicine in Jefferson Medical College. Oct. 19, 1848.

1849   Dr. Dunglison's Printed Views on Cleansing the City, in the Expectation of Cholera, are in report of the subcommittee of cleansing the city etc. Philadelphia, 1849.

1854   Charge to the graduates of Jefferson Medical College of Philadelphia; delivered Mar. 11, 1854. 8vo. Philadelphia, 1854 (mentions the number of volumes of Dr. Dunglison's works sold up to that time).

1854   On the Blind and Institutions for the Blind in Europe. A letter to the President of the Board of Managers of the Pennsylvania Institution for the Instruction of the Blind. 8vo. Philad. 1854.

1854   Recollections of Europe in 1854. An introductory lecture to the class of Institutes of Medicine etc. in the Jefferson Medical College, delivered Oct. 9, 1854. 8vo. Philad. 1854.

1858   Preliminary Remarks to an Introductory Lecture to the Course of Institutes of Medicine, delivered Oct. 11, 1858 (on the occasion of the death of Prof. J. K. Mitchell, and the appointment of Prof. S. H. Dickson). 8vo. Philad. 1858.

1860   Introductory Lecture to the Course of Lectures in Jefferson Medical College, delivered Oct. 9, 1860, 8vo. Philad. 1860.

1862   Obituary Notice of Geo. W. Bethune, D.D., read before the American Philos. Society, Oct. 3, 1862.[8]

70–85, 1862.

Obituary Notice of Prof. George Tucker, read before the American Philos. Society, Oct 3, 1862.[9]

[An "Obituary Notice of John K. Mitchell" by Robley Dunglison was also published in the Proceedings of the American Philosophical Society, VI: 340.]

1864   Exhortation to the Graduates of the Jefferson Medical College of Philadelphia. Delivered on the day of the Commencement. March the 10th, 1864. 8vo. Philad. 1864.

[1873   Chiudoku Rioho (The Treatment of Poisoning. Translated by Shingu Rioyen). Tokio.]

VII. BOYHOOD LETTERS, AND A LETTER TO HIS SON RICHARD

First letter by me, preserved by my revered mother. R.D.[1]

Brisco Hill [2]
Augt. 24th 1803.
Dear Mother:

I am happy to say that I am now able to open a correspondence with you, and to inform you that I am very well and that I have been so since you left me. I have attended school regularly since I had the pleasure of seeing you, and have made no bad progress, for after the vacation, which will commence the sixth Sepr. I shall begin the Latin testament.

I went down to Carlisle to see the Judge and spent an agreeable day with William Nicholson, who returned with me and stayed a few days.

Mr. Scott has got a letter from my uncle at London. Mr. and Mrs. S. both join me in love. I am, dear mother,

Yours affectionately
Robley Dunglison

Another boyhood letter.[3]

Green Row.
Decr. 22nd. 1810
Dear Mother

I seize upon the occasion with pleasure to write you a few lines with Jos. Rookin as the masters are not in at present. I am in want of a waiscoat, as I only brought two with me at Midsummer; a red one and a black one and the red one is quite worn out, therefore I would be much obliged to you if you would send me a new one as I am forced to wear my best every day, and also two neckcloths, as I have worn one ever since I wrote to you about them.

I am learning at present Mechanics and Mr. Saul says that learning Book-keeping at school is so much time lost as they keep their accounts quite different at every counting-house and therefore I think it would be better to defer it till after.

[9] Published ibid., pp. 64–70.
[1] This and the next two letters were loosely inserted in the manuscript by Dr. Dunglison.
[2] Brisco Hill is near Wigton, in Cumberlandshire, England. Apparently he was not quite six years of age when this letter was written. Unless he had the usual solicitous assistance of his teacher in penning this dutiful letter, it gives evidence of marked precocity.
[3] Lonsdale (ibid., p. 264) states that as a schoolboy, Robley Dunglison was noted as a lively, amiable and clever youth, of very studious habits.

[7] Joshua I. Cohen (1801–1870), M.D. 1823, University of Maryland, early American otologist, President of the Medical and Chirurgical Faculty of Maryland 1857–58, with whom Dunglison carried on a correspondence as early as 1828 (A.L.S. Philadelphia, April 1828, Robley Dunglison introducing Dr. Brown-Sequard, Historical Collections Yale Medical Library, courtesy of Miss Madeline E. Stanton).
[8] Proceedings of the American Philosophical Society 9: 70–85, 1862.

We have entered the new school but the walls are very damp as yet but Mr. Saul has got a stove put up in the adjacent house, the pipe of which runs from one end of the school to the other and which throws some heat on both sides.

I would be very glad to hear from my sister Mary as I have not heard from her this long time.

Give my best respects to my brother etc.

I would be sorry to impose upon your goodness by unnecessary demands but I hope you will not think me extravagant if I should ask you to send me some cash and a small Christmas treat.

I am, dear mother,

Your dutiful son
Robley Dunglison

A LETTER TO HIS SON RICHARD

Philadelphia,
July 31, 1853.

My dear boy.

You are now, doubtless, in the midst of your affectionate relatives, and in the home of your ancestors, which can not fail to be full of interesting associations to you; in a country, too, which for beauty of scenery, according to my experience, which has not been limited, can compare favorably with any on either side of the Atlantic. The arrival of the Asia has just instructed us of the arrival of the City of Glasgow at Liverpool, which sailed a few days after you. So favorable a passage leads me to hope that yours may have been propitious also. I did not write to you last week, nor did Willie. We found that but little had transpired since I wrote to you before, and I thought that little would be conveyed through the newspapers I have sent you.

You would see by them the death of Mr. Wetherill, commonly called *Price Wetherill*.[4] It created no little sensation, his peculiarities were forgotten; his objectionable qualities unnoticed, and the favorable points of character brought out in full relief—exaggerated, indeed, as they are apt to be on such occasions. Price was certainly a useful man, and all are disposed to award him that character. I know not, that, in his sphere, he could well be ambitious of a higher repute than that of a correct and useful man. There are so many who live for themselves alone, without any desire of serving or benefiting others, that the eminently useful will richly merit eulogy. My feelings on this subject are expressed in the motto *"Non mihi sed toti,"*[5] and I do not wish a higher encomium, when I have passed away, than that I have been a learned and useful man.

Talking of learning, the tenth edition of the "Dictionary" is now about to be issued, and I see that the ninth edition is noticed in the *British and Foreign Medico-Chirurgical* in most exalted terms. The reviewer says that now the work is finished, there will be no necessity for another dictionary for centuries provided it be kept up in successive editions. This task, dear Dick, will, I hope, devolve hereafter upon you, and if the reviewer be right may pass down even to your children's children. You will be able to see the number of the Review for July—if not before—(one of the medical gentlemen of Keswick may

have it)—at Churchills, 46 Prince Street, Soho. It may be worth your while, indeed, when you are in London to call there and see what new works are appearing, letting him know who you are, or at Highley's, 32 Fleet Street. They are publishers of the Review. I should like you also to report to me the appearance the English copies of the "Dictionary" present. I forgot to ask the names of the London Publishers; but you will easily learn them of the gentlemen I have mentioned, where you can see a copy if they do not possess one. When you are there—at any of those places, purchase for me a copy of a new work on Materia Medica, which is advertized as about to appear, by Dr. Garrod.[6]

Our town has been reported everywhere as affected with yellow fever. A vessel arrived at the South Street wharf from South America. She was examined and they say, purified, but still cases of fever occurred on board: and others in the neighborhood became affected. About half a dozen deaths are said to have occurred but this is not sanctioned by the Bills of Mortality which do not report so many. And Dr. Bell[7] told me that he had seen one of the cases, which looked very much like yellow fever. The Board of Health has published a card in which they say that no new case has occurred since Sunday; and that the disease was a high grade of bilious remittent fever. They considered the affair therefore ended. It may be so. I have no great idea, that the disease will become epidemic here, although there is a great disposition to it in the South. New Orleans is now losing sixty or seventy persons a day by it. It is upwards of thirty years since it has appeared here; and our sanitary regulations are on the whole so respectable—although far from being what they ought to be; that not much fear need, I think, be entertained, that it will become rife.[8] In Philadelphia little has been said or thought of it. The locality is at the foot of South Street wharf, which is a mile and a half from us, you know; but the New York papers, which are anxious to divert the fall trade to that city from this, are making the most of it; and I have no doubt some timid merchants will be afraid to tarry to make their purchases with us.

Mr. Styles, who sold his house on Broad Street to Mr. Davis, died on the same day as Mr. Wetherill of dysentery. At this season of the year, as you are aware, bowel affections are prevalent and this year, both here and all over the country, they prevail to a greater extent perhaps than common.

At home, everything is as usual; we miss you much, dear Dick, your absence renders my distressing bereavement still more poignant, but I console myself with the reflection that it is for your benefit. Tom is an energetic proof reader. We last night read thirty-two pages of the *Materia Medica and Therapeutics* which has all at once received an impulse from my suggestion that we were not going on with sufficient rapidity; and that it was possible I might myself have to detain them, should I again leave

---

[4] John Price Wetherill (1794–1853), Quaker scion of a chemical manufacturing family, member of the American Philosophical Society and the Academy of Natural Sciences of Philadelphia, for many years an influential Whig member of Philadelphia City Council.

[5] The last word of this motto is rather difficult to decipher in the original manuscript and may be "alius."

[6] Alfred Baring Garrod (1819–1907): *Essentials of Materia Medica, Therapeutics and the Pharmocopeias*, London, 1855, a student textbook.

[7] See chapter II, fn. 98.

[8] For the first time since 1820, yellow fever broke out in Philadelphia on July 17, 1853, and during the next three months there were 170 cases with 128 deaths. Dr. Wilson Jewell, President of the Board of Health, attributed the outbreak to the sewers opening into the Delaware River at South Street Wharf and to surface drainage, unsanitary alleys and damp cellars, while others traced it to putrescent bilge water in the hold of a Cuban ship docked at the South Street Wharf (R. LaRoche, *Yellow Fever* 1: 111–113, 1855).

town, of which, however, I have, at this time, no fixed intention. We have reached page 384 of the first volume.

Many books have arrived but no one interferes with them, and I apprehend they will load the tables until you return and assign them their appropriate places. Emma returned to town on Thursday in rude health. She had passed a fortnight most agreeably with Judge Kane. I dined with her there a day or two after I wrote to you. The Kanes are very pleasantly placed there; and did all in their power to render Emma's sojourn agreeable, in which they succeeded. Sometime this week she will probably go to Dr. Pancoast's. I feel somewhat fidgety, however, about the country at this season; having no doubt, that the town is much more healthy; and certainly, I think, less subject to dysentery, and kindred affections.

I saw a letter recently from Judge King, who was pleased with Vienna where he then was. He designs returning in October. He talked, when he was at Keswick, of visiting there again on his return; but this I should suppose, is doubtful. The last news from Drs. Wood and Bache was, that they were at Munich en route for Moscow, St. Petersburg and Stockholm; I believe Mr. Lea and family are at the baths of Schwalbach. It is doubtful whether they will return this year.

Dr. Henry K. Green of Macon, Georgia, has just paid me a visit. He is on his way, with his sister, to Saratoga; and was desirous of consulting me. He was always an invalid and now looked wretchedly; but I could not detect any other sign of tuberculosis than interrupted respiration —the *Respiration Saccadée* of the French writers.[9] He will call again on his return.

You ask about Tom and the cider. That speculation has turned out disastrously. In one day he found several bottles burst; and many without their corks; so that about two dozen were *hors de combat*.

By the way, my eye was caught by an advertisement in the Ledger of the "Keswick Institute"—a boarding school— at Norristown by Mr. and Mrs. Charles Lancaster. Ask

---

[9] Respiration saccadee was equated by Dunglison in his Dictionary (11th edition, p. 750, 1854) with "Respiration, jerking" and defined as follows: "When the murmur of inspiration, in place of being continuous, is interrupted as by starts, it is called 'jerkings.' It is a concomitant of incipient pleurisy, pleurodynia, spasmodic asthma, and tuberculosis of the lung with corresponding pleuritic adhesion." In modern times it is more apt to be called "Cogwheel respiration," although these finely differentiated physical signs in pulmonary diseases are fast being forgotten with growing dependence on x-ray diagnosis.

Aunt Mary if she knows who they are. The whole thing looks Cumbrian.

Our last arrivals mentioned the death of Mr. Pendarves. This is probably no news to you. We have not heard from the Trists since; I wonder what the old lady will adopt in regard to Browse. If Mr. Trist had given you a letter to her, you could not have presented it. He evidently felt uncomfortable in not having done so, for which there was no necessity; as I should have greatly preferred your not presenting it if he had given you one under other circumstances.

In regard to the letter to Mr. Ingersoll and Mr. Buchanan,[10] who may both be in London when you are there, it may be well for you to call, and if they are not in, or not visible, leave your letters with your mourning card, with or without your address, as you may think fit. It would be better, on the whole—as you are so young—not to leave your address, as they might be put to the inconvenience of calling upon you, which would be out of the question. If they can offer you any facilities for visiting places of interest to you, avail yourself of them. Should they be absent, state to the Secretary of Legation who you are, and leave the letter with him—for Mr. Ingersoll or for Mr. Buchanan as the case may be.

Willie's busy season has commenced. For the last week, we have only had his society late in the evening; and on one or two occasions, he did not return home until ten or eleven o'clock. He is as energetic as ever, and, although he has not left town, is in good bodily health and vigour.

The house has been upset for the last week by the putting in of a new furnace, which is evidently most complete; and will warm the house effectively.

The print of the Philadelphia Art Union has just been delivered. Those of the London Art Union will not be here for some time. Mr. Wood has just dined with us. He visited Roland Stephenson last Sunday with Dr. W. Rush, at Bristol. You know I have nothing to do with Roland. Ask Aunt Mary what his offense was for which he left England.

All join me in kindest regards to you, to my dear mother, and sister,

and believe me
my dear Dick
your affectionate father,
Robley Dunglison.

---

[10] James Buchanan (1791-1868), elected President of the United States in 1856, was Minister to Great Britain from 1853 to 1855.

INDEX

Abbey Holme, 7
Abbott, Mr. William, 125
Aberdeen University, 81
Abernethy, John (1764–1831), 17
Académie Royale of Marseilles, 11, 195
Academy of Natural Sciences of Philadelphia, 148, 195
Academy of Science of St. Louis, 196
Acid in the gastric juice, 50, 51
Acland, Henry Wentworth (1815–1900), 176–177
Actors of the 19th century, see: Wood, William B.
Adams, John (1735–1826), 191
Adams, John Quincy (1767–1848), 20
Adelon, N. P. (1780–1852), 49
Adelphia Hotel, Liverpool, 168
Aesculapian Society of the University of Virginia, 195
Agnew, D. Hayes (1818–1892), 124, 125
Agricultural Society of Philadelphia, 16
Agriculture, Randolph's improvement of, 32
Ague, 41
Alabama, 98
Albemarle County, Virginia, 21, 55
Alcock, Thomas A. (1784–1833), 13, 197
Alexandria, Virginia, 51
Alison, William Pulteney (1790–1859), 111, 112, 115
Allibone, Samuel Austin (1816–1889), 67, 68
Allumette Box, 184
Almshouse, Philadelphia, see Philadelphia General Hospital
Alpinus, Prosper (1553–1617), 13
Ambleside, 171
American Academy of Arts and Sciences, 154
American Association for the Advancement of Sciences, 195
American Cyclopaedia of Medicine and Surgery, 76, 198
American Journal of the Medical Sciences, 4, 13, 25, 51, 74–76, 86, 107, 120, 138, 139, 143, 183, 184, 198
American Medical Association, 25, 61, 71, 72, 143, 144, 148, 156, 196
American Medical Intelligencer, 53, 75, 81, 86, 88, 89, 91, 92, 101–104, 106, 109, 110, 119, 121, 125, 133, 135, 143, 145, 159, 196–198
American Medical Library and Intelligencer, see: American Medical Intelligencer
American Medical Recorder, 89
American, newspaper of Baltimore, 82
American Philosophical Society, 4, 26–28, 35, 38, 43, 50, 51, 53, 61, 64, 71, 75, 79, 82, 84, 85, 92, 96, 98, 107, 110, 118 119, 121, 124, 125, 130–133, 139, 140, 148–153, 168, 170–174, 177, 179–183, 186, 194–196, 198
American Quarterly Review, 15, 29, 36, 43, 177, 178, 198
American Statistical Association, 195
Amia Calva, 153

Ana, quoted, 35
Anaesthetics, 136
Anatomical Building at University of Virginia, 9, 23, 24, 36
Anatomical Hall at University of Virginia, 9, 23, 24, 36
Anatomical law of New York State, 81
Anatomy, teaching of, 36, 70, 113
Andalusia, 87
Anderson, Edwin A, (1816–1894?), 121
Andral, Gabriel (1797–1876), 197
Animal magnetism, 114; see also Mesmerism
Annals of Natural History, 117
Annals of Philosophy, 12, 198
Apjohn, James, 197
Apoplexy, 12
Apothecaries Act of 1815, 7
Apothecaries, associated, and surgeon—apothecaries of England and Wales, 11, 194
Apothecaries' Hall, 162
Apothecaries, Society of, London, 7, 11, 13, 79, 194
Arachnitis Cerebelli, article on by Dunglison, 12, 198
Armstrong, John (1784–1829), 13, 151, 197
Artimisia Moxa, 12
Ashmead, John W., 109
Asphyxia, 76, 197, 198
Association of Medical Superintendents of Insane Asylums of America, 110
Astor, Mr. & Mrs. John Jacob, of New York, 190
Astor Library, 190
Asylum for the Insane Poor of Pennsylvania, 108; see also Insane poor
Athenian Institute of Philadelphia, 107, 149, 195, 198
Athenaeum of London, 145
Athenaeum of Philadelphia, 128, 150; see also Athenian Institute of Philadelphia
Atkinson, James, 7, 87
Atkinson, John Richard, 87, 89, 99
Atkinson, Sara Ann, 169, 170
Audubon, John James (1780–1851), 123
Australia, 81
Autopsy of daughter, 158–159; of wife, 166–167
Azores, 19

Babington, Benjamin Guy (1794–1866), 197
Babington, William (1756–1833), 160
Bache, Alexander Dallas (1806–1867), 131, 132, 149, 173, 180, 183
Bache, Franklin (1792–1864), 72, 92, 110, 130, 136, 162, 197, 201
Bacot, John (1781–1870), 13
Baird, Henry Carey, 133
Baker, G. N., 108
Baker, Samuel (1785–1835), 41, 57, 58, 59, 83
Baker, William Nelson (1811–1841), 70, 71

Baltimore, 36–40, 54, 56, 59, 60, 65, 68, 69–83; status of medical profession, 70–71; see also Monday Evening Club
Baltimore Almshouse, 38, 70
Baltimore *American,* 82
Baltimore Infirmary, 70, 83, 195
Baltimore Lyceum, 59
Baltimore Medical and Surgical Journal and Review, 76, 198
Baltimore, Medico Chirurgical Society of, 83, 195
Baly, William (1814–1861), 112
Bancroft, Mr., 48, 179
Bank of the United States, 73
Bankhead, Mrs. Anne Cary Randolph, 34
Barbour, James (1775–1842), 65–66
Barbour, Philip Pendleton (1783–1841), 66
Barclay, James J., 108
Barlow, George Hilaro (1806–1866), 197
Barlow, Professor Peter, of Woolwich, England (1776–1827), 62
Barry, Major, Postmaster General, 68
Barry, Martin (1802–1855), 116
Bartlett, Elisha (1804–1855), 74, 75, 76, 96, 97
Barton, Benjamin Smith (1766–1815), 171
Barton, John Rhea (1796–1871), 81
Barton, William Paul Crillon (1783–1856), 37
Baxley, Henry Willis (1803–1876), 71
Beadle, E. R., 181
Bean, William B., 2, 5, 133
Beaumont, Samuel, 4, 52
Beaumont, William (1785–1853), 4, 50–53; experimental work with, by Dunglison, 50–53; suggestions to, by Dunglison, 50; opinion of, by Dunglison, 51, 53; report of experiments sent to Dr. Hays by Dunglison, 51, 52; published in *Elements of Hygiene* by Dunglison, 51, 52; requests Dunglison's support for a federal grant, 53
Beck, Carl Joseph (1794–1838), 197
Beck, John Brodhead (1794–1851), 97, 147
Beckman, Dr. & Mrs. M. Robert, 5
Bedford, Gunning S. (1806–1870), 91
Bell, Sir Charles (1774–1842), 49
Bell, John (1796–1872), 43, 60, 200
Belladonna, 11, 198
Bellevue Hospital, 95
Bellevue Medical College, 148
Bennett, James Risdon (1809–1891), 156
Bennett, John Hughes (1812–1875), 169
Berkshire Medical Institute, 96
Bermudas, 20, 24
Bernard, Karl, Duke of Saxe-Weimar-Eisenach, see Saxe-Weimar
Bethune, George Washington (1805–1862), 149, 158, 163, 167, 174, 179–181, 191, 192, 194
Bibby, Dr. 56
Biddle, C. C., 132
Biddle, John Barclay (1815–1879), 185, 198

www.ingramcontent.com/pod-product-compliance
Lightning Source LLC
Chambersburg PA
CBHW081338190326
41458CB00018B/6037